Advancing Project Management

emerald PUBLISHING

ice Publishing

Advancing Project Management:

Frameworks, Digital Transformation, Sustainability and Governance

Mohammed Reyasudin Basir Khan
PhD, CEng, MIET

Published by Emerald Publishing Limited,
Floor 5, Northspring, 21-23 Wellington Street, Leeds LS1 4DL.

ICE Publishing is an imprint of Emerald Publishing Limited

Other ICE Publishing titles:
Advancing Leadership and Collaboration in Project Management: Culture, Knowledge, and Team Dynamics
Khan, Mohammed Reyasudin Basir ISBN 978-1-83742-803-8
Sustainability and Digital Engineering Management: African Case Studies in Sustainable Digital Transformation
Telukdarie, Arnesh ISBN 978-1-83662-645-9
Essentials of Construction Planning and Scheduling
Robertson, Graham ISBN 978-0-7277-6592-5

A catalogue record for this book is available from the British Library

ISBN 978-1-83742-799-4

Cover photo: shutterstock.com/g/worawee

Commissioning Editor: Michael Fenton
Content Development Editor: Ryan Molyneux
Books Production Lead: Emma Sudderick
Project Manager: Rajachitra S

Typeset by: KnowledgeWorks Global Limited
Printed and bound by CPI Group (UK) Ltd, Croydon, CR0 4YY

Contents

Preface xi
Acknowledgement xiv
About the editor xvi
List of contributors xviii
List of abbreviations xx
List of figures and tables xxii

Part I – Methods and frameworks 1

1 .... **Framing innovations in project management 3**
Mohammed Reyasudin Basir Khan
1.1. Introduction 3
1.2. The changing landscape of project management 4
1.3. Theoretical foundations of contemporary project management 5
1.4. Conceptual models for framing project management innovations 5
1.5. Thematic synthesis of the content of this book 9
1.6. Malaysia as a representative global context 10
1.7. Conclusion 10
References 10

2 .... **Adopting project management tools in defence projects: strengthening monitoring, control and governance 11**
Dennis Kueh Abdullah and Mohammed Reyasudin Basir Khan
2.1. Introduction 11
2.2. Methodology 14
2.3. Results and analysis 15
2.4. Practical insights 23
2.5. Conclusion and recommendations 27
References 29

3 .... **Quality management systems in energy projects: the role of standardisation, stakeholders and resources 31**
Nor Aina Mazlan and Mohammed Reyasudin Basir Khan
3.1. Introduction 31
3.2. Methodology 36

3.3. Results and analysis 39
3.4. Practical insights 47
3.5. Conclusion and recommendations 49
References 50

4 .... **Green building certification and its impact on project management practices 51**
Praveen Raj Durai Ragavan and
Mohammed Reyasudin Basir Khan
4.1. Introduction 51
4.2. Methodology 55
4.3. Results and analysis 57
4.4. Practical insights 80
4.5. Conclusion and recommendations 82
References 83

5 .... **Managing projects amid geopolitical challenges: lessons from engineering firms 85**
Nur Dayana Fatin Zulkhurnain and
Mohammed Reyasudin Basir Khan
5.1. Introduction 85
5.2. Methodology 89
5.3. Results and analysis 92
5.4. Practical insights 101
5.5. Conclusion and recommendations 102
References 104

Part II – Digital transformation 107

6 .... **Digital project management systems: adoption, integration and project performance 109**
Suganya Sukhumaran and
Mohammed Reyasudin Basir Khan
6.1. Introduction 109
6.2. Methodology 114
6.3. Results and analysis 116
6.4. Practical insights 119
6.5. Conclusion and recommendations 120
References 121

7 .... **Risk management in infrastructure projects: examining schedule, cost and quality impacts on project success** **123**
Rozlano Muhamad Tobat and
Mohammed Reyasudin Basir Khan
7.1. Introduction 123
7.2. Methodology 128
7.3. Results and analysis 128
7.4. Practical insights 136
7.5. Conclusion and recommendations 136
References 137

8 .... **Impact of Lean Six Sigma on cost reduction in electronics manufacturing service companies** **139**
Khugan Kanagasabapathy and
Mohammed Reyasudin Basir Khan
8.1. Introduction 139
8.2. Methodology 144
8.3. Results and analysis 145
8.4. Practical insights 148
8.5. Conclusion and recommendations 150
References 151

Part III – Sustainability and strategy **153**

9 .... **Engineering management practices for optimising EPC project performance** **155**
Robeatul Adawiah Osman and
Mohammed Reyasudin Basir Khan
9.1. Introduction 155
9.2. Methodology 164
9.3. Results and analysis 166
9.4. Practical insights 175
9.5. Conclusion and recommendations 176
References 177

10 .... **Sustainable project management: balancing triple constraints with environmental goals** **181**
Thong Jin Cheng and
Mohammed Reyasudin Basir Khan
10.1. Introduction 181
10.2. Methodology 185
10.3. Results and analysis 186
10.4. Practical insights 193
10.5. Conclusion and recommendations 194
References 195

11 .... **Project cost performance: economic, organisational and project-level influences** **197**
Zaharullail Abdul Rashid,
Mohammed Reyasudin Basir Khan,
Chong Peng Lean and Ahmad Anwar Zainuddin
11.1. Introduction 197
11.2. Methodology 206
11.3. Results and analysis 208
11.4. Practical insights 215
11.5. Conclusion and recommendations 216
References 217

Part IV – Risk and governance 221

12 .... **The role of project management applications in delivering railway infrastructure success** **223**
Mohd Suhail Kamari and
Mohammed Reyasudin Basir Khan
12.1. Introduction 223
12.2. Methodology 227
12.3. Results and analysis 229
12.4. Practical insights 235
12.5. Conclusion and recommendations 235
References 236

13 .... **Construction challenges and governance realities: a project delivery perspective** **239**
Noorazan Morshidi and
Mohammed Reyasudin Basir Khan
13.1. Introduction 239
13.2. Methodology 244
13.3. Results and analysis 247

13.4. Practical insights 256
13.5. Conclusion and recommendations 256
References 257

14 Diagnosing construction delays: an empirical study of a major commercial development project 259
Rozaimoorsyari Abdullah, Mohammed Reyasudin Basir Khan, Chong Peng Lean and Ahmad Anwar Zainuddin
14.1. Introduction 259
14.2. Methodology 262
14.3. Results and analysis 264
14.4. Practical insights 276
14.5. Conclusion and recommendations 277
References 278

Conclusion 281

15 Future directions in project management innovation 283
Mohammed Reyasudin Basir Khan
15.1. Introduction 283
15.2. From methods to adaptive frameworks 284
15.3. Digital transformation as a capability enabler 285
15.4. Sustainability and strategic value creation in projects 286
15.5. Governance, risk and the foundations of execution quality 287
15.6. Implications for theory, practice and policy 288
15.7. An implementation roadmap for project management innovation 289
15.8. Future research and innovation pathways 292
15.9. Concluding integration 292
References 294

Index 295

Preface

Project management continues to evolve as organisations respond to rapid technological advancement, growing sustainability pressures, global competition and increasingly complex regulatory environments. The discipline that once focused primarily on planning and scheduling has grown into a multidimensional field that integrates digital systems, stakeholder engagement, governance mechanisms and strategic decision-making. This book was developed to contribute to the growing understanding of how project management is being transformed in contemporary environments, particularly in rapidly developing, transitioning contexts.

The content presented in this book is based on a series of empirical studies conducted across a range of Malaysian industries. These include defence, construction, engineering, energy, transportation and manufacturing. Malaysia provides a valuable context for exploring innovative project management practices because it reflects many characteristics of emerging and transitional economies. It has a diverse industrial base, an increasing commitment to digital transformation, a growing emphasis on sustainability and expanding infrastructure demands. These conditions create a natural laboratory for assessing how project management approaches are being reshaped in resource-constrained environments while simultaneously pursuing modernisation agendas.

The development of this book was motivated by the need to close the gap between traditional project management frameworks and the realities of modern project environments. Established methodologies such as the Project Management Body of Knowledge (PMBOK) continue to provide essential foundations for integration, scoping, scheduling and quality control. However, the challenges faced by contemporary organisations require additional frameworks that account for digital adoption, sustainability imperatives, institutional governance, risk complexity and strategic alignment. A growing body of literature highlights the limitations of classical linear approaches and the need for hybrid, adaptive and technology-assisted models to support informed and timely decision-making.

This book aims to provide a consolidated view of project management innovation by combining theoretical perspectives with empirical findings from Malaysian industrial sectors. The intention is to offer insights relevant not only to scholars studying project management theory but also to practitioners leading projects in dynamic environments. Policy makers and organisational leaders will also find the content valuable, as it highlights capability gaps and systemic challenges that influence project outcomes at the national level.

The book is organised into four major parts.

Part 1: Methods and frameworks introduces foundational and emerging tools for project planning, monitoring, procurement, quality management and environmental compliance. The chapters address topics including defence project tools, quality management systems in energy companies, green building certification and navigating geopolitical risk. These chapters highlight the importance of strong methodological foundations in enabling project teams to navigate uncertainty and complexity.

Part 2: Digital transformation reflects the increasing influence of digital tools on project execution. Organisations across industries now rely on cloud platforms, data analytics, automation systems and collaborative software to improve coordination and enhance visibility across project phases. The studies in this section explore topics such as risk analytics in construction, Lean Six Sigma integration in electronics manufacturing and challenges related to digital adoption in project management environments. The findings demonstrate that while digital tools offer significant benefits, successful adoption requires supportive leadership, adequate training and organisational readiness.

Part 3: Sustainability and strategy reflect how sustainability has become a central element of project management. Many organisations now recognise that economic performance must be balanced with environmental and social considerations. The chapters in this section examine engineering management in engineering, procurement and construction (EPC) projects, sustainable project management practices, material selection frameworks and economic modelling for cost optimisation. They emphasise how sustainability considerations influence decision-making throughout the project lifecycle.

Part 4: Risk and governance explores the challenges which continue to play a significant role in determining project outcomes. Many industries face issues related to unclear decision rights, weak accountability structures, insufficient risk mitigation processes and communication breakdowns. The chapters in this section examine railway infrastructure projects, governance constraints in Borneo and causes of delays in Malaysian construction projects. The studies demonstrate that governance quality is strongly linked to project performance and that improvements in governance can significantly reduce risk.

This book is intended for a broad audience. Academics will find value in the theoretical grounding and empirical analyses. Practitioners will benefit from practical recommendations and sector-specific insights that can inform decision-making. Students will gain exposure to current developments in project management, particularly in digital technology and sustainability. Policy makers and organisational leaders will find the content helpful in identifying institutional-level capability gaps and guiding policy formulation.

By integrating empirical studies with established theories such as the technology acceptance model, the resource-based view, contingency theory and sustainability frameworks, this book seeks to bridge the gap between scholarly knowledge and practical relevance. The chapters illustrate how organisational culture, technological readiness, governance structures and external constraints influence project management practices. They also demonstrate that innovation in project management often arises from the need to address systemic challenges rather than solely from technological advancements.

As organisations continue to face increasingly complex environments, project management will remain at the forefront of strategic implementation. It is hoped that the insights presented in this book will encourage further exploration, collaborative research and evidence-informed practice. The content reflects the need for continuous innovation and adaptability as essential components of effective project delivery in the modern world.

Acknowledgement

I could not have completed this book without the support and guidance of many individuals and institutions. I am deeply grateful for their help.

I sincerely thank Universiti Tun Abdul Razak (UNIRAZAK) for providing the academic environment and resources essential to conducting interviews, collecting data and facilitating peer review, which enabled the research, writing and editorial development of this book. The university's commitment to applied research, knowledge creation and professional development directly supported the conceptualisation, data analysis and validation processes in this book.

I am especially grateful to the Chief Executive Officer of UNIRAZAK, Mr Amil Izham Hamzah, for his leadership, guidance and strategic vision in strengthening the university's research agenda. His direction and long-term planning have shaped the research foundation of this book. I also wish to thank Mr Fairuz Mohamed Noor, Chief Operating Officer, for his encouragement and support, and Dr Muhamad Firdaus Jamal, Chief Financial Officer, for guidance in strengthening the sustainability of the institution's research initiatives. Their collective leadership has contributed meaningfully to the growth of UNIRAZAK's research ecosystem.

I sincerely thank the Vice Chancellor, Deputy Vice Chancellors and senior academic leadership of UNIRAZAK for fostering academic excellence and research engagement. I am grateful to my academic colleagues and the UNIRAZAK academic team for their cooperation and collaboration. Their commitment advances teaching and research in project management and related fields.

I want to acknowledge the invaluable contributions of the industry partners, participating organisations and professionals who supported the empirical studies in this book. By providing expert perspectives, access to operational data and detailed industry context, they enriched the quality of the research. These contributions strengthened the practical relevance of the findings across Malaysian industries. I also wish to acknowledge Nadia Aqmal for her valuable copyediting assistance. Her help improved the clarity and presentation of this book.

I thank the postgraduate students whose research shaped chapters of this volume. Their work in data collection, analysis and theory has improved our understanding of project management in Malaysia and has strengthened research in the field. I also thank my peers, mentors and reviewers for their feedback. Their comments helped me refine the structure and raise the scholarly quality of this book.

I thank my late mother, Aysha Banu Abdul Rahman, whose support and values shaped my academic journey and inspired my scholarship. I am also grateful to my wife, Nadirah Abdul Kadir, for her patience and encouragement throughout this process.

Thank you to the readers of this volume. I hope these insights will support academic inquiry, professional practice and policy development and advance project management as a discipline.

About the editor

Mohammed Reyasudin Basir Khan, Tun Razak Graduate School, Universiti Tun Abdul Razak, Kuala Lumpur, Malaysia

Mohammed Reyasudin Basir Khan is a Professor of Artificial Intelligence and project management specialist with over 14 years of experience at the intersection of academia, research and industry. While his technical expertise spans artificial intelligence (AI), data analytics, computing and engineering, his career is equally distinguished by his work in project management, organisational performance and leadership of large-scale initiatives. He has successfully directed complex, multidisciplinary projects across domains such as healthcare, energy analytics, environmental, social and governance (ESG) and business transformation, combining technology adoption with proven project management methodologies.

As Deputy Vice Chancellor (Research) at Universiti Tun Abdul Razak (UNIRAZAK), Prof Reyasudin leads the strategic enhancement of the university's research and project delivery ecosystem. His responsibilities include building research capacity, increasing the number of funded projects and publications, and cultivating strong industry partnerships. He also serves as Senior Director of URIMPACT, a business unit that leverages AI and project management frameworks to execute AI and digital transformation initiatives. In this role, he oversees programmes in research, innovation, digital apps, business analytics and stakeholder engagement, ensuring effective delivery through governance structures and project management office (PMO)-style practices.

At the Tun Razak Graduate School, he teaches and supervises in AI, data analytics, business analytics and project management, equipping postgraduate students and executives with both technical and managerial skills. His earlier roles included academic leadership and programme development, where he contributed to curriculum design, accreditation and compliance with bodies such as the Malaysian Qualification Agency (MQA), the Engineering Accreditation Council (EAC), the Engineering Technology Accreditation Council (ETAC) and the Technology and Technical Accreditation Council (TTAC).

Prof Reyasudin holds a BEng and PhD in Electrical and Computer Engineering from Universiti Tenaga Nasional (UNITEN). He is a Professional Engineer (Board of Engineers, Malaysia), Chartered

Engineer (Engineering Council, UK), Chartered Scientist (Science Council, UK), European Engineer (FEANI) and Professional Technologist (Malaysian Board of Technologists). His professional credentials also include Registered Electrical Energy Manager (Energy Commission, Malaysia), Certified Energy Manager (ASEAN Centre of Energy), Professional Data Analytics (Google) and Data Science Specialist (IBM).

He has published widely in international journals and at conferences, and has served as an editor, reviewer and chair for academic and professional events. With deep experience in project governance, accreditation management and delivery of high-impact initiatives, he combines the rigour of project management with the innovation of emerging technologies to drive measurable outcomes in both academia and industry.

List of contributors

Mohammed Reyasudin Basir Khan: Tun Razak Graduate School, Universiti Tun Abdul Razak (UNIRAZAK), Kuala Lumpur, Malaysia.

Dennis Kueh Abdullah: Tun Razak Graduate School, Universiti Tun Abdul Razak (UNIRAZAK), Kuala Lumpur, Malaysia.

Nor Aina Mazlan: Tun Razak Graduate School, Universiti Tun Abdul Razak (UNIRAZAK), Kuala Lumpur, Malaysia.

Praveen Raj Durai Ragavan: Tun Razak Graduate School, Universiti Tun Abdul Razak (UNIRAZAK), Kuala Lumpur, Malaysia.

Nur Dayana Fatin Zulkhurnain: Tun Razak Graduate School, Universiti Tun Abdul Razak (UNIRAZAK), Kuala Lumpur, Malaysia.

Suganya Sukhumaran: Tun Razak Graduate School, Universiti Tun Abdul Razak (UNIRAZAK), Kuala Lumpur, Malaysia.

Rozlano Muhamad Tobat: Tun Razak Graduate School, Universiti Tun Abdul Razak (UNIRAZAK), Kuala Lumpur, Malaysia.

Khugan Kanagasabapathy: Tun Razak Graduate School, Universiti Tun Abdul Razak (UNIRAZAK), Kuala Lumpur, Malaysia.

Robeatul Adawiah Osman: Tun Razak Graduate School, Universiti Tun Abdul Razak (UNIRAZAK), Kuala Lumpur, Malaysia.

Thong Jin Cheng: Tun Razak Graduate School, Universiti Tun Abdul Razak (UNIRAZAK), Kuala Lumpur, Malaysia.

Zaharullail Abdul Rashid: Tun Razak Graduate School, Universiti Tun Abdul Razak (UNIRAZAK), Kuala Lumpur, Malaysia.

Chong Peng Lean: MILA University, Nilai, Negeri Sembilan, Malaysia.

Ahmad Anwar Zainuddin: Kulliyyah of Information and Communication Technology, IIUM, Gombak, Malaysia.

Mohd Suhail Kamari: Tun Razak Graduate School, Universiti Tun Abdul Razak (UNIRAZAK), Kuala Lumpur, Malaysia.

Noorazan Morshidi: Tun Razak Graduate School, Universiti Tun Abdul Razak (UNIRAZAK), Kuala Lumpur, Malaysia.

Rozaimoorsyari Abdullah: Tun Razak Graduate School, Universiti Tun Abdul Razak (UNIRAZAK), Kuala Lumpur, Malaysia.

List of abbreviations

AAGR	average annual growth rate
AI	artificial intelligence
ANOVA	analysis of variance
BIM	building information modelling
BREEAM	Building Research Establishment Environmental Assessment Method
CIDB	Construction Industry Development Board
CPM	critical path method
CSFs	critical success factors
DECRIS	detail engineering completion rating index system
DMAIC	define, measure, analyse, improve and control
E&E	electronics and electrical
EFA	exploratory factor analysis
EMS	electronics manufacturing service
EOT	extension of time
EPC	engineering, procurement and construction
ERP	enterprise resource planning
ESG	environmental, social and governance
FEED	front-end engineering design
FIDIC	International Federation of Consulting Engineers (standardised contract frameworks)
GBC	green building certification
GBI	green building index
GDP	gross domestic product
GTFS	green technology financing scheme
HRC	human resource challenges
IBS	industrialised building systems
IoT	Internet of Things
ISO	International Organization for Standardization
KMO	Kaiser-Meyer-Olkin
LCA	lifecycle assessments
LEED	Leadership in Energy and Environmental Design

LSS	Lean Six Sigma
ML	machine learning
MNC	multinational corporation
NIMP	new industrial master plan
NRO	national rail operator
PEM	project execution models
PEOU	perceived ease of use
PERT	program evaluation and review technique
PMBOK	project management body of knowledge
PMIS	project management information systems
PMRE	project monitoring and reporting efficiency
PMS	project management software
PMT	project management tools
PSM	project sustainability management
PU	perceived usefulness
QA	quality assurance
QC	quality control
QMS	quality management system(s)
RBV	resource-based view
SBSC	sustainability balanced scorecard
SCD	supply chain disruption
SCM	supply chain management
SEM	structural equation modelling
SME	small and medium-sized enterprises
SOP	standard operating procedure
SPSS	statistical package for the social sciences
SWOT	strengths, weaknesses, opportunities and threats
TAM	technology acceptance model
TBL	triple bottom line
TQM	total quality management
UNIRAZAK	Universiti Tun Abdul Razak
UTAUT	unified theory of acceptance and use of technology
VIF	variance inflation factor

List of figures and tables

Figures

Figure 1.1	Project management innovation ecosystem	6
Figure 1.2	Digital transformation adoption model	7
Figure 1.3	Governance and risk performance alignment model	8
Figure 2.1	Conceptual framework	13
Figure 2.2	Technology adoption	23
Figure 2.3	Leadership support and training	24
Figure 2.4	Project complexity and risk	24
Figure 2.5	Resource availability	25
Figure 2.6	Project monitoring and reporting efficiency	25
Figure 3.1	Conceptual framework	35
Figure 3.2	Percentage of respondents by job role	39
Figure 4.1	Conceptual framework	55
Figure 4.2	Percentage of respondents by role in the organisation	58
Figure 4.3	Histogram, Q-Q plots and box plot for cost management	66
Figure 4.4	Histogram, Q-Q plots and box plot for time management	67
Figure 4.5	Histogram, Q-Q plots and box plot for quality management	68
Figure 4.6	Histogram, Q-Q plots and box plot for stakeholder engagement	69
Figure 4.7	Histogram, Q-Q plots and box plot for impact of GBC practices	70
Figure 5.1	Conceptual framework	89
Figure 5.2	Respondent career field	93
Figure 5.3	Respondents by company size	94
Figure 5.4	Histogram of normality	95
Figure 5.5	Q-Q plot	96
Figure 6.1	Conceptual framework	113
Figure 7.1	Conceptual framework	127
Figure 7.2	Percentage of respondents by age group	129
Figure 7.3	Distributions of the dependent and independent variables	131
Figure 7.4	Scatter plots for the dependent and independent variables	132
Figure 7.5	Q-Q plots for DV and IVs	132
Figure 8.1	Conceptual framework	143
Figure 8.2	LSS tool usage in EMS companies	146

Figure 8.3	Cost reduction observed due to LSS	146
Figure 8.4	Areas of cost reduction	147
Figure 8.5	Challenges in LSS implementation	147
Figure 8.6	Measures to improve LSS effectiveness	148
Figure 9.1	Conceptual framework	163
Figure 9.2	Geographic location of EPC projects executed by the participants	167
Figure 9.3	Scatter plots illustrating the correlation between engineering management practices and schedule performance of EPC projects	171
Figure 10.1	Conceptual framework	184
Figure 10.2	Q-Q plot for corporate policies	189
Figure 10.3	Q-Q plot for resource management	189
Figure 10.4	Q-Q plot for stakeholder engagement	190
Figure 10.5	Q-Q plot for organisational learning	190
Figure 10.6	Likert distribution across resource management variable	191
Figure 11.1	Conceptual framework	206
Figure 11.2	Industry sector	209
Figure 11.3	Fluctuation in exchange rates	211
Figure 11.4	Government regulations and policies	211
Figure 11.5	Dependent variable	212
Figure 11.6	Normal P-P plot of regression standardised residual	213
Figure 12.1	Conceptual framework	226
Figure 12.2	Years of experience in project management	230
Figure 12.3	Q-Q plot of correlation coefficients	233
Figure 13.1	Conceptual framework	243
Figure 13.2	Respondents' construction site location	248
Figure 13.3	Q-Q plot of Pearson correlation coefficients	254
Figure 14.1	Conceptual framework	262
Figure 14.2	Demographic profile of project stakeholders	265
Figure 14.3	Regression standardised residuals	269
Figure 14.4	Normal Q-Q plot of project delay	270
Figure 14.5	Normal Q-Q plot of financial constraints	270
Figure 14.6	Normal Q-Q plot of contractual changes	271
Figure 14.7	Normal Q-Q plot of site-related matters	271
Figure 14.8	Normal Q-Q plot of communication issues	272
Figure 14.9	Regression standardised residual against regression standardised predicted value	276
Figure 15.1	Integrated pathways for project management innovation	293

Tables

Table 2.1	Demographic profile	16
Table 2.2	Reliability assessment	16
Table 2.3	Normality tests	18
Table 2.4	Survey items, mean and standard deviation for all constructs (n = 175)	19
Table 2.5	Spearman's rank correlation matrix (n = 175)	21
Table 2.6	Standardised beta coefficients	21
Table 2.7	Summary of hypothesis testing (n = 175)	22
Table 3.1	Descriptive statistics analysis for all the variables	41
Table 3.2	Cronbach's alpha values for variables	41
Table 3.3	Pearson correlation analysis for all the variables	42
Table 3.4	Spearman correlation analysis for all the variables	43
Table 3.5	Simple linear regression analysis by coefficient for process standardisation and impact on QMS	44
Table 3.6	Simple linear regression analysis by ANOVA for process standardisation and impact on QMS	44
Table 3.7	Simple linear regression analysis by coefficient for stakeholder management and impact on QMS	45
Table 3.8	Simple linear regression analysis by ANOVA for stakeholder management and impact on QMS	45
Table 3.9	Simple linear regression analysis by coefficient for training and impact on QMS	45
Table 3.10	Simple linear regression analysis by ANOVA for training and impact on QMS	45
Table 3.11	Simple linear regression analysis by coefficient for time management and impact on QMS	46
Table 3.12	Simple linear regression analysis by ANOVA for time management and impact on QMS	46
Table 3.13	Simple linear regression analysis by coefficient for resource allocation and impact on QMS	46
Table 3.14	Simple linear regression analysis by ANOVA for resource allocation and impact on QMS	46

Table 3.15	Multiple linear regression analysis by coefficient for process standardisation, stakeholder management, training, time management, resource allocation and impact on QMS	47
Table 3.16	Multiple linear regression analysis by ANOVA for process standardisation, stakeholder management, training, time management, resource allocation and impact on QMS	47
Table 4.1	Cost management descriptive results	59
Table 4.2	Time management descriptive results	60
Table 4.3	Quality management descriptive results	61
Table 4.4	Stakeholder engagement descriptive results	62
Table 4.5	Impact of GBC on project management practices	64
Table 4.6	Kolmogorov–Smirnov and Shapiro–Wilk test results for cost management	65
Table 4.7	Kolmogorov–Smirnov and Shapiro–Wilk test results for time management	66
Table 4.8	Kolmogorov–Smirnov and Shapiro–Wilk test results for quality management	67
Table 4.9	Kolmogorov–Smirnov and Shapiro–Wilk test results for stakeholder engagement	69
Table 4.10	Kolmogorov–Smirnov and Shapiro–Wilk test results for impact of GBC	70
Table 4.11	Cronbach's alpha for each variable	71
Table 4.12	Spearman's correlation analysis (impact of GBC against cost management)	72
Table 4.13	Spearman's correlation analysis (impact of GBC against time management)	73
Table 4.14	Spearman's correlation analysis (impact of GBC against quality management)	73
Table 4.15	Spearman's correlation analysis (impact of GBC against stakeholder engagement)	74
Table 4.16	Pearson's correlation analysis (impact of GBC against cost management)	75
Table 4.17	Pearson's correlation analysis (impact of GBC against time management)	75
Table 4.18	Pearson's correlation analysis (impact of GBC against quality management)	76

Table 4.19	Pearson's correlation analysis (impact of GBC against stakeholder engagement)	77
Table 4.20	Regression model summary	77
Table 4.21	ANOVA summary for regression analysis on impact of GBC on project management	78
Table 4.22	Regression coefficients for cost management, quality management and stakeholder engagement	79
Table 4.23	Collinearity diagnostics for regression predictors	81
Table 5.1	Reliability test	95
Table 5.2	Descriptive analysis of project performance	96
Table 5.3	Skewness and kurtosis	97
Table 5.4	Pearson correlation coefficient	98
Table 5.5	Model summary	98
Table 5.6	ANOVA	99
Table 5.7	Coefficient regression	100
Table 5.8	Summary of hypotheses results	100
Table 6.1	Summary of operationalisation and measurement	115
Table 6.2	Reliability analysis	117
Table 6.3	Pearson's correlation analysis	118
Table 6.4	Regression coefficients for predicting project outcomes	119
Table 7.1	Summary of reliability analysis findings	130
Table 7.2	Descriptive statistics of variables	130
Table 7.3	Findings of Pearson's correlation	133
Table 7.4	Multiple regression model summary	134
Table 7.5	Coefficients	134
Table 7.6	Analysis of variance (ANOVA)	135
Table 7.7	Multiple regression and correlation analysis findings	135
Table 8.1	Pearson's correlation analysis	148
Table 8.2	Multiple regression analysis	149
Table 9.1	Summary of reliability test	168
Table 9.2	Summary of validity test results using EFA	168
Table 9.3	Descriptive analysis for research variables ($n = 139$)	169
Table 9.4	Pearson's correlation test between engineering management practices and schedule performance of EPC projects ($n = 139$)	170

Table 9.5	ANOVA	172
Table 9.6	Coefficients and interpretation of hypotheses	172
Table 9.7	Model summary of moderation analysis using hierarchical multiple regression	173
Table 9.8	Coefficients of moderation analysis	173
Table 9.9	Summary of hypothesis testing	174
Table 10.1	Distribution of questionnaire sections	185
Table 10.2	Respondents' demographic profile	187
Table 10.3	Cronbach's alpha across variables	188
Table 10.4	Normality analysis	188
Table 10.5	Descriptive statistics of key constructs ($n = 200$)	191
Table 10.6	Summary of Pearson correlation coefficient results	192
Table 10.7	Summary of hypothesis testing	192
Table 11.1	Demographic	208
Table 11.2	Summary of reliability statistics	209
Table 11.3	Statistics: macroenvironmental	210
Table 11.4	ANOVA	212
Table 11.5	Hypothesis testing	213
Table 12.1	Questionnaire sections	228
Table 12.2	Mean, standard deviation, minimum and maximum	230
Table 12.3	Cronbach's alpha	231
Table 12.4	Pearson's correlation	232
Table 12.5	Multiple regression analysis	233
Table 12.6	ANOVA	234
Table 12.7	Coefficients	234
Table 13.1	Reliability analysis	249
Table 13.2	Descriptive statistics of project environment	249
Table 13.3	Descriptive statistics of the project team	250
Table 13.4	Descriptive statistics of the project management process	250
Table 13.5	Descriptive statistics of project risks	251
Table 13.6	Descriptive statistics of project performance	251
Table 13.7	Findings of Pearson correlation	253
Table 13.8	Model summary	254
Table 13.9	Coefficients	255
Table 13.10	ANOVA	255
Table 14.1	Reliability assessment of project delays	265
Table 14.2	Item total statistics – financial constraints	266

Table 14.3	Item total statistics – contractual changes	267
Table 14.4	Item total statistics – site-related matters	267
Table 14.5	Item total statistics – communication issues	268
Table 14.6	Descriptive statistics of factor delays	269
Table 14.7	Normality test results for study variables	272
Table 14.8	Test for mean difference	272
Table 14.9	Relationship between potential factors and project delays	273
Table 14.10	ANOVA results for the multiple regression model	274
Table 14.11	Regression coefficients for factors contributing to project delay	275

Part I

Methods and frameworks

Mohammed Reyasudin Basir Khan
ISBN 978-1-83742-799-4
https://doi.org/10.1108/978-1-83742-798-720261001

Chapter 1
Framing innovations in project management

Mohammed Reyasudin Basir Khan

Tun Razak Graduate School, Universiti Tun Abdul Razak (UNIRAZAK), Kuala Lumpur, Malaysia

1.1. Introduction

Project management has transitioned from a specialised operational function into a strategic organisational capability that supports competitiveness, innovation and long-term sustainability. Organisations now rely on projects to implement digital transformation, expand infrastructure, enhance operational efficiency and respond to complex geopolitical and market pressures. As projects grow in scale and complexity, the knowledge, tools and governance mechanisms required to manage them effectively also expand. Scholars highlight that modern project environments require more than the traditional triple constraint of time, cost and scope. They demand adaptability, resilience, data-driven decision-making and collaborative structures that bridge organisational boundaries (Kerzner, 2017; PMI, 2021).

This book responds to this broader evolution of the discipline by examining how organisations apply project management theories and tools to improve decision quality, strengthen communication, optimise resource allocation, reduce operational inefficiencies and enhance governance practices. While empirical studies draw heavily on Malaysian industries such as construction, defence, energy, transportation and manufacturing, the insights are not country-specific. Instead, the book positions Malaysia as an empirical landscape that reflects challenges commonly observed in global project environments.

This chapter provides a foundation for the book by discussing the evolution of project management, outlining the theoretical frameworks that underpin the research, examining the relevance of the subsequent chapters and illustrating how innovations in project management can be interpreted within three major domains: frameworks and methodological foundations; digital transformation; and governance and risk.

1.1.1 Background and context

Increasing project complexity, coupled with changing economic and technological conditions, has reshaped expectations of project managers and organisations. Global supply chains, regulatory systems and stakeholder networks have created an environment in which effective project management requires integration of diverse competencies. These include strategic planning, communication, governance, innovation management and digital fluency.

1.1.2 Purpose of the book

The purpose of this book is to explore innovations in project management across multiple sectors and contexts. The chapters highlight the adoption of project frameworks, digital tools, methodological improvements, governance mechanisms and risk management strategies. This chapter provides the foundation by introducing the theoretical perspectives and conceptual structures that inform the analyses presented.

1.1.3 Structure of the chapter

This chapter discusses the evolution of project management, presents theoretical foundations, introduces conceptual frameworks and synthesises the thematic contributions of subsequent chapters. It concludes with a discussion of Malaysia as a representative empirical context and outlines the relevance of these themes for global project environments.

1.2. The changing landscape of project management

1.2.1 Rising complexity and interdependence

Global organisations operate within interconnected supply chains, regulatory environments, technological disruption and high-stakes expectations. Projects often involve multiple partners and require coordination across diverse knowledge domains. Construction projects must manage contractors and environmental assessments. Defence and engineering projects require strict compliance and documentation. Infrastructure projects involve collaboration across public and private sectors. Communication breakdowns, stakeholder misalignment and unclear responsibilities are recurring challenges that influence project performance (Schwalbe, 2015).

1.2.2 Technological transformation

Digital transformation has reshaped project management practices. Cloud platforms, integrated project management tools, progress monitoring dashboards, automation, analytics and mobile applications support real-time communication and decision-making. The technology acceptance model highlights that perceived usefulness and ease of use influence the adoption of these tools (Davis, 1989). Studies in this book show varying levels of digital maturity across sectors. Some industries have adopted advanced digital systems, while others still rely on manual communication and documentation.

1.2.3 Sustainability and environmental responsibility

Sustainability increasingly influences project decisions. The triple bottom line framework encourages organisations to consider environmental, social and economic impacts (Elkington, 1997). Construction and engineering projects now integrate environmental assessments, material optimisation, waste reduction and green certification. Infrastructure sectors increasingly emphasise long-term resource efficiency and environmental stewardship.

1.2.4 Governance and accountability expectations

Governance has become central to project success. Effective governance ensures clear decision rights, transparent reporting and alignment between project goals and organisational strategies. Governance failures often manifest as delays, cost overruns or stakeholder dissatisfaction. Chapters in this book highlight governance dynamics in construction, rail transport, engineering and defence. These studies support broader literature showing that governance structures significantly influence project outcomes (Too and Weaver, 2014).

1.3. Theoretical foundations of contemporary project management

1.3.1 Project management body of knowledge

The project management body of knowledge (PMBOK) outlines knowledge areas and processes that support integration, scope management, schedule development, cost control, quality assurance and stakeholder engagement (PMI, 2021). Many chapters in this book evaluate how organisations apply these processes or adapt them to unique sectoral needs.

1.3.2 Technology acceptance model

The technology acceptance model (TAM) explains how users adopt technologies based on perceived usefulness and perceived ease of use (Davis, 1989). Several chapters analyse digital project management tools, communication platforms and monitoring systems using TAM principles.

1.3.3 Resource-based view

The resource-based view suggests that organisational capabilities create competitive advantage when they are valuable, rare, inimitable and non-substitutable (Barney, 1991). Relevant capabilities include human skills, digital infrastructure and analytical processes that support project performance.

1.3.4 Contingency theory

Contingency theory argues that project management practices must be adapted to fit the specific characteristics of each project environment. Different sectors face unique challenges related to regulatory systems, risk profiles and resource availability.

1.3.5 Sustainability models

Sustainability frameworks guide project decisions related to environmental impact, social responsibility and long-term economic value. Several chapters draw on these models to evaluate green certification, sustainable engineering and resource efficiency.

1.4. Conceptual models for framing project management innovations

Innovation in project management arises from the interaction between environmental forces, organisational capabilities and the practices applied in project settings. To understand how these interactions shape contemporary project delivery, this section presents an integrated set of conceptual models. Together, these models provide a structured analytical lens for interpreting the themes across the chapters in this book. They demonstrate that project management innovation is not simply the adoption of new tools, but the outcome of a complex system shaped by context, capability and governance. The following subsections describe three interrelated conceptual perspectives: the project management innovation ecosystem;

digital transformation adoption dynamics; and the alignment between governance, risk and performance.

1.4.1 Project management innovation ecosystem

The first model conceptualises the ecosystem within which project management innovation develops. As illustrated in Figure 1.1, innovation begins with external forces such as globalisation, technological change, regulatory requirements and sustainability expectations. These forces shape the organisational capabilities that organisations must build, including human expertise, cultural alignment and digital infrastructure. When these capabilities interact with operational demands, they generate innovation across three domains: frameworks and methods; digital transformation; and governance and risk. These domains, in turn, influence key project outcomes such as time, cost, quality, sustainability and stakeholder satisfaction.

Figure 1.1 provides a holistic view of how project management practices evolve. It shows that innovation emerges when organisations align internal capabilities with external expectations. The model also underscores that the innovation domains identified in this book do not

Figure 1.1 Project management innovation ecosystem

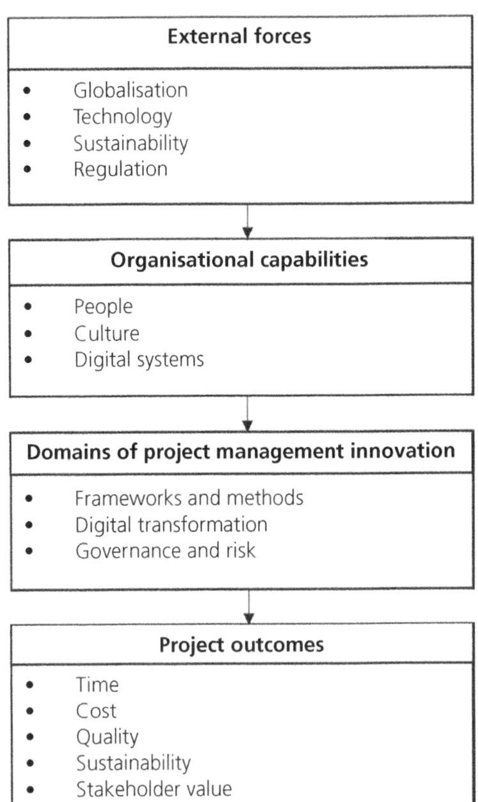

operate independently of one another. Instead, they function as interconnected components of a larger ecosystem that determines project performance.

1.4.2 Digital transformation adoption model

Digital transformation is a key area of innovation in contemporary project management. However, its impact depends not only on the availability of digital tools but also on user perceptions and organisational readiness. As shown in Figure 1.2, digital adoption begins with the establishment of digital infrastructure such as project management software, communication platforms and integrated data systems. User perceptions about usefulness and ease of use influence their acceptance of these tools, consistent with the technology acceptance model. Once the tools are accepted, their practical application enhances coordination, transparency and decision quality, ultimately improving project outcomes.

Figure 1.2 Digital transformation adoption model

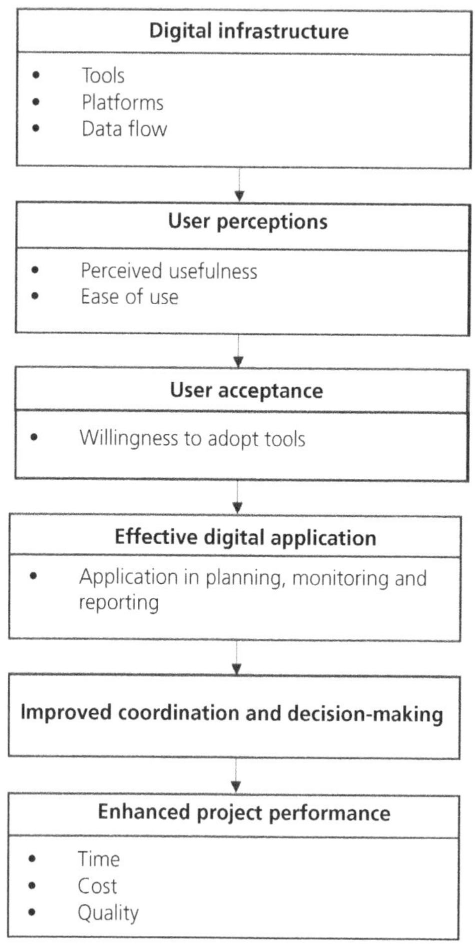

This model reinforces that digital transformation involves behavioural, organisational and cultural dimensions. As illustrated in Figure 1.2, technology becomes transformative only when users adopt it willingly and organisations integrate it into routine project workflows. The model also aligns with global findings that digital tools create the greatest value when supported by training, leadership commitment and structured implementation.

1.4.3 Governance and risk performance model

The third model highlights the central role of governance and risk management in shaping execution quality and project outcomes. As illustrated in Figure 1.3, governance structures establish decision rights, accountability mechanisms and reporting systems. These structures form the foundation for effective risk management practices, including risk identification, assessment, mitigation and monitoring. Strong governance and risk processes enable consistent and disciplined execution, which improves communication,

Figure 1.3 Governance and risk performance alignment model

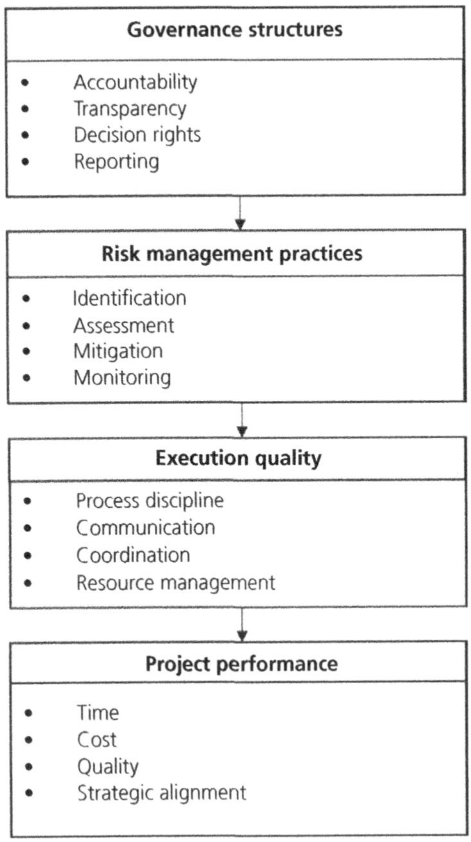

resource coordination and procedural integrity. These improvements collectively enhance project performance.

Figure 1.3 illustrates that governance is not merely an administrative function. Instead, it is a structural driver of performance that shapes how project teams interpret priorities, manage risk and execute work processes. When governance structures are aligned with risk practices and organisational capabilities, they enhance the consistency and reliability of project delivery.

1.5. Thematic synthesis of the content of this book

Although the chapters cover defence, construction, engineering, energy, manufacturing and rail infrastructure, they share common themes that reflect global project management trends.

1.5.1 Frameworks and methodological foundations

Several chapters examine structured methodologies that support planning, monitoring and performance. Examples include

- Lean Six Sigma as applied to manufacturing for waste reduction and process improvement
- communication frameworks for construction project stakeholder alignment
- engineering management models in engineering, procurement and construction (EPC) environments
- quality management standards in the energy sector.

These studies demonstrate that strong methodological foundations remain essential.

1.5.2 Digital transformation and technology adoption

Multiple chapters emphasise the role of digital tools in enhancing project visibility, communication and decision-making. Examples include

- digital project management applications in rail infrastructure projects
- information systems supporting construction documentation and communication
- data-driven tools for manufacturing optimisation.

These findings align with global research showing the need for digital readiness to improve project performance.

1.5.3 Governance, sectoral realities and risk

Governance and risk management form the third primary thematic domain. Examples include

- governance constraints in East Malaysian construction environments
- risk factors associated with urban construction delays
- compliance and documentation requirements in defence projects
- organisational and communication risks in large infrastructure projects.

These chapters illustrate that project governance plays a central role in project outcomes across sectors.

1.6. Malaysia as a representative global context

Although the case studies originate in Malaysia, the issues they explore mirror global challenges. These include

- communication barriers
- digital adoption obstacles
- governance weaknesses
- sustainability pressures
- sector-specific risks in construction, transportation, manufacturing and engineering.

Malaysia serves as a microcosm for examining universal project management issues that apply across diverse geographic and industrial contexts.

1.7. Conclusion

Project management continues to evolve as organisations navigate complexity, technological advancement, governance expectations and sustainability imperatives. This book provides insights into methodological innovations, digital tools, sustainability practices and governance mechanisms that shape project outcomes. This chapter has outlined the theoretical foundations and thematic perspectives that guide the book. By drawing on diverse empirical settings, the book contributes to a deeper understanding of how organisations worldwide can strengthen project management capability, embrace innovation and achieve performance excellence.

REFERENCES

Barney J (1991) Firm resources and sustained competitive advantage. *Journal of Management* **17(1)**: 99–120.

Davis FD (1989) Perceived usefulness, perceived ease of use and user acceptance of information technology. *MIS Quarterly* **13(3)**: 319–340.

Elkington J (1997) *Cannibals with Forks*. Capstone, Oxford, UK.

Kerzner H (2017) *Project Management*. Wiley, Hoboken, NJ, USA.

PMI (Project Management Institute) (2021) *A Guide to the Project Management Body of Knowledge (PMBOK Guide)*, 7th edn. PMI, Newtown Square, PA, USA.

Schwalbe K (2015) *Information Technology Project Management*, 8th edn. Cengage Learning, Boston, MA, USA.

Too EG and Weaver P (2014) The management of project management. *International Journal of Project Management* **32(8)**: 1382–1394.

Mohammed Reyasudin Basir Khan
ISBN 978-1-83742-799-4
https://doi.org/10.1108/978-1-83742-798-720261002

Chapter 2
Adopting project management tools in defence projects: strengthening monitoring, control and governance

Dennis Kueh Abdullah and Mohammed Reyasudin Basir Khan

Tun Razak Graduate School, Universiti Tun Abdul Razak (UNIRAZAK), Kuala Lumpur, Malaysia

2.1. Introduction

A national defence organisation is a critical pillar in safeguarding national security, enhancing operational capabilities and ensuring readiness in both peacetime and crises. Managing defence projects within such an organisation is inherently complex due to the advanced technology associated with modern defence systems, extensive infrastructure requirements and multifaceted coordination across departments. These projects often involve intricate tasks such as planning, resource allocation, task delegation, risk management and reporting. Consequently, effective project management practices are essential to ensure that projects are delivered on schedule, within budget and in accordance with technical and regulatory requirements endorsed by relevant procurement and capability governance bodies (Schwalbe, 2015).

Despite the increasing sophistication of defence projects, many national defence organisations continue to rely heavily on traditional project management methods, including manual spreadsheets, presentation slides, email and telephone-based tracking systems. Such practices are time-consuming, error-prone and unable to provide real-time visibility into projects (Kerzner, 2013). In contrast, defence organisations globally have increasingly adopted digital project management tools that support data integration, communication and real-time reporting, thereby enhancing efficiency and decision-making (National Audit Office, 2021). Nevertheless, defence organisations face several challenges in adopting these technologies, including data security concerns, system compatibility issues and difficulties in scheduling training without disrupting operational commitments (PMI, 2021).

Against this backdrop, this chapter explores how adopting modern project management tools can enhance the efficiency and effectiveness of project monitoring and reporting within a national defence organisation. Although defence projects are becoming increasingly complex and resource-intensive, reliance on fragmented and outdated management systems persists, limiting the ability to respond effectively to high-stakes operational demands. The continued use of manual methods contributes to communication inefficiencies, reporting delays, cost overruns and limited transparency among project stakeholders (Kerzner, 2013; National Audit Office, 2018). Accordingly, the research presented in this chapter investigates how digital project management tools can improve monitoring and reporting efficiency and strengthen overall project performance. The chapter further examines challenges related to technology adoption, particularly perceived usefulness and ease of use, and analyses the role of leadership support and staff training in facilitating successful implementation.

This research holds both theoretical and practical significance. Theoretically, it contributes to the technology adoption literature by applying the technology acceptance model (TAM) in the context of a security-oriented, hierarchical defence organisation. It extends understanding of how organisational culture, security protocols and structural rigidity influence digital transformation initiatives. Practically, the findings offer valuable insights for project managers, information technology specialists and defence leaders by identifying strategies for integrating project management tools, promoting user acceptance and maintaining operational continuity.

From a policy perspective, the chapter provides evidence-based recommendations to support digital transformation initiatives within defence organisations. These include targeted training frameworks, incentive mechanisms and resource allocation strategies that facilitate the transition towards technology-enabled project environments. Enhanced project management practices also contribute to broader public accountability by improving transparency in defence expenditure, strengthening governance and supporting operational readiness.

Finally, this chapter lays the foundation for future research on emerging technologies, including artificial intelligence-enabled project management, blockchain-based security solutions and interagency interoperability systems. By doing so, it offers transferable insights into how technology adoption in complex, risk-averse environments can be extended to other public-sector and critical-infrastructure organisations.

TAM was developed by Davis (1989) and provides the theoretical basis for understanding how personnel within a national defence organisation perceive and adopt digital project management tools. According to TAM, behavioural intention to use technology is influenced by perceived usefulness and perceived ease of use. In defence environments, these perceptions are particularly important, as users must trust that digital tools enhance efficiency while integrating seamlessly with existing workflows and security requirements. Technology adoption is therefore treated as a key independent variable influencing project monitoring and reporting efficiency.

Leadership support and training are grounded in contingency theory (Lawrence and Lorsch, 1967), which suggests that organisational effectiveness depends on alignment between managerial actions and contextual conditions. In defence project environments characterised by uncertainty and complexity, leadership support through resource provision, guidance and motivation is essential, while structured training enhances user competence and confidence (PMI, 2021). Hierarchical command structures further amplify the importance of leadership commitment in enabling technological change and organisational learning.

Project lifecycle theory and the critical success factors framework (PMI, 2021; Rockart, 1979) explain how project complexity and risk influence performance across project phases. Defence projects typically involve multiple stakeholders, stringent security controls and technical interdependencies, increasing exposure to risk and coordination challenges. Managing these risks effectively is therefore essential to sustaining project control and performance.

The resource-based view (RBV) (Barney, 1991) emphasises that organisational performance depends on the effective deployment of valuable financial, technical and human resources. Within defence organisations, resource availability directly affects the adoption and sustainability of project management tools. Adequate budgets, information systems infrastructure and skilled personnel enable effective monitoring and reporting, thereby enhancing project outcomes.

Based on these perspectives, the proposed framework links four independent variables – technology adoption, leadership support and training, project complexity and risk, and resource availability – to the dependent variable, project monitoring and reporting efficiency (PMRE). This framework integrates theoretical foundations with practical relevance in a national defence organisational context. Figure 2.1 presents the conceptual framework illustrating these relationships.

Building on the conceptual framework and literature review, this section develops hypotheses that examine the relationships between the independent variables (technology adoption, leadership support and training, project complexity and risk, and resource availability) and the dependent variable (project monitoring and reporting efficiency (PMRE)) within a national defence organisational context. These hypotheses are grounded in established theoretical perspectives, including TAM, contingency theory, project lifecycle, critical success factors (CSF) theory and RBV.

The first hypothesis proposes a positive relationship between technology adoption and PMRE. According to TAM, when project management tools are perceived as useful and easy

Figure 2.1 Conceptual framework

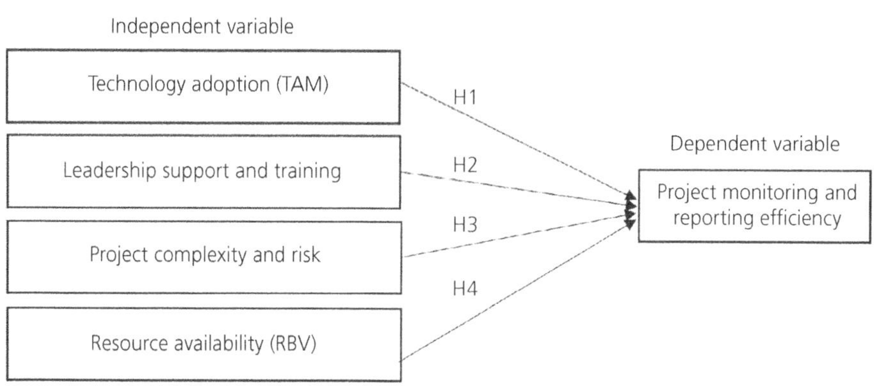

to use, personnel are more likely to adopt them, leading to improved coordination, real-time communication and more effective decision-making.

The second hypothesis suggests a positive association between leadership support, training and PMRE. Drawing on contingency theory, this hypothesis posits that leadership endorsement and structured training programmes enhance user confidence and commitment, thereby facilitating the successful implementation of digital project management tools. Within hierarchical defence organisations, leadership involvement is particularly critical for fostering innovation, compliance and accountability.

The third hypothesis proposes a positive association between project complexity and risk (PCR) and PMRE. Based on project lifecycle and CSF theories, increasing project size, technical complexity, and security requirements tend to expand the scope and depth of monitoring and reporting needs. When supported by advanced project management systems, higher complexity can drive greater investment in structured reporting mechanisms, ultimately enhancing the accuracy and timeliness of PMRE.

Finally, the fourth hypothesis predicts a positive relationship between resource availability and PMRE, consistent with RBV. Adequate financial, human and technical resources support effective tool deployment, sustain system functionality and enhance overall project monitoring and reporting efficiency.

2.2. Methodology

2.2.1 Research design

This study employs a quantitative survey research design, selected for its suitability in examining relationships between measurable constructs such as technology adoption, leadership support, project complexity and resource availability. The quantitative approach enables standardised data collection and supports statistical analysis to identify patterns and associations across organisational levels (Fowler, 2014). Survey methods are particularly appropriate for large, structured defence-sector organisations, as they enable efficient, confidential data collection from personnel across multiple departments and functional roles. The use of surveys is further justified by their ability to capture insights from a broad cross-section of respondents, including project managers, engineers, administrative staff and senior leaders, within a relatively short timeframe. Additionally, the survey approach allows respondents to complete the instrument privately, which is especially important in security-sensitive environments. This helps ensure data integrity, reduce response bias and maintain confidentiality while enabling robust quantitative analysis (PMI, 2021).

2.2.2 Population and sampling procedures

The study population includes personnel directly or indirectly involved in project management, encompassing project officers, department heads, technical staff and administrative support personnel. This population was chosen because these individuals possess firsthand experience in planning, executing and monitoring defence-related projects. To ensure representation across different hierarchical levels and departments, stratified random sampling was adopted. The sample was stratified by rank, department and project role, ensuring proportional representation of perspectives from leadership, management and operational levels (Sekaran and Bougie, 2016). Within each stratum, random selection ensures equal opportunity to participate while reducing bias.

2.2.3 Data collection method

Primary data were collected through a structured electronic questionnaire distributed through the secure intranet or official email channels. The questionnaire consists mainly of closed-ended Likert-scale items ranging from 1 (strongly disagree) to 5 (strongly agree) and was designed to measure attitudes and perceptions regarding technology adoption, leadership support, project complexity, resource adequacy and reporting efficiency.

2.2.4 Operationalisation and measurement

Each construct in the conceptual framework was operationalised using validated scales drawn from established theoretical models. Technology adoption was measured using TAM, which assesses perceived usefulness and ease of use (Davis, 1989). Leadership support and training, grounded in contingency theory, were evaluated through measures of managerial encouragement and training adequacy. Project complexity and risk were conceptualised using the project lifecycle and CSF frameworks, capturing the degree of departmental involvement and risk exposure. Resource availability, informed by RBV, was measured through indicators of budget sufficiency, technical infrastructure and workforce competency. The dependent variable, PMRE, was assessed using items that reflect the timeliness, accuracy and overall quality of project reports. Each item was rated on a five-point Likert scale, ensuring consistency and comparability across constructs.

Data analysis was performed using the statistical package for the social sciences (SPSS). Before hypothesis testing, the data underwent reliability and validity assessments, including Cronbach's alpha for internal consistency and exploratory factor analysis, to verify construct validity. Given the ordinal nature of the data, Spearman's rank correlation was employed to test the strength and direction of associations between each independent variable and project monitoring and reporting efficiency. The four key hypotheses test the relationships predicted by TAM, contingency theory, project lifecycle and RBV. When necessary, multiple regression analysis was conducted to examine the relative influence of each independent variable on PMRE, provided that the statistical assumptions were met.

2.3. Results and analysis

2.3.1 Response rate

Out of 175 distributed questionnaires, all were fully completed and returned, yielding a 100% usable response rate. Although such a rate is unusually high in general research contexts, it is considered plausible within military organisations, where command endorsement and institutional hierarchy can significantly enhance participation.

2.3.2 Demographic profile

The demographic profile of respondents provides essential context for interpreting the findings. The sample comprised a balanced mix of officers and technical staff, representing a broad cross-section of departments including procurement, logistics, engineering, finance and administration. Most respondents had more than 6 years of project-related experience, and a substantial proportion held mid- to senior-level positions with decision-making authority. This diversity enhances the validity of insights by reflecting perspectives from both the strategic and operational levels of the project management structure. Table 2.1 summarises the respondents' demographic characteristics and professional backgrounds.

2.3.3 Reliability analysis

The internal consistency of all constructs is presented in Table 2.2. The reliability analysis confirmed strong internal consistency across all constructs, with Cronbach's alpha (α) values exceeding the recommended 0.70 threshold, indicating that the measurement items

Table 2.1 Demographic profile

Dimension	Category	Frequency	Percentage
Rank/position	Junior officers/staff (Captain and below)	85	48.6%
	Mid-level managers (Major to Colonel)	69	39.4%
	Senior officers and civilians (specialised roles)	21	12.0%
Departmental affiliation	Operations	103	58.9%
	IT/engineering/communication	31	17.7%
	Logistics, planning, training	41	23.4%
Project management experience	6 years	37	21.1%
	6–10 years	85	48.6%
	> 10 years	53	30.3%

Table 2.2 Reliability assessment (continued on next page)

Construct	Number of items	Cronbach's alpha, α	Acceptable range	Interpretation
Technology adoption (TA)	6	~ 0.867	$\alpha \geq 0.70$	Strong internal consistency; items on ease of use, usefulness and system integration cohere well
Leadership support and training (LST)	6	~ 0.862	$\alpha \geq 0.70$	Leadership endorsement and training-related items cluster effectively around a unified construct
Project complexity and risk (PCR)	6	~ 0.896	$\alpha \geq 0.70$	Indicates respondents share a consistent perception of complexity, risk factors and mitigation

Table 2.2 Continued

Construct	Number of items	Cronbach's alpha, α	Acceptable range	Interpretation
Resource availability (RA)	5	~ 0.794	$\alpha \geq 0.70$	Slightly lower but still acceptable; departmental variations in budgets and workforce may contribute
Project monitoring and reporting efficiency (PMRE)	6	~ 0.903	$\alpha \geq 0.70$	Highest alpha, signifying a shared, well-defined concept of efficient reporting among participants

were statistically sound and dependable. The project monitoring and reporting efficiency (PMRE) construct achieved the highest reliability ($\alpha = 0.903$), indicating a shared understanding among respondents about the importance of timely, accurate and detailed reporting. Technology adoption ($\alpha = 0.867$) also showed strong reliability, reflecting coherent perceptions of usefulness, ease of use and system integration. However, slight variability may stem from differing levels of experience with project management software. Similarly, leadership support and training ($\alpha = 0.862$) showed that leadership endorsement and structured training are tightly linked in command-driven settings. Project complexity and risk ($\alpha = 0.896$) also produced a high reliability score, suggesting a consistent view that complexity encompasses multi-stakeholder coordination, technical sophistication and risk intensity. Lastly, resource availability ($\alpha = 0.794$) was slightly lower but still met acceptable reliability standards, reflecting minor variations across departments in budgeting and staffing exposure. Overall, these results confirm that all constructs were measured robustly and were suitable for subsequent statistical analyses.

2.3.4 Normality analysis

Table 2.3 reports the results of the normality tests for all study constructs. The normality analysis using the Kolmogorov–Smirnov (KS) and Shapiro–Wilk (SW) tests indicated that all constructs – technology adoption, leadership support and training, project complexity and risk, resource availability, and project monitoring and reporting efficiency – had p-values below 0.05, suggesting minor deviations from normality. However, given the large sample size ($n = 175$), these deviations are not practically meaningful, as normality tests become increasingly sensitive with larger datasets, often flagging minor departures that do not affect parametric analyses. The skewness and kurtosis values, ranging from approximately 0.0 to 0.3, indicate that the data are only slightly skewed and well within acceptable limits for standard distributional assumptions. In line with the central limit theorem, the sample size is sufficiently large to justify the use of parametric statistical methods, such as Pearson's correlation and linear regression, for hypothesis testing. Non-parametric methods, such as Spearman's rho (ρ), were also used to assess consistency, confirming that despite minor distributional irregularities, the dataset is robust and suitable for advanced statistical analyses.

Table 2.3 Normality tests

Construct	KS, *p*	SW, *p*	Kurtosis	Interpretation
Technology adoption (TA)	< 0.05	< 0.05	~ 0.12	Minor deviation from normal but not extreme; large *n* (175) justifies parametric use
Leadership support and training (LST)	< 0.05	< 0.05	~ 0.15	Slightly skewed, but Q-Q plots show near-normal shape; parametric methods are robust at this sample size
Project complexity and risk (PCR)	< 0.05	< 0.05	~ 0.20	Some mild skewness consistent with complexity distribution; still suitable for parametric analysis
Resource availability (RA)	< 0.05	< 0.05	~ 0.07	Noticeable but not severe skew; large sample size reduces impact on correlation and regression
Project monitoring and reporting efficiency (PMRE)	< 0.05	< 0.05	~ 0.08	Near-normal with minor skewness; remains appropriate for parametric tests

2.3.5 Descriptive analysis

The descriptive analysis revealed a generally positive perception of the use of project management tools across all constructs, although with notable areas for improvement. Table 2.4 presents the descriptive statistics, including mean and standard deviation (SD), for all survey items. Technology adoption (mean ≈ 3.7) scored favourably, particularly in enhancing monitoring and decision-making (TA3, TA4), although moderate scores for ease of learning (TA2) and tool complexity (TA6) suggest a learning curve among users. Leadership support and training (mean ≈ 3.5) reflected strong endorsement from senior leaders but inconsistent training opportunities, implying a gap between advocacy and structured implementation. Project complexity and risk (mean ≈ 3.7) indicated that respondents acknowledged the challenges posed by high security requirements and multi-stakeholder coordination, yet recognised that project tools effectively mitigate risks (PCR5). Resource availability (mean ≈ 3.5) emerged as a limiting factor, with respondents highlighting inadequate budgets, IT support and personnel, although many agreed that additional investments could significantly improve efficiency (RA4, RA5). Finally, project monitoring and reporting efficiency (mean ≈ 3.8) demonstrated overall satisfaction with timeliness, accuracy and transparency, emphasising the importance of real-time metrics (PMRE5) for better decision-making. Collectively, the findings suggest that while the organisation's personnel generally value digital project management tools, success depends on improved training, greater resource allocation and continued leadership commitment to realise their potential fully.

Table 2.4 Survey items, mean and standard deviation for all constructs (n = 175) (continued on next page)

Survey item			
Item code	Technology adoption (TA)	Mean	SD
TA1	The project management tools are easy to navigate	3.70	0.82
TA2	Learning to use the project management tools does not require much effort	3.48	0.89
TA3	Using project management tools enhances my ability to monitor and report on project progress	3.98	0.86
TA4	Project management tools help improve decision-making quality in my projects	3.93	0.84
TA5	The project management tools integrate well with our existing workflows	3.78	0.82
TA6 (R)	The complexity of these tools makes them difficult to implement effectively	3.27	0.96
	(Cronbach's alpha for TA = 0.867)		
Leadership support and training (LST)			
Item code	Survey item	Mean	SD
LST1	Senior leadership in my unit actively endorses the use of project management tools	3.74	0.91
LST2	Leadership provides sufficient guidance and encouragement to adopt these tools	3.64	0.90
LST3	I have received adequate training to use project management tools effectively	3.20	1.01
LST4	Training sessions are relevant and help me perform my tasks better	3.73	0.97
LST5	There is a well-structured, ongoing training programme on project management tools	3.13	1.14
LST6 (R)	A lack of training prevents me from using these tools effectively in my work	3.74	0.92
	(Cronbach's alpha for LST = 0.862)		
Project complexity and risk (PCR)			
Item code	Survey item	Mean	SD
PCR1	The projects I work on involve multiple stakeholders, making coordination more challenging	3.42	1.02
PCR2	High security requirements affect how easily I can monitor and report on project updates	3.65	0.87

Table 2.4 Continued

PCR3	The complexity of our projects makes real-time reporting more difficult	3.70	0.87
PCR4	Unforeseen risks frequently affect project timelines and data accuracy	3.79	0.87
PCR5	Our project management tools help mitigate risks and improve overall efficiency	3.85	0.85
PCR6 (R)	Security concerns limit the full implementation of digital project management solutions	3.64	0.80
	(Cronbach's alpha for PCR = 0.896)		
	Resource availability (RA)		
Item code	Survey item	Mean	SD
RA1	My department has sufficient budget to invest in project management tools	3.13	1.03
RA2	Adequate IT support is available to ensure these tools operate smoothly	3.35	1.10
RA3	We have enough trained personnel to fully utilise modern project management software	3.02	1.08
RA4	Additional investments in these tools would significantly improve reporting efficiency	3.98	0.88
RA5 (R)	A lack of resources is a major barrier to improved project monitoring and reporting in my unit	3.95	0.88
	(Cronbach's alpha for RA = 0.794)		
	Project monitoring and reporting efficiency (PMRE)		
Item code	Survey item	Mean	SD
PMRE1	Project reports are generated on time for decision-making	3.81	0.81
PMRE2	The reports produced are detailed and accurate	3.76	0.76
PMRE3	These tools have improved the transparency and accountability of project reporting	3.83	0.81
PMRE4	I am satisfied with the reporting process in my department	3.61	0.91
PMRE5	Real-time metrics allow for faster, more effective risk management and decision-making	3.95	0.76
PMRE6	Using digital project tools has reduced errors in my reports	3.81	0.84
	(Cronbach's alpha for PMRE = 0.903)		

Note: R, reverse-scored item.

Table 2.5 Spearman's rank correlation matrix ($n = 175$)

Variable	PMRE	TA	LST	PCR	RA
PMRE	1	0.670	0.676	0.615	0.716
TA	0.670	1	0.549	0.619	0.597
LST	0.676	0.549	1	0.403	0.714
PCR	0.615	0.619	0.403	1	0.441
RA	0.716	0.597	0.714	0.441	1

Note: all correlations are significant at $p < 0.01$ level (two-tailed).

2.3.6 Spearman's rank correlation

The Spearman's rank correlation results between all study variables are shown in Table 2.5. The Spearman's rank correlation analysis revealed strong and statistically significant positive associations between all independent variables and PMRE at $p < 0.01$. The highest correlation was observed between resource availability ($\rho = 0.716$) and PMRE, followed closely by leadership support and training ($\rho = 0.676$) and technology adoption ($\rho = 0.670$), indicating that sufficient resources, strong leadership backing and effective technological integration are the most influential factors in enhancing monitoring and reporting efficiency. Project complexity and risk ($\rho = 0.615$) also showed a meaningful positive relationship, suggesting that as project complexity increases, structured supervision and risk management become more crucial for maintaining efficiency. Furthermore, the positive inter-construct correlations between all variables (e.g. between leadership support, technology adoption and resource availability) demonstrate that these factors work synergistically, reinforcing one another to improve overall project monitoring and reporting performance.

2.3.7 Regression analysis

Table 2.6 reports the standardised beta (β) coefficients from the regression analysis, indicating the relative influence of each predictor. The multiple regression analysis demonstrated a strong predictive relationship between the independent variables, technology adoption (TA), leadership support and training (LST), project complexity and risk (PCR) and resource availability (RA) and PMRE, with the model explaining approximately 72% of the variance

Table 2.6 Standardised beta coefficients

Predictor	Standardised beta, β	Sig., p
Technology adoption (TA)	0.255	0.000
Leadership support and training (LST)	0.292	0.000
Project complexity and risk (PCR)	0.228	0.000
Resource availability (RA)	0.229	0.001

($R^2 = 0.717, p < 0.01$). All predictors were statistically significant, confirming their collective impact on project monitoring performance. Among them, leadership support and training ($\beta = 0.292$) emerged as the strongest predictor, underscoring the importance of consistent managerial endorsement and structured training for improving reporting efficiency. Technology adoption ($\beta = 0.255$) and resource availability ($\beta = 0.229$) also contributed substantially, supporting the TAM and RBV frameworks, which highlight the importance of user acceptance and adequate resources. Meanwhile, project complexity and risk ($\beta = 0.228$) showed that as project intricacy increases, formal monitoring systems become more critical for maintaining control. Overall, these findings confirm that technology, leadership, risk management and resources function interdependently, aligning with contingency theory and PMBOK principles to drive project efficiency.

2.3.8 Hypothesis testing

Table 2.7 summarises the hypothesis-testing results and provides statistical support for each proposed relationship. Hypothesis testing confirmed that all four hypotheses were strongly supported, validating the study's conceptual framework. Technology adoption (H1) showed a significant positive relationship with PMRE ($\rho = 0.670$; $\beta = 0.255, p < 0.001$), indicating that effective use of digital tools enhances accuracy and coordination. Leadership support and training (H2) was the strongest predictor ($\rho = 0.676$; $\beta = 0.292, p < 0.001$), underscoring the importance of managerial endorsement and structured training for successful technology adoption. Project complexity and risk (H3) also exhibited a positive association ($\rho = 0.615$; $\beta = 0.228, p < 0.001$), suggesting that greater complexity drives the need for stronger monitoring and risk control mechanisms. Resource availability (H4) showed the strongest correlation ($\rho = 0.716$; $\beta = 0.229, p < 0.001$), indicating that adequate financial, technical and human resources are essential for efficient project reporting. Overall, these findings confirm that the use of technology, leadership, resource adequacy and complexity management significantly

Table 2.7 Summary of hypothesis testing ($n = 175$)

Hypothesis	Statement	Key statistical evidence	Result
H1	Technology adoption (TA) is positively associated with PMRE	Correlation with PMRE: $\rho = 0.670$ ($p < 0.01$) regression ($\beta = 0.255$, $p < 0.001$)	Supported
H2	Leadership support and training (LST) is positively associated with PMRE	Correlation with PMRE: $\rho = 0.676$ ($p < 0.01$) regression ($\beta = 0.292$, $p < 0.001$)	Supported
H3	Project complexity and risk (PCR) is positively associated with PMRE	Correlation with PMRE: $\rho = 0.615$ ($p < 0.01$) regression ($\beta = 0.228$, $p < 0.001$)	Supported
H4	Resource availability (RA) is positively associated with PMRE	Correlation with PMRE: $\rho = 0.716$ ($p < 0.01$) regression ($\beta = 0.229$, $p < 0.001$)	Supported

Figure 2.2 Technology adoption

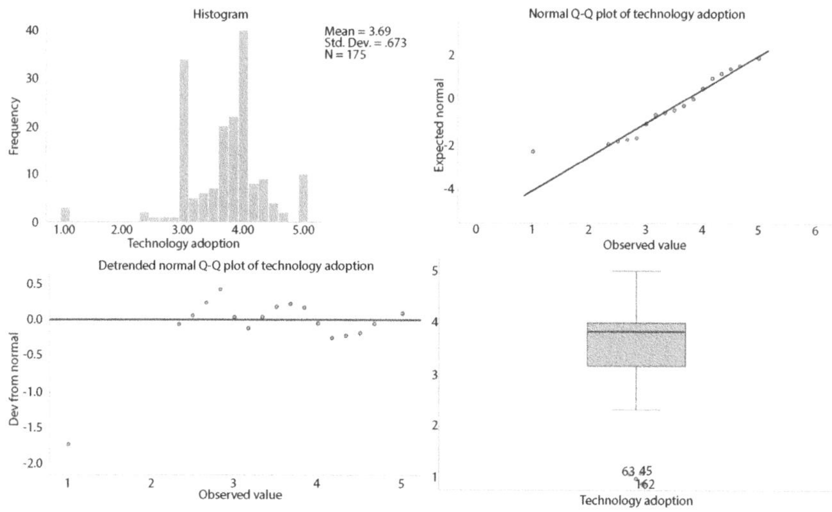

strengthens monitoring and reporting efficiency in defence projects. Figure 2.2 illustrates respondents' perceptions of technology adoption and its relationship to project monitoring and reporting efficiency.

As shown in Figure 2.3, leadership support and training are critical to effective use of project management tools. Building on this, Figure 2.4 illustrates how project complexity and risk influence the efficiency of monitoring and reporting processes. In addition, Figure 2.5 presents respondents' perceptions of resource availability and its effect on project monitoring and reporting capabilities. Collectively, these factors contribute to overall system performance, as summarised in Figure 2.6, which provides an overview of respondents' assessments of project monitoring and reporting efficiency.

2.4. Practical insights

The empirical investigation provides several important insights for practitioners operating within defence and security-oriented project environments, particularly those characterised by hierarchical command structures, high operational risk and strict security requirements. The findings demonstrate that effective project monitoring and reporting in defence projects does not depend on a single intervention, but instead on the coordinated integration of technology, leadership, risk management and resource capability.

First, the results show that digital project management tools significantly enhance monitoring and reporting efficiency when they are perceived as valuable and compatible with operational workflows. For defence practitioners, this highlights the importance of selecting tools that support real-time visibility, structured reporting and decision traceability while complying

Figure 2.3 Leadership support and training

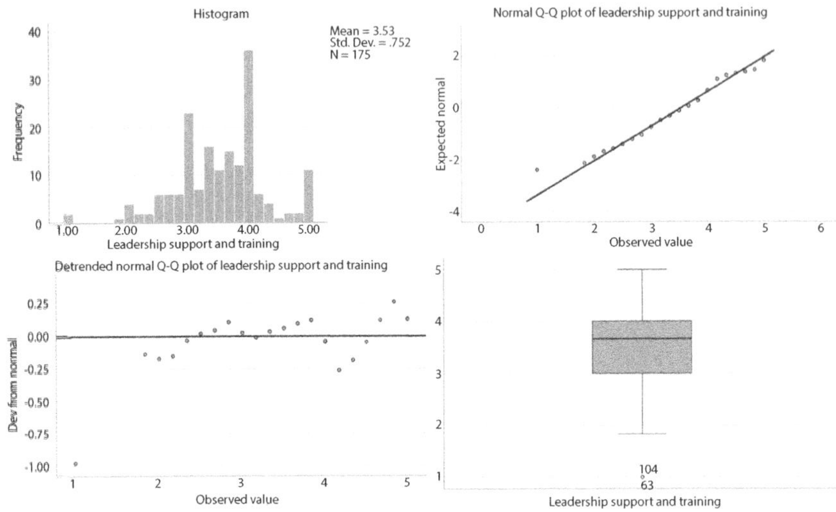

Figure 2.4 Project complexity and risk

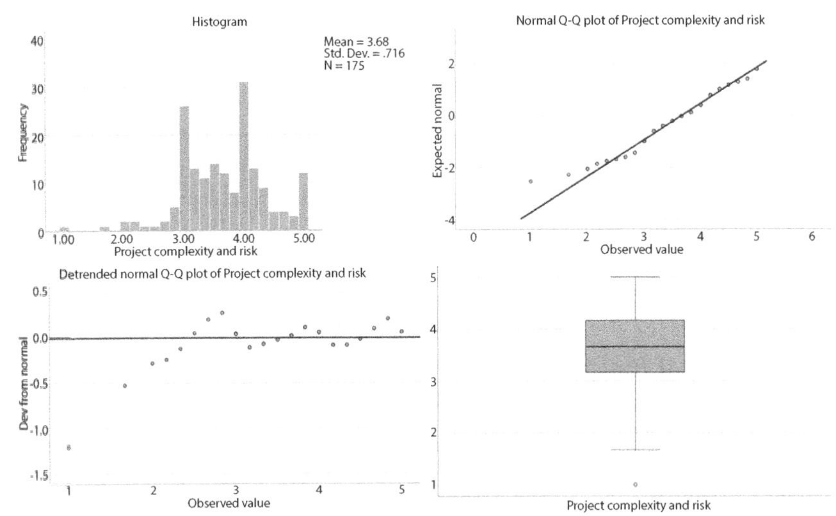

with security protocols. The positive relationship between technology adoption and project monitoring efficiency indicates that manual systems, spreadsheets and fragmented reporting channels are increasingly inadequate for managing complex defence projects. However, the findings also suggest that tool complexity and learning effort can limit adoption. Practitioners should therefore prioritise systems with intuitive interfaces, role-based access controls and precise alignment with defence operating procedures.

Figure 2.5 Resource availability

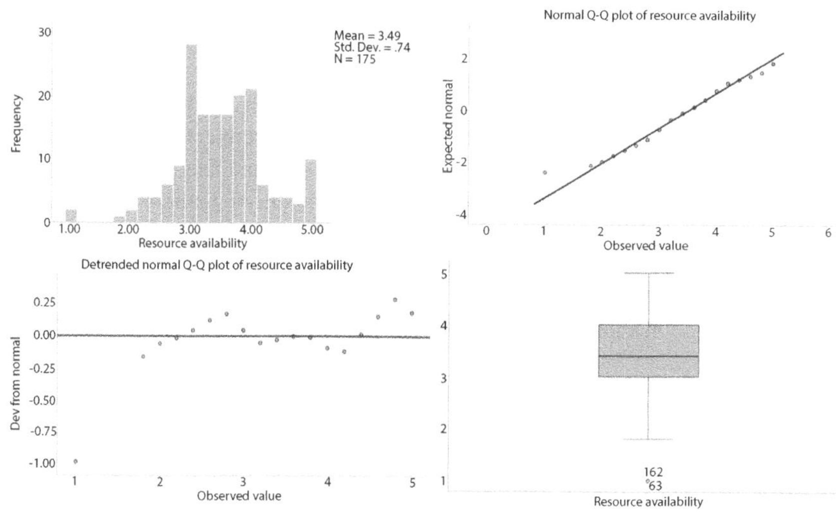

Figure 2.6 Project monitoring and reporting efficiency

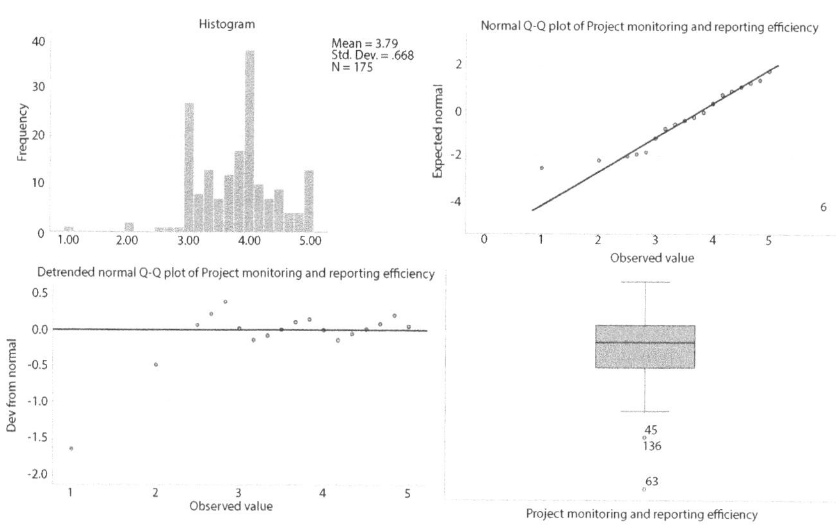

Second, the evidence clearly demonstrates that leadership support and structured training are the strongest drivers of successful tool adoption within the defence context. In command-driven organisations, leadership endorsement plays a decisive role in shaping behaviour. When senior officers actively reference digital dashboards, require system-generated reports and model technology usage, adoption becomes institutionalised rather than optional.

Practitioners should recognise that issuing directives alone is insufficient. Structured, continuous training programmes are essential for building confidence, competence and consistency across ranks and departments. Training should be operationally relevant, modular and scheduled to minimise disruption to mission readiness.

Third, the study highlights that project complexity and risk intensify the need for structured monitoring systems rather than weaken their relevance. Defence projects often involve multiple stakeholders, advanced technologies, classified information and long procurement cycles. The positive association between project complexity and monitoring efficiency suggests that as complexity increases, the value of formal project management tools becomes more pronounced. Practitioners should therefore avoid scaling back monitoring practices in complex or sensitive projects. Instead, they should strengthen governance routines, risk registers and reporting discipline to maintain control and situational awareness.

Fourth, the findings underscore that resource availability is a critical enabler of effective project monitoring and reporting. Adequate financial allocations, IT infrastructure and skilled personnel are essential for sustaining digital project management systems. Defence organisations frequently operate under budgetary constraints and competing priorities. The evidence indicates that underinvestment in resources undermines even well-designed systems and leadership initiatives. Practitioners should advocate for dedicated funding lines for project management tools, training and system maintenance, framing these investments as contributors to accountability, auditability and operational readiness rather than administrative overhead.

Fifth, the results emphasise the synergies between technology, leadership, risk management and resources. No single factor independently guarantees improved project performance. Instead, effectiveness emerges when these elements are aligned. For defence practitioners, this implies that piecemeal reforms are unlikely to succeed. Implementing digital tools without leadership commitment or mandating their use without adequate resources creates resistance and inefficiency. A balanced and coordinated approach is required to institutionalise best practices in project monitoring and reporting.

From a governance perspective, the evidence suggests that digital project management systems strengthen transparency, accountability and decision traceability, which are increasingly crucial in defence procurement and public accountability contexts. Practitioners should leverage these systems to improve oversight, support audits and enhance confidence in defence spending decisions. At the same time, security considerations must remain paramount. Systems should incorporate encryption, access controls and compliance with defence cybersecurity standards to maintain trust and operational integrity.

Overall, the practical implications indicate that defence organisations seeking to improve project performance should view digital project management not as a technical upgrade, but as an organisational transformation initiative. By aligning leadership behaviour, training frameworks, resource planning and risk governance with appropriate digital tools, defence practitioners can significantly enhance monitoring and reporting efficiency, strengthen governance and improve mission outcomes in complex and high-stakes project environments.

2.5. Conclusion and recommendations

The study revealed that PMRE is influenced by four interrelated factors: technology adoption; leadership support and training; project complexity and risk; and resource availability. Technology adoption showed a strong positive relationship with PMRE ($\rho = 0.670$; $\beta = 0.255$), indicating that modern digital platforms improve the efficiency of real-time monitoring, communication and decision-making. However, challenges remain regarding user familiarity and technical complexity. According to TAM, perceived usefulness and ease of use are vital to broader acceptance, underscoring the need for ongoing training.

Leadership support and training emerged as the strongest predictor ($\beta = 0.292$), emphasising that leadership endorsement and structured training programmes drive organisational compliance and tool utilisation. Anchored in contingency theory, effective leadership should adapt to the situational realities by balancing command authority with participative decision-making to build user confidence.

Project complexity and risk, with a positive association ($\beta = 0.228$), indicated that as defence projects grow in scale and sensitivity, more stringent oversight and structured risk management become necessary. The PMBOK framework reinforces this by emphasising integrated monitoring systems, risk registers and periodic reviews.

Resource availability showed the strongest correlation ($\rho = 0.716$; $\beta = 0.229$), indicating that financial, human and technical resources are crucial for successful implementation. RBV theory aligns with this finding, demonstrating that internal capabilities, both tangible (infrastructure, funding) and intangible (skills, culture), collectively determine performance outcomes.

These findings affirm that no single factor guarantees project success; rather, integrating technology, leadership, risk management and resource optimisation enables effective monitoring and reporting across the organisation.

2.5.1 Implications and strategic considerations

The study's implications are twofold: leadership-driven transformation and the institutionalisation of digital project management practices. Command-level endorsement is essential for embedding project management tools within standard operating procedures. When leaders model technology use by referencing dashboards or requiring digital updates, they normalise adoption and accelerate behavioural change. Furthermore, a cultural shift toward data-driven decision-making should be cultivated. Senior officers can demonstrate commitment by participating in training and rewarding accurate, timely and system-generated reporting. Leadership development programmes must also include exposure to analytics, risk management and integrated resource planning, aligning technical literacy with strategic leadership capability.

2.5.2 Training, technology and resource management

Training programmes must be modular and progressive, allowing staff to transition from basic data entry to advanced project analysis. The train-the-trainer model can identify 'power users' who mentor their units, promoting sustainability and reducing reliance on external trainers. Regular refresher courses ensure adaptation to new software features, preventing skill stagnation.

Technology integration requires user-friendly interfaces and interoperability across systems. Simple, visual dashboards improve usability, reinforcing TAM's ease of use principle. Integration with existing databases (finance, logistics, maintenance) ensures real-time synchronisation, while adherence to security protocols (encryption, multi-factor authentication, role-based access) safeguards sensitive information and fosters trust.

Resource management emerged as a critical enabler. The organisation should conduct comprehensive resource audits to identify gaps in budget, infrastructure and workforce. Allocating funds specifically for software, training and IT infrastructure will sustain performance improvements. Human resource strategies should promote cross-functional rotations to allow officers to gain both operational and analytical expertise. Moreover, consistent infrastructure upgrades, such as high-speed connectivity and reliable hardware, are vital to ensuring the full accessibility and utility of project management tools.

2.5.3 Theoretical contributions

This study contributes to academic discourse by extending established theories to the defence context. PMBOK integration confirms that communication, resource and risk management must function in harmony to achieve project success ($R^2 = 0.717$). TAM remains applicable even in security-oriented environments, highlighting that ease of use and perceived usefulness continue to drive technology adoption when leadership and training are strong. RBV emphasises that organisational performance stems from leveraging internal assets – financial, technological and human – while contingency theory underscores that adaptive leadership is essential for aligning project goals with contextual realities. Collectively, these frameworks explain how the institutional capacity can evolve into a more agile, data-informed and accountable system of project governance.

2.5.4 Limitations

The single-service focus on the organisation limits generalisability across other military branches or civilian agencies. The self-reported nature of survey data introduces potential bias, and the cross-sectional design captures perceptions at only one point in time. Additionally, unmeasured variables such as organisational culture, motivation and bureaucracy may influence the results. Future studies could integrate qualitative methods, such as interviews or focus groups, to capture more profound insights into the behavioural and cultural dynamics that influence technology adoption.

2.5.5 Recommendations and future research

- *Leadership and accountability*: incorporate project management reporting performance into promotion criteria and appoint 'monitoring champions' responsible for tool usage compliance and standard operating procedure (SOP) updates.
- *Revitalised training*: establish phased training aligned with real projects and partner with universities or external experts to integrate advanced analytics, predictive modelling and cybersecurity into the curriculum.
- *Holistic resource planning*: develop a dedicated project management support fund for software licensing, training and hardware upgrades. Align resource audits with budgetary cycles to ensure sustainable funding.

- *Progressive technology deployment*: pilot new project management tools on small-scale projects before expanding to critical missions. Incremental adoption encourages learning, feedback and a smoother transition.
- *Cross-functional collaboration*: form inter-departmental task forces to capture and share best practices, fostering a culture of knowledge exchange and reducing siloed inefficiencies.

For future research, longitudinal studies could examine how technological and leadership interventions evolve and how they sustain their impact on the effectiveness of project monitoring and reporting. Comparative analyses across different branches or units within the defence sector could further reveal the broader applicability of the findings across diverse operational contexts. Advanced analytical techniques, such as structural equation modelling (SEM) and partial least squares (PLS), may be employed to test mediating and moderating relationships between key variables. In addition, qualitative research methods could provide deeper insights into personnel's lived experiences of adapting to digital transformation in security-sensitive environments. Future studies may also explore the adoption of emerging technologies, including artificial intelligence, blockchain and the Internet of Things (IoT), to enhance predictive project oversight, data security and decision-making capabilities in defence and other high-reliability organisations.

REFERENCES

Barney J (1991) Firm resources and sustained competitive advantage. *Journal of Management* **17(1)**: 99–120.

Davis FD (1989) Perceived usefulness, perceived ease of use and user acceptance of information technology. *MIS Quarterly* **13(3)**: 319–340.

Fowler FJ (2014) *Survey Research Methods*, 5th edn. Sage, Thousand Oaks, CA, USA.

Kerzner H (2013) *Project Management: A Systems Approach to Planning, Scheduling, and Controlling*, 11th edn. Wiley, Hoboken, NJ, USA.

Lawrence PR and Lorsch JW (1967) *Organization and Environment: Managing Differentiation and Integration*. Harvard University Press, Boston, MA, USA.

National Audit Office (2018) *Improving Government's Planning and Spending Framework*. National Audit Office, London, UK. https://www.nao.org.uk (accessed 08/10/2025).

National Audit Office (2021) *Improving Operational Delivery in Government*. National Audit Office, London, UK. https://www.nao.org.uk (accessed 08/10/2025).

PMI (Project Management Institute) (2021) *A Guide to the Project Management Body of Knowledge (PMBOK® Guide)*, 7th edn. Project Management Institute, Newtown Square, PA, USA.

Rockart JF (1979) Chief executives define their own data needs. *Harvard Business Review* **57(2)**: 81–93.

Schwalbe K (2015) *Information Technology Project Management*, 8th edn. Cengage Learning, Boston, MA, USA.

Sekaran U and Bougie R (2016) *Research Methods for Business: A Skill-Building Approach*, 7th edn. Wiley, Chichester, UK.

Advancing Project Management: Frameworks, Digital Transformation, Sustainability and Governance

emerald PUBLISHING ice Publishing

Mohammed Reyasudin Basir Khan
ISBN 978-1-83742-799-4
https://doi.org/10.1108/978-1-83742-798-720261004

Chapter 3
Quality management systems in energy projects: the role of standardisation, stakeholders and resources

Nor Aina Mazlan and Mohammed Reyasudin Basir Khan

Tun Razak Graduate School, Universiti Tun Abdul Razak (UNIRAZAK), Kuala Lumpur, Malaysia

3.1. Introduction

The Malaysian energy sector plays a crucial role in driving the nation's economic growth, contributing substantially to national income, employment generation and long-term development. As the sector continues to evolve in response to rising demands for efficiency, sustainability and regulatory compliance, organisations face increasing pressure to uphold high quality standards in project delivery. In this context, implementing an effective quality management system (QMS) has become a strategic priority, especially given the complexity and dynamic nature of energy-related projects. Quality management ensures that projects meet stakeholder expectations, comply with regulatory requirements and optimise resource use. However, despite the recognised importance of QMS, many organisations in Malaysia continue to face challenges in integrating quality practices into daily project operations. These shortcomings often result in inconsistent outcomes and reduced stakeholder satisfaction. Structured planning, monitoring and control mechanisms form the foundation of modern project management systems (Kerzner, 2017). In addition, digital transformation is increasingly reshaping how organisations design, monitor and govern project processes, influencing communication flows, data transparency and performance control mechanisms across industrial sectors (Vial, 2021).

A growing body of literature highlights the importance of project management practices, particularly process standardisation, stakeholder management, training, time management and resource allocation, in enhancing QMS outcomes. Process standardisation promotes uniform procedures that reduce variability and improve overall efficiency, while effective stakeholder management fosters communication and collaboration, which are essential for meeting diverse expectations. Training strengthens project personnel's skills and competencies, providing the knowledge base needed to implement quality practices effectively.

Time management ensures timely project completion and minimises errors that could compromise quality. Similarly, adequate resource allocation guarantees that financial, human and physical resources are optimally utilised to support project requirements. Although these factors have been acknowledged in prior studies, their specific influence on QMS effectiveness within the Malaysian energy sector remains underexplored. Existing research often treats these factors independently, overlooking their interrelationships and combined impact on quality performance.

This chapter addresses these gaps by examining how the identified project management practices collectively contribute to the effectiveness of QMS in Malaysian energy projects. The problem arises mainly because organisations frequently depend on fragmented or generalised quality strategies that do not adequately reflect the sector's operational realities. Without empirical evidence tailored to the energy context, organisations risk implementing ineffective or misaligned quality initiatives, leading to wasted resources, delays and diminished project outcomes. The research presented in this chapter generated data-driven insights to guide industry stakeholders in designing more robust, context-specific quality management frameworks.

The research also recognises that existing gaps in the literature extend beyond a lack of sector-specific focus. Earlier studies have not sufficiently examined how individual variables interact to influence QMS, nor have they assessed how factors such as educational background or competency levels alter perceptions of quality practices. Notably, emerging evidence suggests that training may not always have a significant impact on QMS, an unexpected finding that warrants deeper exploration into the relevance, quality and alignment of current training programmes with actual industry needs. Understanding these nuances is critical because the effectiveness of QMS depends not only on organisational procedures but also on human factors, team composition and resource environments.

Six research questions guided this investigation, focusing on whether process standardisation, stakeholder management, training, time management and resource allocation, individually and collectively, affect QMS performance. These questions support the main research objectives, which include assessing the impacts of each variable, understanding their interrelationships and offering practical recommendations based on empirical findings. The study employed a quantitative methodology, using structured questionnaires distributed to project management professionals in the Malaysian energy sector. This approach enabled systematic examination of relationships between variables and provided measurable insights into how project management practices contribute to quality performance.

The significance of this study is twofold. Academically, it fills a clear gap in the literature by offering sector-specific empirical evidence on the interplay between project management practices and QMS effectiveness. This area has been largely overlooked in Malaysian energy research. The findings will enrich scholarly discussions and provide a foundation for future research to deepen understanding of quality management in complex industrial environments. In practice, the study offers valuable implications for industry practitioners, project leaders and policy makers. Organisations can apply the findings to refine their project management strategies, improve training programmes, strengthen stakeholder engagement, optimise resource allocation and ultimately enhance the overall quality of project outcomes. As Malaysia continues to expand its energy infrastructure and align with global standards, the insights from this research will be essential to promoting efficiency, reliability and competitiveness in the sector.

The theoretical foundation of this study rests on several key frameworks that collectively explain how project management practices influence the effectiveness of QMS in the Malaysian energy sector. Central among these is ISO 9001 (ISO, 2015), which provides a structured approach for designing, implementing and maintaining a QMS. Built on seven quality management principles – customer focus, leadership, people engagement, process approach, improvement, evidence-based decision-making and relationship management – the standard emphasises systematic, data-driven and stakeholder-oriented practices. In the energy sector, where projects are complex, capital-intensive and highly regulated, these principles are especially relevant.

Customer focus and relationship management directly relate to stakeholder management in this study. Energy projects involve multiple stakeholders, including regulators, customers, suppliers and local communities, whose expectations concerning safety, reliability and environmental performance must be consistently met. Effective stakeholder engagement, as reflected in ISO 9001, helps align project activities with these expectations and contributes to higher quality outcomes. Empirical evidence further supports the importance of stakeholder salience and proactive engagement in complex project environments (Aaltonen, 2010). Leadership and employee engagement, meanwhile, underscore the role of management and employee involvement in creating a quality-oriented culture. They strongly connect to training, highlighting the need to equip project teams with the relevant skills and competencies to implement standardised processes and uphold QMS requirements. The principles of process approach and improvement reinforce the importance of process standardisation and continuous refinement of procedures to reduce variability, prevent errors and enhance efficiency. Finally, evidence-based decision-making and effective resource use are closely linked to time management and resource allocation, as organisations rely on performance data to optimise schedules, allocate resources wisely and sustain QMS performance across projects.

Complementing ISO 9001, total quality management (TQM) offers a broader philosophical underpinning that views quality as a responsibility shared by everyone in the organisation. TQM emphasises customer satisfaction, continuous improvement, employee involvement and systematic problem-solving. In the Malaysian energy sector, TQM principles underscore that formal procedures alone are insufficient; there must be a pervasive culture that supports quality throughout project delivery. Process standardisation under TQM provides the foundation for consistent performance, while continuous improvement encourages organisations to continually evaluate and refine these processes. Training is central to TQM, as employees must be continuously developed to handle new technologies, regulatory requirements and project complexities. Time management and resource allocation similarly flow from TQM's emphasis on efficiency and effectiveness: projects delivered late, over budget or with misallocated resources risk undermining quality and stakeholder trust, regardless of technical success.

Project management theory and the project management body of knowledge (PMBOK: PMI, 2021) further strengthen this foundation by providing a practical framework that covers the process groups – initiating, planning, executing, monitoring and controlling, and closing – and key knowledge areas. Contemporary empirical research confirms that alignment between project management methodologies and organisational context significantly enhances project success (Joslin and Müller, 2015). Within this view, process standardisation is embedded in structured planning and execution, where clear procedures, templates and checklists guide project activities. Stakeholder management is recognised as a distinct knowledge area that requires the systematic identification, analysis and engagement of stakeholders to ensure their expectations are incorporated into project decisions. PMBOK's focus on human resource and competency development aligns with the training dimension of this study, emphasising that capable, well-trained teams are a precondition for successful project performance and quality. Its specific knowledge areas on time management and resource management also provide conceptual grounding for understanding how scheduling, sequencing, budgeting and allocation decisions impact project quality and, by extension, QMS effectiveness.

Stakeholder theory adds a relational and ethical dimension to these predominantly process-oriented frameworks. It posits that organisations should consider and balance the interests of all parties affected by their activities, not just shareholders. This foundational perspective was originally articulated in Freeman's stakeholder framework (Freeman, 1984). In the energy sector, where projects often attract public scrutiny and regulatory oversight, the theory highlights the importance of recognising stakeholders' power, legitimacy and urgency. Effective stakeholder management through communication, consultation and collaboration can enhance trust, reduce conflict and support smoother project implementation. Recent studies further demonstrate that structured stakeholder engagement practices significantly improve transparency, trust-building and long-term project performance outcomes (Mok *et al.*, 2015).

Visual mapping and structured identification of stakeholder influence further improve transparency and the effectiveness of engagement (Bourne and Walker, 2006). This, in turn, contributes positively to QMS performance by minimising rework, delays and resistance that may arise from neglected stakeholder concerns. Stakeholder theory also intersects with training and resource allocation: project teams must be equipped with the skills to manage complex stakeholder relationships, and sufficient resources must be allocated to engagement activities to make these efforts effective.

On this theoretical foundation, empirical research plays a crucial role in validating and quantifying the relationships proposed. In project management and quality management literature, numerous studies have used surveys, case studies and statistical analyses to explore how process standardisation, stakeholder engagement, training, time management and resource utilisation influence project success and QMS outcomes. Empirical investigations, particularly those using quantitative methods such as regression and correlation analysis, provide evidence of which factors are most influential and how they interact in different industrial contexts. Prior research in construction and infrastructure contexts confirms that structured management practices significantly influence quality performance outcomes (Jha and Iyer, 2006). In the energy sector, such research is especially important because projects are high-risk and resource-intensive; decisions about process design, training investment and stakeholder strategies must therefore be grounded in reliable data rather than assumptions. This study aligns with that tradition by employing structured questionnaires and statistical techniques to examine how the identified independent variables affect QMS effectiveness among project management teams in the Malaysian energy industry.

Based on these theoretical perspectives and empirical traditions, a conceptual framework is proposed in which QMS effectiveness is the dependent variable, while process standardisation, stakeholder engagement (or stakeholder management), training, time management and resource allocation are the independent variables. The model assumes that each of these project management factors directly influences QMS performance. Process standardisation was expected to enhance quality by establishing consistent, documented procedures that reduce variability. Stakeholder engagement was anticipated to improve QMS outcomes by ensuring that quality objectives align with stakeholder expectations and that potential conflicts are proactively managed. Training was proposed to strengthen QMS implementation by improving the competencies, awareness and commitment of project personnel. Time management is critical because schedule adherence and timely execution can prevent quality deterioration from rushed work or extended delays. Resource allocation is essential to ensure teams have the financial, human and material resources needed to meet quality requirements. The conceptual relationships between the independent and dependent variables are illustrated in Figure 3.1.

The study further refines this framework by proposing six hypotheses, each of which can be empirically tested. The first hypothesis proposes a positive relationship between process standardisation and QMS effectiveness, suggesting that more standardised processes are associated with better quality outcomes. The second hypothesis argues that effective stakeholder management improves project quality and, in turn, enhances QMS, especially in a sector where regulatory and public expectations are high. The third hypothesis states that training

Figure 3.1 Conceptual framework

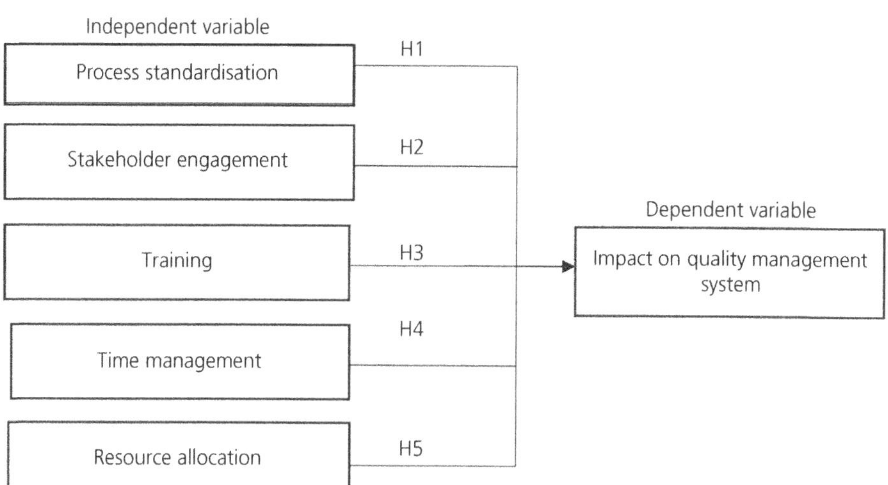

has a significant positive correlation with project team performance and, consequently, with QMS effectiveness, reflecting the view that a skilled workforce is fundamental to quality success. The fourth hypothesis links effective time management to higher quality deliverables, proposing that teams that plan, schedule and monitor activities efficiently are more likely to meet quality standards. The fifth hypothesis asserts that adequate resource allocation significantly improves QMS effectiveness by ensuring that projects are properly supported and executed under favourable conditions. The sixth and final hypothesis takes a more integrative view, positing that the combined effects of process standardisation, stakeholder management, training, time management and resource allocation will significantly enhance overall QMS performance. This recognises that these variables do not operate in isolation but interact to shape the quality environment within project-based organisations.

Together, these theoretical, empirical and conceptual elements provide a coherent foundation for the study. They justify the selection of variables, guide the formulation of research questions and hypotheses, and provide a robust lens for interpreting findings on QMS effectiveness in the Malaysian energy sector.

3.2. Methodology

3.2.1 Research design

This study adopted a quantitative research design to examine how project management factors influence the effectiveness of QMS among project management teams in the Malaysian energy sector. Quantitative survey designs are particularly suitable for examining relationships between measurable constructs and testing theoretically derived hypotheses (Field, 2018). The design was grounded in the need to collect standardised, quantifiable data that can be statistically analysed to test the proposed hypotheses and validate the conceptual framework. The research instrument, a structured questionnaire, was developed specifically to measure five independent constructs: process standardisation, stakeholder management, training, time management and resource allocation, as well as the dependent construct: the impact on QMS.

The development of the questionnaire drew heavily from established literature to ensure that each construct was properly operationalised. Items measuring process standardisation were adapted from studies emphasising the role of standard operating procedures and formalised processes in enhancing quality. Stakeholder management items were based on prior work that highlighted stakeholder engagement, feedback integration and communication quality. Training items captured perceptions of the availability, relevance and effectiveness of training programmes in supporting project-related tasks. Time management was reflected in items related to clarity of timelines, adherence to deadlines and the use of planning or scheduling tools. Resource allocation was assessed through statements on the adequacy, fairness and monitoring of personnel, budget and material resources. All items were rated on a five-point Likert scale ranging from 'strongly disagree' to 'strongly agree', enabling the quantification of attitudes and perceptions and providing a consistent metric across all constructs.

The questionnaire was directly aligned with the conceptual framework and research objectives, ensuring that each section captured data relevant to specific variables and hypotheses. Data analysis was conducted using the statistical package for the social sciences (SPSS), using descriptive and inferential statistics, including frequency distribution, mean, correlation and regression analysis. The use of a structured instrument, supported by established scales and analytical tools, ensures that the research design is systematic, replicable and capable of yielding robust empirical evidence.

3.2.2 Population and sampling procedures

The study population comprised project management professionals working in the Malaysian energy sector, a critical industry encompassing oil and gas, renewable energy and utilities. This population includes individuals occupying roles such as project manager, coordinator, engineer and other team members directly involved in planning and executing projects. Their perspectives are essential for understanding how quality management practices are implemented in real project environments and how they affect project outcomes. Given the estimated population of approximately 1000 project management professionals across energy-related organisations, an appropriate sample size was determined using the Krejcie and Morgan approach (Jimoh *et al.*, 2025). For a population of this size, a sample of around 278 respondents is sufficient to achieve a 95% confidence level with a 5% margin of error. This sample size is considered statistically adequate for generalising findings to the broader population of project management professionals in the sector.

To ensure the sample reflected the population's diversity, a stratified random sampling technique was employed. The population was divided into relevant strata based on characteristics such as role (e.g. project manager or team member) and subsector (e.g. oil and gas or renewables). Respondents were then randomly selected from each stratum. This approach increases representativeness and enables more nuanced analysis of how quality management factors may vary across roles and subsectors. Stratified random sampling also enhances the reliability and generalisability of the results, supporting more credible conclusions about quality management practices in the Malaysian energy industry.

3.2.3 Data collection method

Data were gathered using a self-administered online questionnaire hosted on Google Forms. This platform was chosen for its accessibility, ease of distribution and efficient data management capabilities. Respondents could complete the survey at their convenience, which is particularly important for busy professionals working on time-sensitive energy projects. The questionnaire included an introductory section that explains the purpose of the study, assures confidentiality and seeks informed consent in line with ethical guidelines.

Before large-scale data collection, a pilot study was conducted with a small subset of project management professionals. The pilot aimed to assess the clarity, relevance and reliability of the items. Feedback from pilot respondents was used to refine wording, adjust ambiguous questions and ensure that the instrument effectively captures the intended constructs. This process strengthens the questionnaire's content validity and reliability.

Once finalised, the survey link was distributed to the selected sample through email or professional networks. Ethical considerations were strictly observed: participation was voluntary, responses were anonymous and data were used solely for academic purposes. Completed Google Forms responses were exported to SPSS for analysis, ensuring a streamlined and accurate transfer of data from collection to statistical processing.

3.2.4 Operationalisation and measurement

To enable statistical analysis, both independent and dependent variables were clearly defined and operationalised. The independent variables – process standardisation, stakeholder management, training, time management and resource allocation – were each measured using multiple Likert-scale items that reflect their conceptual definitions.

Process standardisation was operationalised through items assessing the clarity, documentation and consistent application of procedures and guidelines within project teams. Stakeholder management was measured by items that capture stakeholder involvement in decision-making, the effectiveness of communication channels and the extent to which stakeholder feedback is considered. Training was operationalised through perceptions of the availability, relevance and impact of training programmes on project-related skills and performance. Time management was reflected in planning, scheduling, task prioritisation and adherence to project timelines. Resource allocation was operationalised through perceptions of the adequacy, fairness and monitoring of human, financial and material resources for project tasks.

The dependent variable – the impact on QMS – was also measured using multiple indicators. These include perceptions of project performance (such as adherence to budget, schedule and quality standards), reduction in errors and rework, stakeholder satisfaction with deliverables, evidence of continuous improvement and alignment of project outcomes with organisational goals. Respondents evaluated the extent to which QMS practices have contributed to these dimensions, again using a five-point Likert scale. This structured operationalisation enabled quantitative assessment of how strongly QMS is perceived to influence project success.

By translating abstract concepts into measurable indicators, the study ensured that variables were captured consistently and could be analysed statistically. This operationalisation framework contributes to the study's reliability and validity and supports a meaningful examination of relationships between the constructs.

3.2.5 Data analysis techniques

Data analysis relies on both descriptive and inferential statistical techniques. Descriptive analysis was used to summarise respondents' demographic characteristics and to provide an overview of their perceptions of each construct. Measures such as mean, standard deviation (SD) and frequency distribution were computed for all key variables. These statistics offer a general picture of the extent to which project management practices and QMS-related outcomes are perceived as effective across the sample.

Inferential analysis was used to test the study's hypotheses and to explore relationships between variables. Correlation analysis (e.g. Pearson's correlation) assessed the strength and direction of the relationship between each independent variable and the dependent variable. Multiple regression analysis was used to examine the predictive power of process standardisation, stakeholder management, training, time management and resource allocation on QMS impact, both individually and jointly. Multiple regression analysis is appropriate for assessing the influence of multiple independent variables on a single dependent variable in behavioural and management research (Hair *et al.*, 2019). This indicates which factors make the strongest contributions to QMS effectiveness when other variables are controlled.

Where relevant, additional inferential techniques, such as analysis of variance (ANOVA), were used to compare groups (for example, across different levels of training or subsectors) and to explore associations between categorical variables.

Together, these analytical approaches enabled the study to move beyond simple description to draw statistically supported conclusions about the impact of project management factors on QMS effectiveness in the Malaysian energy sector. The results provide evidence-based insights to guide organisations in refining their quality management strategies and enhancing project outcomes.

3.3. Results and analysis

3.3.1 Demographic profile

This section begins by presenting the respondents' demographic profile, as this context is essential for interpreting the subsequent statistical findings on QMS. Four key demographic dimensions were analysed: job role, years of experience, highest level of education and involvement in the energy sector.

In terms of job role, engineers formed the largest group of respondents, reflecting the strong technical base of the energy sector and the centrality of engineering functions in project delivery. Team members also accounted for a substantial portion of the sample, underscoring the importance of collaborative work structures in achieving project and quality objectives. Only a small number of respondents selected 'other' roles, indicating that most participants are in well-defined project-related positions. The distribution of respondents by job role is presented in Figure 3.2.

Figure 3.2 Percentage of respondents by job role

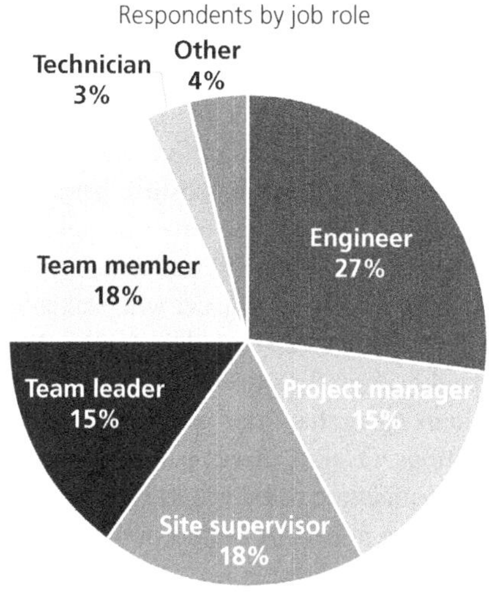

Regarding years of experience, the sample was well balanced between highly experienced professionals and relatively new entrants. The largest proportion of respondents had more than 10 years of experience, suggesting a strong presence of seasoned practitioners with deep knowledge of project and quality practices. A comparable share had between 1 and 3 years of experience, representing newer professionals who bring fresh perspectives but may have less exposure to established QMS practices. Only a small fraction had less than 1 year of experience, indicating limited representation of entry-level staff.

The educational profile of respondents indicates that the workforce was generally well qualified. A substantial proportion held at least a Bachelor's degree and a notable share held a Master's degree. This reflects a strong academic foundation that is likely to support understanding and implementation of structured project and quality management practices. Very few respondents reported having no formal education, indicating that education is a prominent feature of employment in this sector.

Regarding involvement in the energy sector, most respondents were from the energy transmission and distribution segment, underscoring the critical role of infrastructure and grid-related activities in the Malaysian energy landscape. Energy services also accounted for a significant portion of the sample, indicating a growing emphasis on efficiency, optimisation and sustainability. In contrast, the oil and gas sector was the least represented, suggesting a relative decline in dominance compared to more modern or sustainable energy-related activities, or at least a shift in survey representation. Collectively, these demographic patterns provide a nuanced backdrop for understanding variations in perceptions of QMS across roles, experience levels, education levels and subsectors.

Following the demographic analysis, statistical analysis was conducted in SPSS to examine the relationships between the independent variables (process standardisation, stakeholder management, training, time management and resource allocation) and the dependent variable (impact on QMS). Descriptive statistics were first computed to summarise the central tendencies and dispersion of all variables, providing an initial overview of how respondents generally perceived project and quality management practices.

3.3.2 Descriptive analysis

Based on responses from 341 participants, descriptive statistics for all variables show consistently high mean scores, indicating a generally positive perception of project management practices and QMS effectiveness. All constructs (process standardisation, stakeholder management, training, time management, resource allocation and impact on QMS) had recorded mean values above 4.14 on a five-point Likert scale, with stakeholder management (mean = 4.2135) and impact on QMS (mean = 4.2264) slightly higher than the others. The standard deviations (SD), ranging roughly from 0.58 to 0.67, indicate moderate variability, suggesting that while most respondents agree that these practices are present and effective, there is still some diversity in perceptions. Overall, these results indicate a context in which quality-related project management practices are perceived as well established and in which QMS is viewed as having a strong, positive impact on project performance. The descriptive statistics for all key variables are summarised in Table 3.1.

Table 3.1 Descriptive statistics analysis for all the variables

	Descriptive statistics					
	n	Minimum	Maximum	Mean	SD	Variance
Process standardisation	341	1.00	5.00	4.1677	0.64170	0.412
Stakeholder management	341	1.00	5.00	4.2135	0.60502	0.366
Training	341	1.00	5.00	4.1507	0.66503	0.442
Time management	341	1.00	5.00	4.1566	0.64771	0.420
Resource allocation	341	1.00	5.00	4.1478	0.60893	0.371
Impact on QMS	341	1.00	5.00	4.2264	0.58401	0.341

3.3.3 Reliability analysis

To ensure the measurement instrument's robustness, reliability was assessed using Cronbach's alpha (α) for all constructs. Cronbach's alpha remains one of the most widely applied reliability indicators in social science and management research (Pallant, 2020). The dependent variable, impact on QMS, achieved an alpha of 0.751, indicating acceptable internal consistency. Among the independent variables, process standardisation ($\alpha = 0.812$), training ($\alpha = 0.843$) and time management ($\alpha = 0.811$) demonstrated strong reliability, while stakeholder management ($\alpha = 0.777$) and resource allocation ($\alpha = 0.778$) showed adequate reliability. All values exceeded the commonly accepted threshold of 0.70, confirming that the items within each scale consistently measure their intended constructs. According to George and Mallery (2003), Cronbach's alpha values above 0.70 indicate acceptable internal consistency for research instruments. This supports the validity of using these variables in subsequent correlation and regression analyses. The internal consistency reliability results for each construct are presented in Table 3.2.

3.3.4 Correlation analysis

Correlation analysis using Pearson's r revealed very strong, positive relationships between each independent variable and impact on QMS. Correlation coefficients ranged from 0.792

Table 3.2 Cronbach's alpha values for variables

Type of variable	Variable	α	No. of Items	Reliability analysis
Dependent	Impact on QMS	0.751	5	Adequately correlated
	Process standardisation	0.812	5	Strong consistency
	Stakeholder management	0.777	5	Adequately correlated
Independent	Training	0.843	5	Strong consistency
	Time management	0.811	5	Strong consistency
	Resource allocation	0.778	5	Adequately correlated

Table 3.3 Pearson correlation analysis for all the variables

	Process standardisation	Stakeholder management	Training	Time management	Resource allocation	Impact on QMS
Process standardisation	1	0.699[a]	0.795[a]	0.794[a]	0.800[a]	0.792[a]
Stakeholder management	0.699[a]	1	0.769[a]	0.772[a]	0.769[a]	0.827[a]
Training	0.795[a]	0.769[a]	1	0.833[a]	0.803[a]	0.792[a]
Time management	0.794[a]	0.772[a]	0.833[a]	1	0.772[a]	0.812[a]
Resource allocation	0.800[a]	0.769[a]	0.803[a]	0.772[a]	1	0.837[a]
Impact on QMS	0.792[a]	0.827[a]	0.792[a]	0.812[a]	0.837[a]	1

[a]Correlation is significant at the 0.01 level (2-tailed).

to 0.837, all significant at the 0.01 level. Resource allocation showed the strongest correlation with impact on QMS ($r = 0.837$), followed closely by stakeholder management ($r = 0.827$) and time management ($r = 0.812$). Process standardisation and training both showed correlations of 0.792 with the dependent variable, underscoring their relevance to QMS effectiveness. The independent variables were also highly intercorrelated; for instance, training and time management ($r = 0.833$) and resource allocation and process standardisation ($r = 0.800$) showed strong correlations. These patterns suggest that quality-related project management practices tend to reinforce one another and jointly contribute to stronger QMS outcomes. All correlations were significant at the 0.01 level (2-tailed), with p-values below 0.001 across every pair. The Pearson correlation coefficients among all variables are shown in Table 3.3.

3.3.5 Spearman's rank correlation

Spearman's rho (ρ) was also employed to confirm the robustness of these relationships when using ranked data. The results were consistent with Pearson's r, again showing strong positive associations between each independent variable and impact on QMS. In this analysis, time management ($\rho = 0.866$), training ($\rho = 0.862$) and resource allocation ($\rho = 0.864$) emerged as particularly influential, with process standardisation ($\rho = 0.837$) and stakeholder management ($\rho = 0.694$) also showing substantial positive associations. The convergence of Pearson and Spearman findings reinforces the conclusion that the five project management factors are all positively and significantly associated with QMS effectiveness, regardless of the specific statistical assumption set. To confirm robustness, Spearman's rank correlation coefficients are presented in Table 3.4. All Spearman correlations were significant at the 0.01 level (2-tailed), with p-values below 0.001 across every pair.

Table 3.4 Spearman correlation analysis for all the variables

	Process standardisation	Stakeholder management	Training	Time management	Resource allocation	Impact on QMS
Process standardisation	1.000	0.617[a]	0.838[a]	0.872[a]	0.831[a]	0.837[a]
Stakeholder management	0.617[a]	1.000	0.660[a]	0.650[a]	0.680[a]	0.694[a]
Training	0.838[a]	0.660[a]	1.000	0.860[a]	0.860[a]	0.862[a]
Time management	0.872[a]	0.650[a]	0.860[a]	1.000	0.855[a]	0.866[a]
Resource allocation	0.831[a]	0.680[a]	0.860[a]	0.855[a]	1.000	0.864[a]
Impact on QMS	0.837[a]	0.694[a]	0.862[a]	0.866[a]	0.864[a]	1.000

[a]Correlation is significant at the 0.01 level (2-tailed).

3.3.6 Simple linear regression analysis

Simple linear regression analyses were then conducted to examine the individual predictive power of each independent variable on the impact on QMS. In the regression output that follows, B denotes the unstandardised regression coefficient (the change in QMS impact per one-unit change in the predictor), t is the test statistic for the coefficient (B divided by its standard error) and F is the overall model fit statistic from the ANOVA test of the regression. For each model, the independent variable significantly predicted QMS, with all p-values < 0.001 and high F-statistics in the associated ANOVA tests. Process standardisation produced an unstandardised coefficient of 0.721 ($\beta = 0.792$), indicating that higher levels of standardisation are associated with a greater perceived impact on QMS. Stakeholder management showed an even stronger effect ($B = 0.798$, $\beta = 0.827$), affirming the centrality of stakeholder engagement to quality performance. Training, time management and resource allocation also exhibited substantial predictive strength, with coefficients of 0.696 ($\beta = 0.792$), 0.732 ($\beta = 0.812$) and 0.803 ($\beta = 0.837$), respectively. In each case, the model ANOVA returned a highly significant F-value, indicating that each factor alone explains a substantial portion of the variance in QMS impact. Taken together, these models confirm that each of the five project management practices has a strong, direct and statistically significant positive effect on perceived QMS effectiveness. The regression coefficients for process standardisation predicting QMS impact are reported in Table 3.5.

3.3.7 Multiple linear regression analysis

A more comprehensive understanding of these relationships was obtained through multiple linear regression, in which all five independent variables were entered simultaneously. The overall model was highly significant ($F = 292.465$, $p < 0.001$), accounting for a substantial

Table 3.5 Simple linear regression analysis by coefficient for process standardisation and impact on QMS

	Unstandardised coefficient, *B*	Std. error	Standardised coefficient, *β*	*t*	Sig.
(Constant)	1.222	0.127		9.604	0.000
Process standardisation	0.721	0.030	0.792	23.895	0.000

Note: dependent variable: impact on QMS.

share of the variance in QMS (as indicated by the large regression sum of squares). When controlling for all other variables, stakeholder management ($B = 0.316$, $\beta = 0.327$), resource allocation ($B = 0.283$, $\beta = 0.295$), time management ($B = 0.168$, $\beta = 0.186$) and process standardisation ($B = 0.152$, $\beta = 0.167$) remained significant predictors (all $p < 0.001$). Training, however, lost its significance in the combined model ($B = 0.014$, $p = 0.748$), suggesting that its apparent effect on QMS in the simple regressions may be largely mediated or shared with other factors – especially time management and resource allocation, with which it is strongly correlated. This indicates that while training is important, its influence on QMS may be indirect in this context, operating through improvements in other project management practices rather than as a standalone driver. The ANOVA results for the initial regression model are presented in Table 3.6. For stakeholder management, the regression coefficients and corresponding ANOVA results are reported in Tables 3.7 and 3.8, respectively. Similarly, the regression and ANOVA results for training are displayed in Tables 3.9 and 3.10. The time management analysis follows the same structure, with regression coefficients shown in Table 3.11 and ANOVA results in Table 3.12. For resource allocation, the regression coefficients and ANOVA outcomes are provided in Tables 3.13 and 3.14. Finally, the results of the multiple linear regression model are presented in Table 3.15, while the overall model ANOVA summary is reported in Table 3.16.

Table 3.6 Simple linear regression analysis by ANOVA for process standardisation and impact on QMS

Model	Sum of squares	Degrees of freedom, *df*	Mean square	*F*	Sig.
Regression	72.761	1	72.761	570.955	0.000
Residual	43.201	339	0.127		
Total	115.962	340			

Note: dependent variable: impact on QMS; predictors: (constant), process standardisation.

Table 3.7 Simple linear regression analysis by coefficient for stakeholder management and impact on QMS

	Unstandardised coefficient, *B*	Std. error	Standardised coefficient, *β*	*t*	Sig.
(Constant)	0.864	0.126		6.882	0.000
Stakeholder management	0.798	0.029	0.827	27.055	0.000

Note: dependent variable: impact on QMS.

Table 3.8 Simple linear regression analysis by ANOVA for stakeholder management and impact on QMS

Model	Sum of squares	*df*	Mean square	*F*	Sig.
Regression	79.257	1	79.257	731.997	0.000
Residual	36.705	339	0.108		
Total	115.962	340			

Note: dependent variable: impact on QMS; predictors: (constant), stakeholder management.

Table 3.9 Simple linear regression analysis by coefficient for training and impact on QMS

	Unstandardised coefficient, *B*	Std. error	Standardised coefficient, *β*	*t*	Sig.
(Constant)	1.338	0.122		10.941	0.000
Training	0.696	0.029	0.792	23.921	0.000

Note: dependent variable: impact on QMS.

Table 3.10 Simple linear regression analysis by ANOVA for training and impact on QMS

Model	Sum of squares	*df*	Mean square	*F*	Sig.
Regression	72.821	1	72.821	572.215	0.000
Residual	43.142	339	0.127		
Total	115.962	340			

Note: dependent variable: impact on QMS; predictors: (constant), training.

Table 3.11 Simple linear regression analysis by coefficient for time management and impact on QMS

	Unstandardised coefficient, *B*	Std. error	Standardised coefficient, *β*	*t*	Sig.
(Constant)	1.182	0.120		9.838	0.000
Time management	0.732	0.029	0.812	25.643	0.000

Note: dependent variable: impact on QMS.

Table 3.12 Simple linear regression analysis by ANOVA for time management and impact on QMS

Model	Sum of squares	*df*	Mean square	*F*	Sig.
Regression	76.516	1	76.516	657.581	0.000
Residual	39.446	339	0.116		
Total	115.962	340			

Note: dependent variable: impact on QMS; predictors: (constant), time management.

Table 3.13 Simple linear regression analysis by coefficient for resource allocation and impact on QMS

	Unstandardised coefficient, *B*	Std. error	Standardised coefficient, *β*	*t*	Sig.
(Constant)	0.897	0.120		7.506	0.000
Resource allocation	0.803	0.029	0.837	28.151	0.000

Note: dependent variable: impact on QMS.

Table 3.14 Simple linear regression analysis by ANOVA for resource allocation and impact on QMS

Model	Sum of squares	*df*	Mean square	*F*	Sig.
Regression	81.219	1	81.219	792.479	0.000
Residual	34.743	339	0.102		
Total	115.962	340			

Note: dependent variable: impact on QMS; predictors: (constant), resource allocation.

Table 3.15 Multiple linear regression analysis by coefficient for process standardisation, stakeholder management, training, time management, resource allocation and impact on QMS

	Unstandardised coefficient, *B*	Std. error	Standardised coefficient, *β*	*t*	Sig.
(Constant)	0.332	0.103		3.218	0.001
Process standardisation	0.152	0.041	0.167	3.658	0.000
Stakeholder management	0.316	0.041	0.327	7.794	0.000
Training	0.014	0.044	0.016	0.322	0.748
Time management	0.168	0.044	0.186	3.805	0.000
Resource allocation	0.283	0.045	0.295	6.263	0.000

Note: dependent variable: impact on QMS.

Table 3.16 Multiple linear regression analysis by ANOVA for process standardisation, stakeholder management, training, time management, resource allocation and impact on QMS

Model	Sum of squares	*df*	Mean square	*F*	Sig.
Regression	94.348	5	18.870	292.465	0.000
Residual	21.614	335	0.065		
Total	115.962	340			

Note: dependent variable: impact on QMS; predictors: (constant), process standardisation, stakeholder management, training, time management, resource allocation.

3.4. Practical insights

The study's empirical evidence provides several important insights for practitioners managing projects and QMS in the energy sector. The findings confirm that isolated quality initiatives do not drive quality performance in energy projects, but that coordinated application of project management practices related to process standardisation, stakeholder engagement, time management and resource allocation does.

First, the results demonstrate that process standardisation is a foundational enabler of effective QMS in energy projects. Practitioners should recognise that standardised procedures, documented workflows and consistent application of operating guidelines significantly reduce variability and improve compliance in complex, highly regulated environments. Energy projects often involve stringent safety, environmental and regulatory requirements, and the evidence indicates that well-defined processes help project teams maintain quality under these constraints. Practically, this means investing in clear standard operating procedures, quality checklists and formalised approval processes that are consistently applied across projects rather than adapted informally at the discretion of individual teams.

Second, the findings show that stakeholder management is one of the strongest predictors of QMS effectiveness. This has direct implications for energy-sector practitioners who work with regulators, suppliers, contractors, local communities and internal technical teams. The evidence suggests that quality outcomes improve when stakeholders are systematically identified, engaged and involved in decision-making. Practitioners should therefore move beyond ad hoc communication and adopt structured stakeholder engagement strategies, including regular consultation meetings, feedback mechanisms and transparent communication channels. Proactive stakeholder management reduces resistance, minimises rework and supports smoother implementation of quality requirements throughout the project lifecycle.

Third, time management emerges as a critical operational lever for quality performance. The statistical analysis indicates that projects with realistic scheduling, disciplined monitoring of timelines and effective prioritisation are more likely to achieve strong QMS outcomes. In practice, this implies that quality deterioration often arises not from technical incompetence but from schedule pressure, rushed execution and inadequate planning. Energy project practitioners should therefore use robust scheduling tools, monitor critical paths closely and treat time overruns as quality risks rather than purely schedule issues. Integrating time management with quality planning helps prevent shortcuts that compromise safety and compliance.

Fourth, the study highlights resource allocation as the most influential factor affecting QMS effectiveness. An adequate allocation of financial, human and material resources enables project teams to implement quality procedures and respond effectively to emerging issues consistently. For practitioners, this underscores the importance of aligning resource planning with quality objectives rather than treating quality as an add-on responsibility. Under-resourced projects may comply with documentation requirements on paper while failing to achieve substantive quality outcomes in practice. Decision makers in the energy sector should therefore ensure that sufficient budgets, skilled personnel and appropriate tools support quality-related activities.

Fifth, while training shows strong individual correlations with QMS effectiveness, it does not emerge as a significant predictor when combined with other project management factors. This finding carries an important practical message. It suggests that training alone is insufficient to improve quality unless standardised processes, effective time management and adequate resources support it. Practitioners should critically evaluate existing training programmes to ensure they are directly aligned with operational needs, project realities and quality objectives. Emphasis should be placed on practical, context-specific training that reinforces how quality standards are applied in real-world projects, rather than on generic or compliance-focused programmes.

The combined results also indicate that project management practices reinforce one another. Strong correlations between the independent variables suggest that improvements in one area, such as process standardisation, are more effective when accompanied by improvements in stakeholder engagement, scheduling discipline and resource planning. For practitioners, this means that fragmented quality initiatives are unlikely to succeed. Instead, organisations should adopt an integrated approach to QMS implementation that aligns processes, people and resources within a coherent project management framework.

From a governance perspective, the findings suggest that quality management should be embedded into routine project governance rather than treated as a separate function. Energy-sector organisations should integrate QMS indicators into regular project reviews, management reporting and decision-making forums. This improves accountability and ensures that quality considerations are addressed proactively rather than reactively.

Overall, the practical implications indicate that energy-sector practitioners seeking to enhance QMS effectiveness should prioritise disciplined process design, structured stakeholder engagement, realistic scheduling and adequate resource support. Quality performance improves most consistently when these elements are implemented together and reinforced through governance mechanisms. By translating these empirical insights into daily project management practices, organisations in the energy sector can strengthen reliability, compliance and long-term project value.

3.5. Conclusion and recommendations

3.5.1 Limitations

Several limitations of the study must be acknowledged. The research used a purely quantitative survey design, which, while suitable for statistical testing, may not fully capture the depth and complexity of real project environments. The reliance on self-reported perceptions and Likert-scale responses may overlook nuanced contextual factors. The sample, although large (341 respondents), is heavily represented by engineers and more experienced professionals, which may skew perspectives and limit generalisability to other roles or junior staff. The study is also geographically and sector-specific, focusing solely on the Malaysian energy industry; thus, findings may not directly transfer to other countries, cultures or sectors. Furthermore, while the analysis revealed that training had a limited direct effect on QMS, it did not explore in depth why this is the case, whether due to training content, delivery methods, relevance or organisational follow-through. Finally, potential mediating or moderating factors, such as organisational culture, leadership style, technological maturity or market conditions, were not modelled, meaning that the pathways through which project management practices influence QMS may be more complex than captured here.

3.5.2 Recommendations and future research

In light of these implications and limitations, several directions for future research are recommended. Methodologically, future studies should adopt mixed-methods designs that combine surveys with interviews, focus groups or case studies to capture richer, context-specific insights. Sampling should be broadened to include more varied roles, sectors and regions, enabling cross-industry and cross-cultural comparisons and to test whether the relationships found here hold in other contexts. The surprising result regarding training suggests the need for focused investigation into training design, relevance and impact, especially practice-based, scenario-driven or mentorship-oriented approaches. Future work should also integrate mediating and moderating factors, such as organisational culture, leadership, digital maturity, sustainability orientation and external regulatory pressures. It may use more advanced modelling techniques, such as structural equation modelling or longitudinal designs. As the energy sector evolves, research could further explore how QMS and project management practices intersect with sustainability goals, digital transformation and sector-specific challenges (e.g. renewable versus fossil-based projects). Ultimately, building on this study, future research can develop a comprehensive, context-sensitive framework that integrates process, people, technology and stakeholder dimensions to guide effective quality management in the energy sector and beyond.

REFERENCES

Aaltonen K (2010) Stakeholder management in international projects. *International Journal of Project Management* **28(6)**: 573–582.

Bourne L and Walker DHT (2006) Visualising and mapping stakeholder influence. *Management Decision* **44(3)**: 329–344.

Field A (2018) *Discovering Statistics Using IBM SPSS Statistics*, 5th edn. SAGE Publications, London, UK.

Freeman RE (1984) *Strategic Management: A Stakeholder Approach*. Pitman, Boston, MA, USA.

George D and Mallery P (2003) *SPSS for Windows Step by Step: A Simple Guide and Reference*, 4th edn. Allyn & Bacon, Boston, MA, USA.

Hair JF, Black WC, Babin BJ and Anderson RE (2019) *Multivariate Data Analysis*, 8th edn. Cengage Learning, Boston, MA, USA.

ISO (International Organization for Standardization) (2015) *ISO 9001: Quality Management Systems – Requirements*. ISO, Geneva, Switzerland.

Jha KN and Iyer KC (2006) Critical factors affecting quality performance in construction projects. *Total Quality Management and Business Excellence* **17(9)**: 1155–1170.

Jimoh AL, Dunmade EO and Salman A (2025) Green talent management and job performance: the significance of open innovation. *Journal of Intellectual Capital* **26(6)**: 1190–1207.

Joslin R and Müller R (2015) Relationships between a project management methodology and project success in different project governance contexts. *International Journal of Project Management* **33(6)**: 1377–1392.

Kerzner H (2017) *Project Management: A Systems Approach to Planning, Scheduling, and Controlling*, 12th edn. Wiley, Hoboken, NJ, USA.

Mok KY, Shen GQ and Yang J (2015) Stakeholder management studies in mega construction projects: A review and future directions. *International journal of project management* **33(2)**, 446–457.

Pallant J (2020) *SPSS Survival Manual*, 7th edn. Open University Press, Maidenhead, UK.

PMI (Project Management Institute) (2021) *A Guide to the Project Management Body of Knowledge (PMBOK Guide)*, 7th edn. PMI, Newtown Square, PA, USA.

Vial G (2021) Understanding digital transformation: a review and a research agenda. In *Managing digital transformation* (Hinterhuber A, Vescovi T and Checchinato F (eds.)). Routledge, Abingdon, pp. 13–66.

Mohammed Reyasudin Basir Khan
ISBN 978-1-83742-799-4
https://doi.org/10.1108/978-1-83742-798-720261005

Chapter 4
Green building certification and its impact on project management practices

Praveen Raj Durai Ragavan and Mohammed Reyasudin Basir Khan

Tun Razak Graduate School, Universiti Tun Abdul Razak (UNIRAZAK), Kuala Lumpur, Malaysia

4.1. Introduction

Malaysia's construction and real estate sector has long supported national economic growth, but rapid development has also resulted in high carbon emissions, heavy energy use and resource depletion. To address these issues, the country has increasingly adopted sustainable construction and green building practices (Mustaffa *et al.*, 2022). A major step was the introduction of the Green Building Index (GBI), developed by Pertubuhan Arkitek Malaysia (PAM) and the Association of Consulting Engineers Malaysia (ACEM) in 2009 to assess buildings across criteria such as energy efficiency, water efficiency, indoor environmental quality, sustainable site planning and management, materials and resources, and innovation (Solla *et al.*, 2022). Malaysia also integrates global standards such as Leadership in Energy and Environmental Design (LEED) and the Building Research Establishment Environmental Assessment Method (BREEAM), aligning local practices with international benchmarks (Solla *et al.*, 2022).

Government policies, including the Green Technology Financing Scheme (GTFS) and various tax incentives, have further supported the uptake of green buildings, alongside growing environmental awareness and corporate sustainability commitments. Green-certified buildings offer benefits such as lower operational costs, better occupant wellbeing and higher asset value.

Nonetheless, green building adoption remains modest at about 3–4% of the market (Zakaria *et al.*, 2024). Challenges persist, particularly perceptions of high cost and implementation complexity. This highlights the crucial role of project management, as project managers must integrate sustainability requirements into traditional project goals, coordinate diverse stakeholders, manage risks and ensure green objectives are embedded throughout the project lifecycle.

Green building certification systems such as GBI, LEED and BREEAM establish clear benchmarks for sustainable construction, promoting energy efficiency, resource conservation and reduced environmental impact. In Malaysia, the number of certified projects continues to grow, supporting national goals for a low-carbon and resource-efficient economy.

However, certification introduces new demands that reshape project management. Budgets must account for higher upfront costs associated with sustainable materials and technologies, even if these lead to long-term savings. Project schedules become more complex due to extended lead times, additional coordination and commissioning requirements. Quality management must incorporate environmental and performance-based standards, while stakeholder management must expand to include regulators, certification bodies, communities and end users.

Digital tools such as building information modelling (BIM), Internet of Things (IoT), artificial intelligence (AI) and data analytics increasingly support collaboration, performance monitoring and resource optimisation (Cao *et al.*, 2022). Nevertheless, the Malaysian construction industry continues to face challenges, including skills shortages, limited experience with green projects, communication gaps and resistance to change (Muthusamy and Chew, 2020).

Although green building certifications are increasingly adopted in Malaysia, many project managers still struggle to integrate sustainability requirements into conventional project management practices. Traditional frameworks are not fully equipped to handle the added complexity of green projects.

Key challenges include the following.

- *Cost management*: sustainable materials and technologies often involve higher upfront costs. Without structured methods such as lifecycle costing, project managers may struggle to justify expenditures or prevent cost overruns (Liu *et al.*, 2022).
- *Time management*: certification requires extensive documentation, compliance checks and multi-stakeholder coordination, which can lead to delays if not properly planned and integrated into the project schedule.
- *Quality management*: green projects must satisfy both traditional quality standards and environmental performance criteria. Many managers lack clear guidance on how to incorporate these additional requirements into quality processes.
- *Stakeholder engagement*: green projects involve a wider range of stakeholders with differing expectations. Without proper engagement frameworks, miscommunication and conflicting priorities can negatively affect project performance.

Despite the growing literature on sustainability, there is limited empirical evidence on how green certifications directly influence project costs, timelines, quality and stakeholder management in Malaysia. As a result, many project managers rely on ad hoc strategies, leading to inconsistent outcomes. This chapter addresses this gap by examining the impact of green building certifications (GBC) on project management and proposing a structured framework tailored to Malaysian green projects. Specifically, this chapter

- examines the effect of green building certifications on project cost management
- evaluates the influence of green building certifications on project time management
- assesses the impact of green building certifications on project quality management
- investigates the role of green building certifications in enhancing stakeholder engagement in construction projects.

Malaysian construction projects face multiple challenges in project management, including financial constraints, technological change, regulatory requirements, workforce issues and communication gaps. Integrating sustainable project management practices offers a pathway to address these issues while aligning with national environmental and socioeconomic goals. The significance of this study is threefold.

- *Theoretical contribution*: by linking green building certifications with core project management dimensions – cost, time, quality and stakeholder engagement – this research contributes to the growing body of knowledge on sustainable project management in developing countries.
- *Practical contribution*: the study provides project managers, developers and policy makers with empirical evidence and a proposed framework to better plan, execute and monitor green projects. This can lead to improved project outcomes, better resource utilisation, reduced environmental impacts and enhanced stakeholder satisfaction.
- *Policy and industry relevance*: insights from this research can support policy refinement, capacity-building programmes and industry guidelines to mainstream green building practices. By strengthening Malaysia's capability to deliver high-performing green projects, the country can enhance its competitiveness, attract investment and position itself as a regional leader in sustainable construction.

Several theoretical perspectives underpin the analysis of green building certifications and their impact on project management.

- *Sustainability theory*: emphasises balancing environmental, economic and social dimensions to safeguard current and future generations. Green buildings operationalise this through the use of renewable energy, resource efficiency and reduced environmental footprints.
- *Stakeholder theory*: highlights the need to engage all parties affected by or influencing a project. In green projects, close collaboration between architects, engineers, contractors, developers, authorities and occupants is critical for aligning expectations and achieving sustainable outcomes (Mahajan *et al.*, 2023).
- *Institutional theory*: explains how norms, regulations and professional standards, such as those embedded in LEED, BREEAM and GBI, shape organisational behaviour and encourage the adoption of sustainable practices (Risi *et al.*, 2023).
- *The resource-based view (RBV)*: suggests that organisations gain a competitive advantage from valuable, rare, inimitable and well-organised resources. In green construction, capabilities in sustainable design, materials selection and environmental management can become strategic assets (Lubis, 2022).
- *Change management theory*: focuses on how organisations manage transitions to new practices. Shifting from conventional to green project management demands training,

new processes and strong leadership to overcome resistance and embed sustainability into everyday routines (Hayes, 2025).
- *Performance measurement theory*: underscores the importance of metrics and benchmarking. Environmental rating systems and green certification criteria provide measurable indicators of building performance and guide continuous improvement (Mio *et al.*, 2022).
- *Systems theory*: views organisations and projects as interconnected systems. Green buildings exemplify this by integrating site planning, structure, services, materials and user behaviour into a coherent whole, reinforcing the need for holistic project management (Zhang *et al.*, 2023).

Empirical studies further support the link between sustainability and project management. Research indicates that green initiatives and green/green IT decisions can enhance project performance, reduce energy and waste costs, and strengthen organisational reputation. Studies also emphasise the role of leadership, culture and cross-disciplinary collaboration in successfully implementing sustainable practices.

In Malaysia, emerging evidence shows growing recognition of the business case for going green and the need for structured models, tools and training to support sustainable initiatives. However, comprehensive analyses focusing specifically on green building certification and project management practices remain limited, particularly for construction projects and developers with substantial green portfolios.

Given the identified gaps, this study proposes a conceptual framework in which green building certification influences four key project management dimensions: cost management, time management, quality management and stakeholder engagement, which together shape overall project success and sustainability outcomes.

Based on the literature and theoretical foundations, the following hypotheses are formulated.

- H1: the integration of green building practices in project management significantly contributes to controlling overall project costs.
- H2: the relationship between the increased time required for documentation and preparation for green building certification and improved project schedules is positive.
- H3: the integration of sustainable construction practices leads to higher quality standards in projects.
- H4: increased stakeholder engagement in projects with advanced sustainability features results in better project outcomes.

The dependent construct – impact of green building certification (GBC) practices on project management practices – captures the overall effect of these integrations across cost, time, quality and stakeholder dimensions. The framework assumes that sustainability is not an optional add-on but a systemic approach that reshapes how projects are conceived, managed and evaluated. Through testing these hypotheses in the Malaysian construction context, this study provides empirical insights and practical guidance to support the mainstreaming of green building practices in project management. The conceptual relationships between the independent and dependent variables are illustrated in Figure 4.1.

Figure 4.1 Conceptual framework

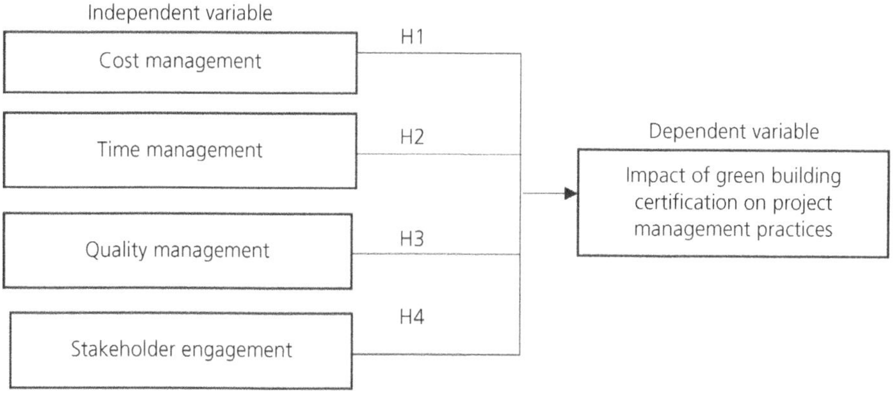

4.2. Methodology

4.2.1 Research design

A quantitative research design was employed to assess how GBC systematically influences project management outcomes. This approach was appropriate as the study aimed to test predefined hypotheses and quantify the relationships between GBC implementation and core project management domains. Data were collected using a structured, closed-ended Likert-scale questionnaire. This allowed responses to be converted into numerical data suitable for statistical analysis. The survey instrument was developed from prior literature on green building and project management and aligned with the study's objectives and hypotheses. The deductive approach enabled hypothesis testing and contributed to the theoretical framework by confirming or refuting assumed relationships between GBC and cost, time, quality and stakeholder engagement.

4.2.2 Study population and sampling procedures

The study population comprised construction professionals in Malaysia with direct experience in GBC-certified projects, including architects, engineers, project managers, contractors and consultants. These respondents were selected because they are directly involved in the planning and execution of green-certified projects and are therefore well-positioned to assess the impact of GBC on project management practices.

A purposive sampling technique was used. This non-probability method allowed the researcher to intentionally target individuals with relevant knowledge and experience in GBC implementation. Although purposive sampling may limit the generalisability of the findings, it was appropriate for this study given its specialised focus.

Participants were recruited through professional networks, industry associations, conferences and social media channels. The sample included respondents aged approximately 20 to 60 years, representing both early-career practitioners and experienced professionals.

This diversity supported a broader understanding of GBC-related challenges and benefits from multiple professional perspectives.

4.2.3 Data collection method

Data were collected using an online questionnaire administered through Google Forms. The online format facilitated efficient access to respondents across Malaysia and enabled timely data collection.

The questionnaire was divided into four main sections.

- *Demographic and professional background*: collected information on respondents' roles, years of experience and level of involvement in green building projects. This contextual data supported subgroup analysis and interpretation of perceptions across different professional profiles.
- *GBC practices and management techniques*: assessed how GBC requirements are integrated into project management processes, particularly in procurement, resource allocation and risk management. Respondents rated the extent to which GBC influences these areas.
- *Influence of GBC on project outcomes*: examined perceived impacts of GBC on cost efficiency, time management and quality of project outcomes. This section presents core data for testing hypotheses about the relationship between GBC and project performance.
- *General perceptions of the GBC process*: captured overall satisfaction with GBC certification, perceived strengths and weaknesses, and practical challenges. While primarily closed-ended limited open-ended items allowed respondents to elaborate on key issues and suggestions.

Most items were closed-ended and measured on a Likert scale to ensure consistency, facilitate coding and enable robust statistical analysis. The survey link was distributed through email and online platforms, and follow-up reminders were sent to improve response rates.

4.2.4 Variables operationalisation

The study's dependent variable was the overall impact of GBC on project management practices. The independent variables corresponded to four core project management domains.

- *Cost management*: operationalised as respondents' perceptions of initial project costs, long-term cost savings and efficiency of resource allocation in GBC-certified projects.
- *Time management*: measured in terms of project schedule performance, including the extent of delays or improvements linked to compliance with GBC requirements, documentation and coordination.
- *Quality management*: assessed through perceptions of material quality, environmental performance and long-term functional performance of GBC projects, including indoor environmental quality and durability.
- *Stakeholder engagement*: defined as the effectiveness of communication, collaboration and coordination among project teams, contractors, consultants, certification bodies, clients and end-users in GBC projects.

Each construct was measured using multiple Likert-scale items, enabling composite scores to be calculated for use in statistical analyses. This operationalisation ensured that abstract concepts such as 'impact on cost' and 'stakeholder engagement' were captured systematically and measurably.

4.2.5 Data analysis techniques

- *Descriptive statistics*: frequency, percentage, mean and standard deviation were used to summarise demographic characteristics and overall response patterns. These statistics provided an overview of respondents' perceptions of GBC's impact on project management.
- *Normality testing*: the Shapiro–Wilk and Kolmogorov–Smirnov tests were applied to assess whether the data approximated a normal distribution. Skewness and kurtosis values were also examined, with acceptable values considered to be between −3 and +3. These checks were important for determining the suitability of parametric tests such as regression.
- *Reliability testing*: Cronbach's alpha was used to evaluate the internal consistency of the scales used for each construct (cost, time, quality, stakeholder engagement and overall GBC impact). A value of 0.70 or higher was considered acceptable, indicating that items within each scale measured the same underlying construct reliably.
- *Correlation and regression analysis*: correlation analysis was conducted to examine the strength and direction of relationships between GBC-related practices and project management outcomes. Subsequently, multiple regression analysis was used to test the study's hypotheses and determine the extent to which GBC (and its associated practices) predicts variations in cost, time, quality and stakeholder engagement outcomes. This approach provided a comprehensive view of the direct and combined effects of GBC on project management.

4.3. Results and analysis

4.3.1 Frequency distribution analysis

The geographical distribution of respondents in this study reflects broad representation across Malaysia, with notable concentrations in economically advanced, industrially active regions. This distribution offers valuable insight into how GBC practices are adopted across diverse regional contexts, each shaped by distinct development priorities and sustainability agendas. Kuala Lumpur, with the largest share of respondents at 24.4%, naturally stands out as the country's central economic hub. As the nation's capital and primary centre for finance, commerce and corporate headquarters, Kuala Lumpur hosts industries that rely heavily on precise, sophisticated forecasting. Organisations within this urban centre operate in highly competitive environments where accuracy and efficiency are vital, and the respondents from this region reflect the advanced forecasting needs of businesses engaged in complex financial and operational decision-making. The distribution of respondents by professional role is shown in Figure 4.2.

The respondents' years of experience, with most having 5 to 10 years in the field, indicate a workforce that is both knowledgeable and adaptable. Their experience equips them to critically assess GBC requirements and sustainability practices, while their relative youth

Figure 4.2 Percentage of respondents by role in the organisation

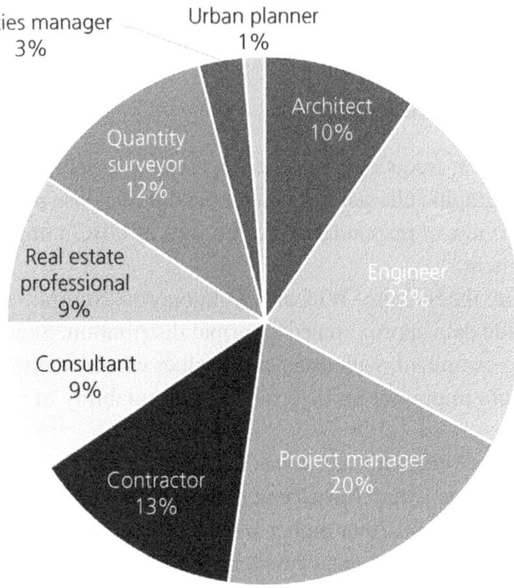

suggests openness to learning and integrating new methods and technologies. This demographic insight provides a basis for understanding how different career stages influence the acceptance, implementation and perceived reliability of various GBC strategies within project management contexts.

Similarly, the relatively high proportion of respondents involved in green building projects highlights the growing importance of sustainability within construction project management practices. When considered alongside data on age, experience and professional roles, this information forms a strong foundation for analysing how GBC adoption affects project management performance across cost, time, quality and stakeholder engagement dimensions. Together, these demographic factors support a deeper exploration of how green building practices can either enhance or hinder organisational effectiveness in a rapidly evolving, sustainability-driven construction environment.

4.3.2 Descriptive statistics

4.3.2.1 Cost management

The descriptive statistics for cost management variables are summarised in Table 4.1. The results show a generally positive perception of sustainability practices in project management, reflected in medium mean scores (around 3.63). Respondents recognise benefits such as improved cost efficiency, quality, time management and stakeholder engagement. However, relatively high standard deviations (SD) (1.30–1.38) indicate wide variation in experience, project types and implementation consistency, suggesting that views on sustainability are not uniform among respondents.

Table 4.1 Cost management descriptive results

	n	Minimum	Maximum	Mean	SD
The integration of sustainability practices in project management helps in controlling overall project costs	208	1	5	3.63	1.335
Adopting energy-efficient technologies leads to long-term cost savings in construction projects	208	1	5	3.47	1.379
The initial financial outlay for implementing advanced green technologies is higher compared to conventional methods	208	1	5	3.63	1.305
Financial incentives such as tax breaks or grants are available for projects incorporating energy-saving measures	208	1	5	3.48	1.383
Investing in high-quality, sustainable materials reduces overall maintenance costs over the lifespan of the project	208	1	5	3.63	1.338
Valid *n* (listwise)	208				

Adoption of energy-efficient technologies shows similar patterns. The mean score of 3.47 suggests acknowledgement of long-term benefits. Still, the high variability in responses indicates uncertainty, likely influenced by cost concerns, differences in technological maturity and varying expectations of return on investment. The initial financial outlay for green technologies also yielded a mean of 3.63, implying cautious acceptance. While many respondents view such investments as worthwhile over time, the variation in responses highlights the need for clearer evidence on long-term payoffs.

Financial incentives, such as tax breaks and grants, received a moderate mean score (3.48) with high variability, suggesting inconsistent access, awareness or effectiveness across regions and project types. Finally, investment in high-quality sustainable materials also scored 3.63. Respondents appear to value their long-term benefits, yet differing project experiences and cost considerations contribute to diverse perceptions.

4.3.2.2 Time management

Table 4.2 presents the descriptive statistics for time management variables. The analysis of time management factors related to green certification shows generally positive yet varied

Table 4.2 Time management descriptive results

	n	Minimum	Maximum	Mean	SD
The process of preparing documentation for green building certification requires additional time compared to standard project requirements	208	1	5	3.66	1.316
Green building certification processes necessitate early and detailed project planning to avoid delays	208	1	5	3.47	1.379
The need to meet specific green building certification milestones positively influences the project timeline by improving efficiency	208	1	5	3.59	1.327
Additional coordination with certifying bodies for green building certification can impact project scheduling and lead to potential delays	208	1	5	3.48	1.383
Incorporating green building practices often requires adjustments in project timelines due to the integration of new, sustainable technologies	208	1	5	3.66	1.316
Valid *n* (listwise)	208				

perceptions among respondents. Documentation requirements received a mean score of 3.66, indicating that certification paperwork tends to extend project timelines, although not severely. The relatively high standard deviation (1.316) reflects diverse experiences, influenced by differences in certification complexity, project compliance demands and team familiarity with documentation procedures. Early and detailed planning is also viewed as important, with a mean of 3.47. However, the high variability (SD = 1.379) suggests that its effectiveness depends heavily on project readiness, team knowledge and the ability to integrate sustainability requirements into existing workflows.

Meeting certification milestones appears to enhance project efficiency (mean = 3.59), but differences in experience (SD = 1.327) indicate that the benefits vary by project scale, management capacity and consistency in applying certification guidelines. Coordination with certifying bodies earned a mean score of 3.48, reflecting moderate agreement that this interaction affects project scheduling. A high standard deviation (1.383) suggests inconsistent responsiveness among certifying agencies, with some providing timely feedback and others causing delays. Lastly, the need to adjust project timelines to incorporate green practices received a relatively positive mean of 3.60. Respondents generally acknowledge that sustainability measures may extend schedules but view the added time as justified. Still, the standard deviation (1.316) indicates diverse experiences, reflecting the complexity of the green measures and team preparedness.

4.3.2.3 Quality management

The analysis of quality management factors in sustainable construction projects shows strong stakeholder confidence in the positive impact of GBC on project outcomes. The descriptive statistics for quality management are shown in Table 4.3. Sustainable construction practices received a high mean score of 3.67, indicating broad agreement that these practices enhance quality standards. Although the standard deviation of 1.319 indicates some variability, it remains relatively modest and likely reflects differences in how sustainability is applied across projects of varying scales and complexities.

Table 4.3 Quality management descriptive results

	n	Minimum	Maximum	Mean	SD
Integrating sustainable construction practices ensures that the project adheres to higher quality standards	208	1	5	3.67	1.319
The use of high-performance materials required for green building projects results in superior overall quality compared to conventional materials	208	1	5	3.48	1.383
The focus on environmental sustainability in project planning enhances the performance outcomes of construction projects	208	1	5	3.59	1.323
The requirements for achieving green building certifications lead to more rigorous quality control measures throughout the project lifecycle	208	1	5	3.48	1.383
Projects that emphasise sustainable practices generally exhibit better durability and longevity due to the quality of materials and methods used	208	1	5	3.67	1.319
Valid *n* (listwise)	208				

High-performance materials scored a mean of 3.48, suggesting moderate agreement regarding their contribution to quality. The higher standard deviation (1.383) indicates diverse experiences influenced by factors such as cost, availability and differing familiarity with advanced materials. Despite these variations, stakeholders generally understand the value of such materials for improving durability and performance. Environmental sustainability also showed a positive influence on outcomes, with a mean of 3.59. While the standard deviation of 1.323 suggests variation in practice, respondents broadly agree that environmentally focused planning enhances project effectiveness and aligns construction with ecological goals.

Green certification processes received a mean of 3.48, reflecting recognition of the rigorous quality control they require. However, the relatively high standard deviation (1.383) indicates significant differences in experience, often linked to the complexity of certification requirements, documentation demands and varying levels of team readiness. Finally, the perception that sustainability-focused projects achieve greater durability and longevity was strongly supported, again with a mean of 3.67. Although stakeholder experiences vary (SD = 1.319), there is widespread agreement that integrating sustainability into materials, design and construction enhances long-term project performance and lifecycle value.

4.3.2.4 Stakeholder engagement

Table 4.4 summarises the descriptive statistics for stakeholder engagement variables. The analysis of stakeholder engagement in green building projects shows generally positive

Table 4.4 Stakeholder engagement descriptive results (continued on next page)

	n	Minimum	Maximum	Mean	SD
The implementation of sustainable practices in projects leads to increased engagement and interest from stakeholders	208	1	5	3.62	1.335
Stakeholders show greater enthusiasm and involvement when projects include advanced environmental sustainability features	208	1	5	3.48	1.383
The collaborative efforts required for adopting green practices foster stronger teamwork and cooperation among project stakeholders	208	1	5	3.63	1.309
Engaging stakeholders early in the planning of sustainability initiatives leads to more effective project outcomes and shared goals	208	1	5	3.47	1.379

Table 4.4 Continued

	n	Minimum	Maximum	Mean	SD
Effective communication strategies for integrating eco-friendly initiatives enhance transparency and trust among project stakeholders	208	1	5	3.62	1.335
Valid *n* (listwise)	208				

perceptions of sustainability's influence on interest, involvement and collaboration. A mean score of 3.62 indicates strong agreement that sustainable practices enhance stakeholder engagement. However, the relatively high standard deviation (1.335) reflects varied experiences shaped by project type, the visibility of sustainability features and stakeholder familiarity with green initiatives. Advanced environmental features received a moderate mean score of 3.48, suggesting that while these elements can boost enthusiasm, this effect is inconsistent. The high variability (SD = 1.383) indicates that enthusiasm depends on factors such as perceived value, visibility of green technologies and stakeholders' understanding of these features.

Collaboration is a comparatively strong area, with a mean score of 3.63 and a lower standard deviation of 1.309. This consistency suggests that sustainable projects often encourage teamwork because of their multidisciplinary nature and the need for coordinated decision-making. Early stakeholder involvement in sustainability planning received a mean score of 3.47, indicating moderate agreement. However, differing views (SD = 1.379) suggest that its effectiveness varies depending on when stakeholders are engaged, their expertise and the clarity of roles within the process. Effective communication received a high mean score of 3.62, emphasising its role in building stakeholder trust. Yet, the standard deviation of 1.335 shows that communication practices vary across projects, influenced by the clarity, frequency and accessibility of sustainability-related information.

4.3.2.5 Impact of GBC on project management practices

The overall descriptive results for the impact of GBC construct are presented in Table 4.5. The assessment of green building certification (GBC) practices shows a moderately positive perception of their influence on project management, with an overall mean of 3.58. However, the relatively high standard deviation (1.20978) indicates substantial variation in experiences, likely due to differences in project scale, scope and the types of sustainability measures implemented. The integration of sustainability initiatives receives strong support, with a mean of 3.68. Stakeholders generally agree that these initiatives enhance cost, time and quality performance, although the SD of 1.316 suggests uneven experiences influenced by organisational readiness and familiarity with sustainability frameworks.

Sustainability-focused practices show moderate agreement (mean = 3.48), with high variability (SD = 1.383). This suggests that while respondents acknowledge their value, implementation differs widely across projects, with some practices proving easier to integrate than

Table 4.5 Impact of GBC on project management practices

	n	Minimum	Maximum	Mean	SD
The integration of sustainability initiatives in projects enhances overall project management effectiveness in terms of cost, time, quality and stakeholder engagement	208	1	5	3.66	1.316
Sustainability-focused practices contribute to improved project outcomes by promoting better planning and execution	208	1	5	3.48	1.383
The adoption of sustainable construction methods leads to more efficient project management through streamlined processes and better resource utilisation	208	1	5	3.59	1.327
Projects that emphasise environmental sustainability generally show higher levels of satisfaction among stakeholders due to improved project outcomes	208	1	5	3.48	1.383
Incorporating advanced sustainable practices into project management results in enhanced overall project performance and success	208	1	5	3.67	1.319
Valid *n* (listwise)	208				

others. Sustainable construction methods score more favourably (mean 3.59), with efficiency and resource use recognised as areas for improvement. Yet the SD of 1.327 indicates that benefits are not uniformly realised, depending on project conditions, available technologies and stakeholder expertise.

Environmental sustainability-oriented projects also receive a medium rating (mean 3.48), with high variability (SD = 1.383). Differences in how environmental goals align with broader project objectives likely contribute to inconsistent perceptions. Finally, advanced sustainability practices earn strong approval (mean 3.67) with a relatively lower SD of 1.319, indicating more consistent recognition of their value. When effectively implemented, these advanced measures appear to offer clear, widely acknowledged benefits for project performance.

4.3.3 Normality analysis

4.3.3.1 Cost management

The results of the normality tests for cost management are presented in Table 4.6. Normality for the green building certification impact study was assessed using the Shapiro–Wilk and Kolmogorov–Smirnov tests. Both tests examine whether the data follow a normal distribution, with the Shapiro–Wilk test providing a sensitive measure for smaller samples and the Kolmogorov–Smirnov test comparing the sample distribution to a theoretical normal curve.

For the cost management variable, both KS and SW tests were significant ($p = 0.000$), indicating non-normality. However, the kurtosis value of −1.058 falls well within the acceptable range of −3 to + 3, indicating no meaningful deviation in the distribution's shape. This discrepancy is expected because normality tests are highly sensitive to large sample sizes; even small deviations become statistically significant. Although both the Kolmogorov–Smirnov and Shapiro–Wilk tests were significant ($p < 0.001$), this is common in large samples ($n > 200$), where minor deviations become statistically significant. Inspection of skewness, kurtosis and graphical plots indicates approximate normality. Given the sample size and use of composite Likert-scale measures, parametric analyses are considered robust and appropriate.

The graphical distribution and normality assessment for cost management are illustrated in Figure 4.3. The histogram shows that responses cluster around values 3 and 5, with few low ratings, reinforcing the higher mean and median reported earlier and indicating generally positive views toward cost-management practices. The normal Q-Q plot shows clear deviations from the diagonal line, particularly in the tails, with the upper tail reflecting a ceiling effect due to many high ratings. The detrended Q-Q plot further highlights these systematic departures from normality and suggests potential response bias. The box plot shows negative skewness, with the median near the top of the box and a longer lower whisker, confirming earlier statistical observations and indicating patterns in the data that merit further consideration.

4.3.3.2 Time management in green building projects

Table 4.7 shows the normality test results for time management. Normality tests using both the Kolmogorov–Smirnov and Shapiro–Wilk methods produced statistically significant results ($p = 0.000$), indicating rejection of the null hypothesis of normality. This suggests that the time management data are not normally distributed, likely due to outliers, negative skewness and variability in response patterns.

Table 4.6 Kolmogorov–Smirnov and Shapiro–Wilk test results for cost management

	Kolmogorov–Smirnov		Shapiro–Wilk	
	Statistic	Degrees of freedom, *df*	Statistic	Sig.
Cost management	0.246	208	0.847	0.000

Figure 4.3 Histogram, Q-Q plots and box plot for cost management

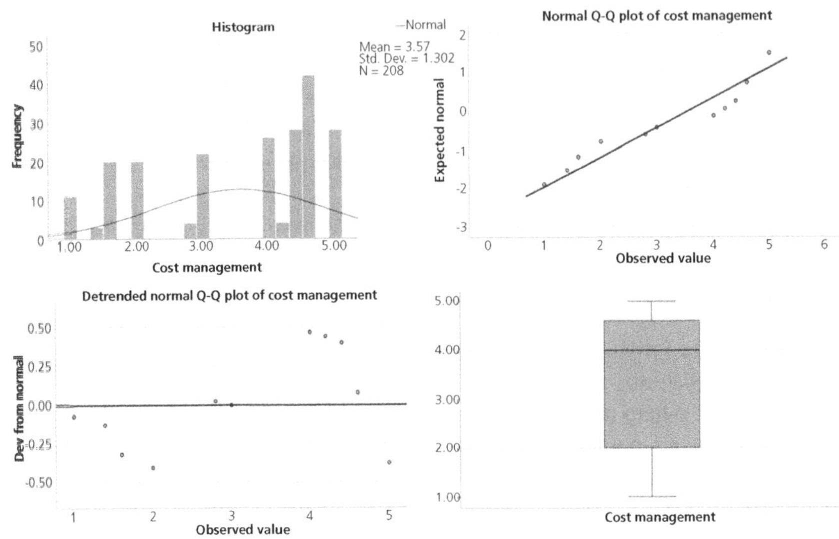

Table 4.7 Kolmogorov–Smirnov and Shapiro–Wilk test results for time management

	Kolmogorov–Smirnov		Shapiro–Wilk	
	Statistic	*df*	Statistic	Sig.
Time management	0.244	208	0.847	0.000

Despite this, the kurtosis value of −1.023 falls within the commonly accepted range of −3 to +3, indicating no substantial deviation from normality. This contradiction is explained by the sensitivity of normality tests, especially with large sample sizes, where even small departures from normality generate significant p-values. Therefore, although the formal tests detect non-normality, the kurtosis measure supports the view that the time management data are approximately normal and can still justify the use of parametric statistical methods.

The distribution and normality diagnostics for time management are displayed in Figure 4.4. The histogram shows a generally bell-shaped distribution concentrated between scores 3.0 and 5.0, with fewer responses at the lower end. This indicates a tendency toward higher time management effectiveness, consistent with descriptive statistics. The normal Q-Q plot shows deviations from the reference line, particularly at the tails, confirming non-normality and consistent with the normality tests. In contrast, the detrended Q-Q plot shows several deviations and possible outliers, illustrating how individual data points deviate from a theoretical normal distribution. The box plot reflects negative skewness, with the median closer to the upper edge of the box and a shorter lower whisker. This provides a clear visual of how time management responses are distributed, with a higher concentration of scores at the upper end.

Figure 4.4 Histogram, Q-Q plots and box plot for time management

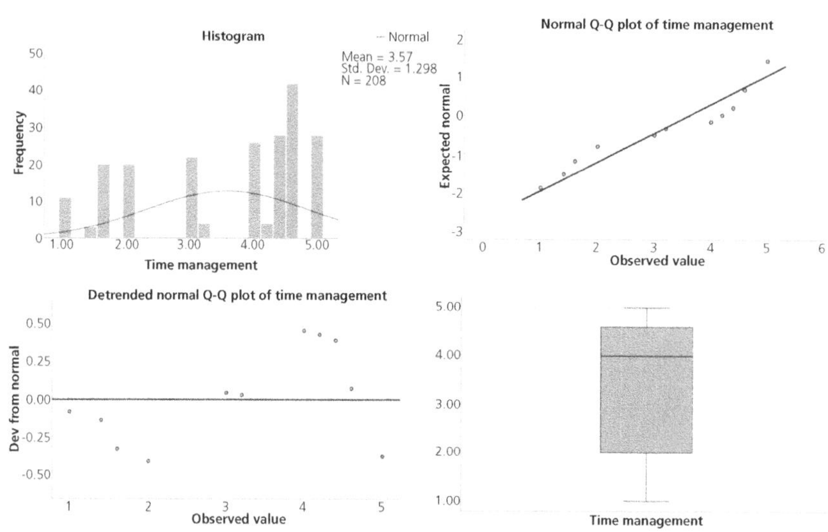

4.3.3.3 Quality management in green building projects

The normality assessment for quality management is summarised in Table 4.8. Both the Kolmogorov–Smirnov and Shapiro–Wilk tests yield statistically significant results ($p = 0.000$), indicating rejection of the null hypothesis of normality. This suggests that the quality management data are not normally distributed, likely due to outliers, negative skewness and uneven response patterns.

However, the kurtosis value of −1.025 falls well within the commonly accepted range of −3 to +3, suggesting no substantial deviation from normality. According to established guidelines in the literature, kurtosis within this range indicates that the data is approximately normal.

The apparent contradiction between the test results and the kurtosis measure can be explained by the sensitivity of normality tests, particularly with large sample sizes, where even small deviations from normality can yield significant results. Thus, although the Kolmogorov–Smirnov and Shapiro–Wilk tests detect non-normality, the kurtosis value indicates that the overall distribution of the quality management variable is sufficiently close to normal to justify the use of parametric statistical methods in subsequent analyses.

Table 4.8 Kolmogorov–Smirnov and Shapiro–Wilk test results for quality management

	Kolmogorov–Smirnov		Shapiro–Wilk	
	Statistic	*df*	Statistic	Sig.
Quality management	0.244	208	0.847	0.000

Figure 4.5 Histogram, Q-Q plots and box plot for quality management

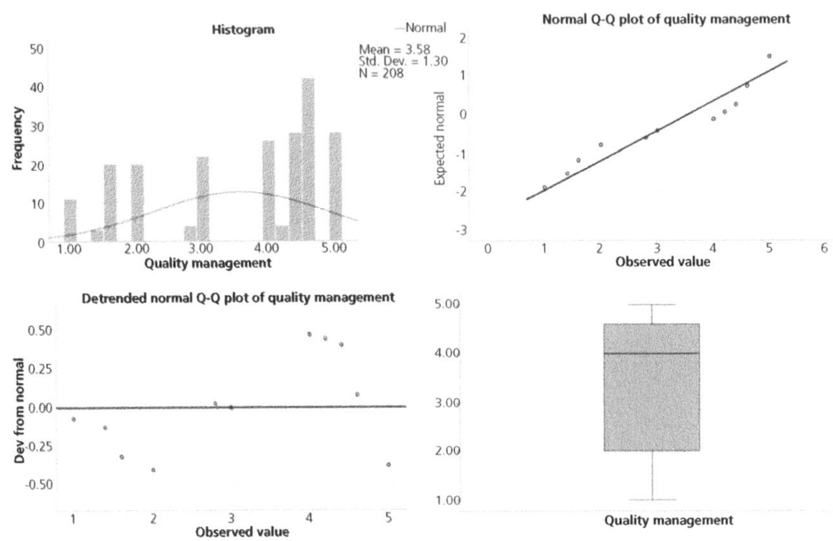

Figure 4.5 illustrates the histogram, Q-Q plots and box plot for quality management. The histogram shows a significant pile-up of responses in the 3.0–5.0 range, tapering off toward fewer responses at lower ratings. This result is consistent with the descriptive statistics, indicating a generally positive view of the influence of sustainability on quality management, although there is some divergence. A normal Q-Q plot shows deviations in the tails (indicating non-normality in statistical tests) and suggests the presence of outliers in the data. The detrended Q-Q plot is a plot of individual points representing deviations from a normal distribution, and it shows that the anomalous data fall at the lower end of the scale. The box plot shows a left-skewed distribution of responses, as indicated by the more prominent left tail, which is centred more closely to the box's fourth quartile, and an extended bottom whisker, suggesting that respondents are more divided when quality management practices are less effective.

4.3.3.4 Stakeholder engagement in green building projects

Table 4.9 presents the results of the normality test for stakeholder engagement. The Kolmogorov–Smirnov and Shapiro–Wilk tests both yield significant results ($p = 0.000$), indicating that the stakeholder engagement data do not meet the assumption of normality. However, the kurtosis value of −1.057 falls well within the acceptable −3 to +3 range, suggesting no substantial deviation from normality in terms of distribution shape. This discrepancy is expected, as normality tests are highly sensitive with larger samples and may flag even minor deviations as significant. Given the acceptable kurtosis value, the distribution can still be considered reasonably normal, supporting the use of parametric statistical methods for further analysis.

Table 4.9 Kolmogorov–Smirnov and Shapiro–Wilk test results for stakeholder engagement

	Kolmogorov–Smirnov		Shapiro–Wilk	
	Statistic	*df*	Statistic	Sig.
Stakeholder engagement	0.242	208	0.849	0.000

Figure 4.6 Histogram, Q-Q plots and box plot for stakeholder engagement

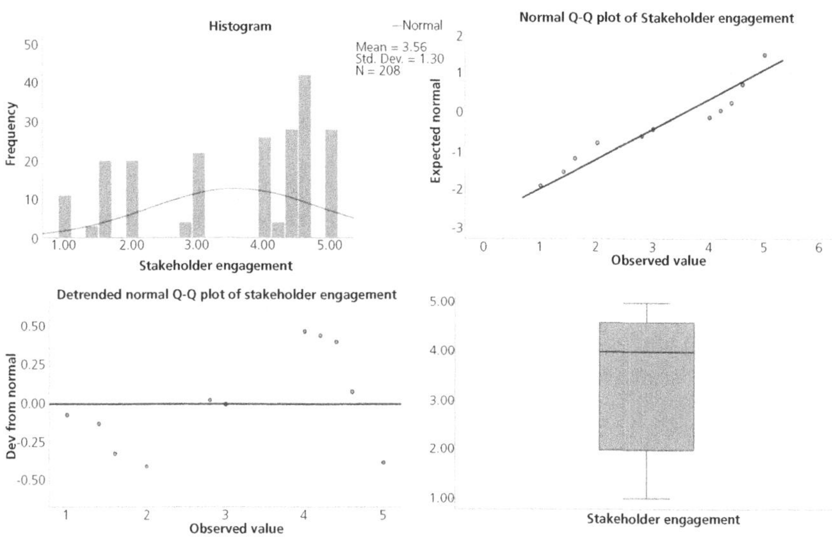

The graphical normality diagnostics for stakeholder engagement are shown in Figure 4.6. The distribution resembles a bell curve centred between 3.0 and 5.0, with more high-end responses, indicating generally favourable perceptions of stakeholder engagement in green building practices. The normal Q-Q plot shows deviations from the reference line at the upper end, suggesting that the highest engagement scores are higher than expected under the normality assumption. The detrended Q-Q plot highlights these deviations more clearly, showing where individual scores depart from an ideal normal distribution. The median's position near the upper quartile and the longer lower whisker indicate a slight left skew. Overall, these visual plots effectively show the distribution of stakeholder engagement scores' central tendency and spread.

4.3.3.5 Impact of GBC

The normality test results for the impact of GBC variable are shown in Table 4.10. Both normality tests (Kolmogorov–Smirnov and Shapiro–Wilk) produced very small *p*-values, indicating that the impact of GBC variable does not follow a normal distribution. This may

Table 4.10 Kolmogorov–Smirnov and Shapiro–Wilk test results for impact of GBC

	Kolmogorov–Smirnov		Shapiro–Wilk	
	Statistic	*df*	Statistic	Sig.
Impact of GBC	0.244	208	0.847	0.000

reflect unusual values or slight skewness in the data. However, the kurtosis value of –1.025 falls within the commonly accepted range of –3 to +3, suggesting no serious deviation from normality (Frost, 2019). The discrepancy arises because normality tests – especially with larger samples – are highly sensitive and may flag even minor deviations. Therefore, despite the significant test results, the kurtosis value indicates that the impact of GBC variable is reasonably close to normal, supporting the use of parametric statistical methods.

Figure 4.7 presents the graphical normality assessment for the impact of GBC variable. The histogram shows a left-skewed distribution, with most responses clustered between 3.0 and 5.0, indicating generally positive perceptions, although fewer respondents provided lower ratings.

For the normal Q-Q plot, deviations from the diagonal line, especially in the tails, indicate non-normality, consistent with the statistical test results. Meanwhile, the detrended Q-Q plot

Figure 4.7 Histogram, Q-Q plots and box plot for impact of GBC practices

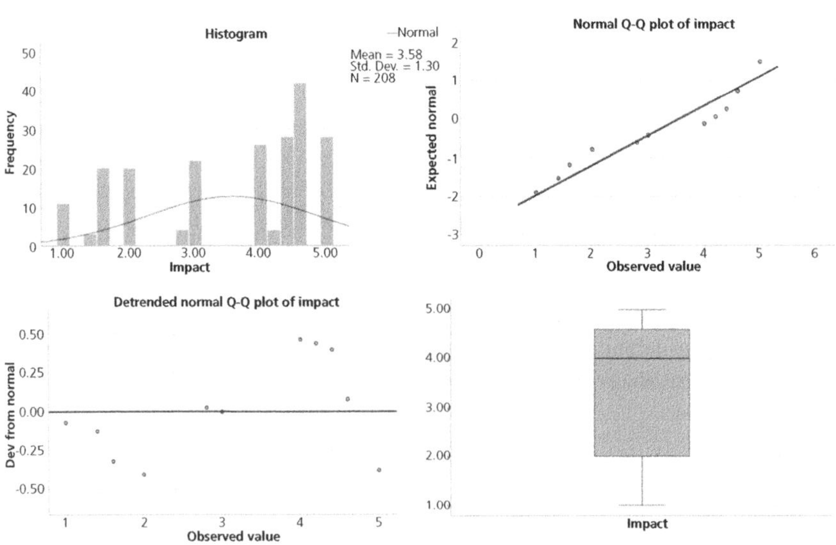

highlights individual deviations from normality and reveals severe outliers contributing to the skew. The box plot shows a longer distance from the median to the upper quartile than to the lower whisker, indicating negative skewness and effectively summarising the distribution's central tendency and spread for the self-reported impact of GBC.

4.3.4 Reliability analysis

Reliability in research refers to the consistency of a measurement tool – its ability to produce stable, repeatable results across time, participants and contexts. When studying the impact of GBC on project management, assessing the instrument's reliability is essential. Cronbach's alpha, which ranges from 0 to 1, is the most commonly used measure of internal consistency. Values above 0.7 are considered acceptable, and those above 0.8 indicate good to excellent reliability (Abolghasemi *et al.*, 2026).

The study evaluated internal consistency for the following variables using Cronbach's alpha (α):

- cost management
- time management
- quality management
- stakeholder engagement
- overall impact of GBC on project management.

The reliability coefficients for each construct are summarised in Table 4.11.

Cronbach's alpha (α) values ranging from 0.849 to 0.918 indicate strong internal consistency, suggesting that items within each construct reliably measure the same underlying dimension. While most constructs showed strong agreement, a few items, particularly those related to time management, were less robust, suggesting that perceived benefits in these areas may vary with project priorities. Overall, the findings show that green certification can support long-term cost savings, improve project quality and strengthen stakeholder collaboration. The strong reliability of the measures enhances confidence in the consistency and stability of these results, offering meaningful insight into the broader advantages of green building certification (Abolghasemi *et al.*, 2026).

Table 4.11 Cronbach's alpha for each variable

Variable	Cronbach's alpha, α	No. of questions
Cost management	0.883	5
Time management	0.849	5
Quality management	0.903	5
Stakeholder engagement	0.892	5
Impact of GBC	0.918	5

4.3.5 Correlation analysis

4.3.5.1 Spearman's correlation analysis

Spearman's correlation is a nonparametric measure used to determine the strength and direction of the association between two ranked variables. The results provide valuable insights into how green building practices relate to different dimensions of project management.

IMPACT OF GBC AGAINST COST MANAGEMENT

The Spearman correlation between impact of GBC and cost management is shown in Table 4.12. The Spearman's rho (ρ) value of 0.680 indicates a strong positive rank-order relationship between impact of GBC and cost management. As the ranked value of one variable increases, the other tends to rise as well. The highly significant p-value (<0.001) confirms that this association is not due to chance. Because Spearman's rho is nonparametric, it is useful when data may be non-normal or influenced by outliers, and it provides insight into relationships that are not strictly linear. This strong coefficient aligns closely with the Pearson correlation, suggesting that the relationship between the two variables is linear and consistent when ranked. The slightly lower value compared to Pearson (0.680 against 0.720) may indicate minor nonlinearities or outlier effects, such as high-impact initiatives occasionally experiencing unexpected cost variations. Overall, the Spearman result reinforces the strong, positive association between impact of GBC and cost management.

IMPACT OF GBC AGAINST TIME MANAGEMENT

Table 4.13 presents the Spearman correlation between impact and time management. The Spearman's rho (ρ) value of 0.710 indicates a strong positive rank-order relationship between impact of GBC and time management. This shows that even if the relationship is not perfectly linear, projects with higher-impact initiatives generally achieve better time management outcomes. The highly significant p-value (<0.001) further supports the strength and reliability of this association. The slight difference between the Pearson (0.750) and Spearman (0.710) coefficients suggests minor nonlinearities or outliers, such as high-impact projects

Table 4.12 Spearman's correlation analysis (impact of GBC against cost management)

Spearman's rho, ρ	Cost management	Impact of GBC
Cost management		
Correlation coefficient	1.000	0.680[a]
Sig. (2-tailed)	–	<0.001
n	208	208
Impact of GBC		
Correlation coefficient	0.680[a]	1.000
Sig. (2-tailed)	<0.001	–
n	208	208

[a]Correlation is significant at the 0.01 level (2-tailed).

Table 4.13 Spearman's correlation analysis (impact of GBC against time management)

Spearman's rho, ρ	Impact of GBC	Time management
Impact of GBC		
Correlation coefficient	1.000	0.710[a]
Sig. (2-tailed)	–	<0.001
n	208	208
Time management		
Correlation coefficient	0.710[a]	1.000
Sig. (2-tailed)	<0.001	–
n	208	208

[a]Correlation is significant at the 0.01 level (2-tailed).

experiencing unexpected delays. Nonetheless, the Spearman result confirms that, across projects, higher-impact efforts consistently correspond to stronger time-management performance, emphasising the practical advantage of focusing on impactful initiatives.

IMPACT OF GBC AGAINST QUALITY MANAGEMENT

The Spearman correlation between impact of GBC and quality management is summarised in Table 4.14. The Spearman's rho (ρ) value of 0.744 indicates a strong positive rank-order relationship between quality management and impact of GBC, meaning that projects with higher quality management rankings also tend to achieve higher impact of GBC rankings. The highly significant p-value (<0.001) confirms that this association is not due to chance. The slight difference between Pearson's r (0.780) and Spearman's ρ (0.744) suggests minor

Table 4.14 Spearman's correlation analysis (impact of GBC against quality management)

Spearman's rho, ρ	Quality management	Impact of GBC
Quality management		
Correlation coefficient	1.000	0.744[a]
Sig. (2-tailed)	–	<0.001
n	208	208
Impact of GBC		
Correlation coefficient	0.744[a]	1.000
Sig. (2-tailed)	<0.001	–
n	208	208

[a]Correlation is significant at the 0.01 level (2-tailed).

nonlinearities or the influence of outliers. In some cases, strong quality practices may not immediately translate into impact due to external challenges, while certain projects may achieve high impact for reasons beyond quality management. Nonetheless, the overall results support a consistently strong and meaningful association between quality management and impact of GBC.

IMPACT OF GBC AGAINST STAKEHOLDER ENGAGEMENT

Table 4.15 displays the Spearman correlation between impact of GBC and stakeholder engagement. The Spearman's rho (ρ) value of 0.720 indicates a strong positive rank-order relationship between impact of GBC and stakeholder engagement. Although slightly lower than the Pearson coefficient, it still reflects a meaningful and consistent association, with the small difference suggesting minor nonlinearities or outliers. The highly significant *p*-value (<0.001) confirms the reliability of this relationship. Overall, the results show that projects with higher levels of stakeholder engagement consistently achieve greater impact, even when the relationship is not perfectly linear.

4.3.5.2 Pearson's correlation analysis

The Pearson correlation, which measures the linear association between two continuous variables, supports and complements the insights obtained from the Spearman analysis.

IMPACT OF GBC AGAINST COST MANAGEMENT

The Pearson correlation between impact of GBC and cost management is presented in Table 4.16. The Pearson correlation coefficient of 0.720 indicates a strong positive linear relationship between impact of GBC and cost management, suggesting that higher-impact initiatives are generally associated with more efficient cost management. The highly significant *p*-value (<0.001) confirms that this relationship is unlikely to be due to chance. Although the association is strong, it is not perfect, meaning that while impactful practices often contribute to better cost outcomes, other factors also influence cost management performance. This coefficient suggests that a substantial portion of the variation in cost management is

Table 4.15 Spearman's correlation analysis (impact of GBC against stakeholder engagement)

Spearman's rho, ρ	Impact of GBC	Stakeholder engagement
Impact of GBC		
Correlation coefficient	1.000	0.720[a]
Sig. (2-tailed)	–	<0.001
n	208	208
Stakeholder engagement		
Correlation coefficient	0.720[a]	1.000
Sig. (2-tailed)	<0.001	–
n	208	208

[a]Correlation is significant at the 0.01 level (2-tailed).

Table 4.16 Pearson's correlation analysis (impact of GBC against cost management)

Pearson correlation	Cost management	Impact of GBC
Cost management		
Pearson correlation	1	0.720[a]
Sig. (2-tailed)	–	<0.001
n	208	208
Impact of GBC		
Pearson correlation	0.720[a]	1
Sig. (2-tailed)	<0.001	–
n	208	208

[a]Correlation is significant at the 0.01 level (2-tailed).

attributable to differences in impact. Still, some variability also arises from external factors such as unexpected expenses or market conditions. Overall, the results highlight that enhancing impactful initiatives can meaningfully support cost efficiency, while acknowledging that additional factors still play a role.

IMPACT OF GBC AGAINST TIME MANAGEMENT

Table 4.17 shows the Pearson correlation between impact of GBC and time management. The Pearson correlation coefficient of 0.750 shows a strong positive linear relationship between impact of GBC and time management, indicating that more effective impact initiatives are generally associated with improved time management performance. The highly significant

Table 4.17 Pearson's correlation analysis (impact of GBC against time management)

Pearson correlation	Impact of GBC	Time management
Impact of GBC		
Pearson correlation	1	0.750[a]
Sig. (2-tailed)	–	<0.001
n	208	208
Time management		
Pearson correlation	0.750[a]	1
Sig. (2-tailed)	<0.001	–
n	208	208

[a]Correlation is significant at the 0.01 level (2-tailed).

p-value (<0.001) confirms that this relationship is statistically reliable and not due to chance. This coefficient suggests that impact initiatives account for about 56% of the variation in time management outcomes, with the remainder attributable to other factors, such as resource constraints, unforeseen events or external disruptions. Overall, the strong association suggests that projects emphasising impactful practices tend to achieve better scheduling accuracy, fewer delays and smoother coordination, even though time management is still affected by real-world complexities.

IMPACT OF GBC AGAINST QUALITY MANAGEMENT

The Pearson correlation between impact of GBC and quality management is reported in Table 4.18. The Pearson correlation coefficient of 0.780 indicates a strong positive linear relationship between quality management and impact of GBC, meaning that better quality management practices are closely associated with higher project impact. The highly significant p-value (<0.001) confirms that this relationship is not due to chance. This coefficient suggests that quality management explains about 60.8% of the variation in impact of GBC, underscoring its crucial role in driving successful outcomes. However, the remaining 39.2% indicates that other factors, such as stakeholder engagement, cost management or external conditions, also contribute to the overall impact of GBC. Thus, while quality is a key driver, it must work alongside other project components to maximise results.

IMPACT OF GBC AGAINST STAKEHOLDER ENGAGEMENT

Table 4.19 presents the Pearson correlation between impact of GBC and stakeholder engagement. The Pearson correlation coefficient of 0.820 indicates a strong positive linear relationship between impact of GBC and stakeholder engagement, meaning that higher levels of stakeholder involvement are closely associated with greater project impact. The highly significant p-value (<0.001) confirms that this relationship is statistically reliable. This coefficient suggests that stakeholder engagement accounts for about 67.2% of the variation in impact of GBC, underscoring the crucial role of active participation, communication and collaboration

Table 4.18 Pearson's correlation analysis (impact of GBC against quality management)

Pearson correlation	Quality management	Impact of GBC
Quality management		
Pearson correlation	1	0.780[a]
Sig. (2-tailed)	–	<0.001
n	208	208
Impact of GBC		
Pearson correlation	0.780[a]	1
Sig. (2-tailed)	<0.001	–
n	208	208

[a]Correlation is significant at the 0.01 level (2-tailed).

Table 4.19 Pearson's correlation analysis (impact of GBC against stakeholder engagement)

Pearson correlation	Impact of GBC	Stakeholder engagement
Impact of GBC		
Pearson correlation	1	0.820[a]
Sig. (2-tailed)	–	<0.001
n	208	208
Stakeholder engagement		
Pearson correlation	0.820[a]	1
Sig. (2-tailed)	<0.001	–
n	208	208

[a]Correlation is significant at the 0.01 level (2-tailed).

in achieving successful outcomes. Although the association is strong, the remaining 32.8% of unexplained variation shows that other factors – such as cost management, quality practices or external influences – also contribute to overall impact.

4.3.6 Multiple regression analysis of green building practices

4.3.6.1 Model summary and statistical fit

The multiple regression analysis evaluates how key management practices influence the outcomes of green building initiatives. The regression model summary is presented in Table 4.20. The model demonstrates a very strong fit, accounting for a substantial proportion of the variance in the dependent variable ($R^2 = 0.884$).

The regression model shows a very strong relationship between the predictors (stakeholder engagement, cost, time and quality) and the dependent variable (impact). The R-value of 0.940 indicates a strong positive correlation, and the R^2 of 0.884 indicates that these four variables explain 88.4% of the variance in impact. This indicates a very large effect size according to conventional standards, suggesting that the selected management practices are dominant predictors of project impact in green building initiatives. The adjusted R^2 (0.877) confirms that the model remains robust even after accounting for the number of predictors,

Table 4.20 Regression model summary

R	R^2	Adj. R^2	Std. error of the estimate	R^2 change	F change	$df1$	$df2$	Sig. F change
0.940[a]	0.884	0.877	0.135	0.884	89.762	4	203	<0.001

[a]Predictors: (constant), stakeholder, cost, time, quality.

indicating minimal risk of overfitting. The low standard error of the estimate (0.135) suggests that predicted values closely match actual outcomes, reflecting strong model accuracy. The F change (89.762) and highly significant sig. F change (<0.001) confirm that the predictors collectively meaningfully explain impact. The large change in R^2 further indicates that these variables are essential components of the model.

Overall, the findings demonstrate that stakeholder engagement, cost management, time management and quality management are major drivers of project impact. While the model explains a large proportion of variance, the remaining unexplained portion indicates that additional external or contextual factors may also influence outcomes. These results highlight the value of strengthening key management practices to enhance the effectiveness and impact of green building initiatives.

4.3.6.2 ANOVA (analysis of variance)

The ANOVA results for the regression model are shown in Table 4.21. The ANOVA table evaluates whether the predictors (stakeholder engagement, cost, time and quality) collectively have a significant effect on impact of GBC. The regression sum of squares (326.500) shows that the model explains a substantial portion of the variance in impact of GBC. In contrast, the much smaller residual sum of squares (42.695) indicates that only a limited amount of variance remains unexplained. This aligns with the high R^2 (0.884), confirming the model's strong explanatory power.

The degrees of freedom ($df = 4$ for regression, 203 for residuals) allow calculation of mean squares, which show an average explained variance of 81.625 and a very low average unexplained variance of 0.018. This highlights the model's precision. The F-statistic of 92.562 demonstrates that the explained variance is significantly greater than the unexplained variance, and the p-value (<0.001) confirms that the model is statistically significant. Overall, the ANOVA results indicate that the four predictors jointly make a meaningful and reliable contribution to explaining impact of GBC.

4.3.6.3 Coefficients analysis

The regression coefficients for each predictor are summarised in Table 4.22. The coefficients table shows how each predictor (cost, time, quality and stakeholder engagement) contributes to impact of GBC. The unstandardised coefficients indicate the expected change in impact of GBC for a one-unit change in each predictor, while the standardised coefficients (β) allow

Table 4.21 ANOVA summary for regression analysis on impact of GBC on project management

Model	Sum of squares	*df*	Mean square	*F* / Sig.
Regression	326.500	4	81.625	$F = 92.562$ / Sig. < 0.001
Residual	42.695	203	0.183	
Total	369.195	207		

Note: dependent variable: impact; predictors: (constant), stakeholder, cost, time, quality.

Table 4.22 Regression coefficients for cost management, quality management and stakeholder engagement

Model	Unstandardised coefficient, B	Std. error	Standardised coefficient, β	t	Sig.	Zero-order	Partial	Part
(Constant)	0.102	0.045		2.267	0.024			
Cost management	0.135	0.053	0.145	2.547	0.012	0.650	0.220	0.180
Time management	−0.082	0.058	−0.089	−1.414	0.159	0.700	−0.090	−0.060
Quality management	0.420	0.067	0.451	6.269	<0.001	0.790	0.400	0.350
Stakeholder engagement	0.305	0.051	0.324	5.980	<0.001	0.760	0.380	0.330

Note: collinearity statistics: cost management (tolerance 0.751, VIF 1.331), time management (tolerance 0.682, VIF 1.466), quality management (tolerance 0.650, VIF 1.538), stakeholder engagement (tolerance 0.702, VIF 1.423).

comparison of their relative importance. Quality management has the strongest effect (β = 0.451), followed by stakeholder engagement (β = 0.324) and cost management (β = 0.145). Time management has a negative but non-significant effect (p = 0.159), suggesting its influence on impact of GBC is limited or indirect.

The t-values (which test whether each predictor's coefficient differs significantly from zero) and p-values confirm that cost management, quality management and stakeholder engagement are significant predictors ($p < 0.05$), whereas time management is not. Zero-order, partial and part correlations further emphasise the importance of quality management and stakeholder engagement in predicting impact. Collinearity diagnostics show low variance inflation factor (VIF) values (1.331–1.538), indicating minimal multicollinearity and confirming the regression model's stability. Overall, the results highlight quality management and stakeholder engagement as the strongest drivers of impact of GBC, with cost management also playing a meaningful role. Time management, however, does not significantly predict impact.

4.3.6.4 Collinearity diagnostics

The collinearity diagnostics for the regression model are presented in Table 4.23. Although condition indices exceed 15 in some dimensions, VIF values remain below 2.0 and tolerance values exceed 0.6, indicating that multicollinearity is not severe and does not materially threaten coefficient stability. The predictors show some shared variance, as expected given the conceptual overlap among project management dimensions, yet the model remains statistically stable.

4.4. Practical insights

The findings of this chapter offer several practical insights for project managers, developers, policy makers and industry stakeholders involved in green building projects in Malaysia.

First, project managers should integrate sustainability requirements early into project planning and budgeting processes. Although green building certification often entails higher upfront costs, empirical results show strong positive relationships with long-term cost efficiency, quality and stakeholder satisfaction. Applying tools such as lifecycle costing and value engineering can help justify initial investments and prevent cost overruns in later project stages.

Second, time management practices must be adapted to accommodate certification requirements. Documentation, coordination with certification bodies and compliance checks increase planning complexity; however, early scheduling, alignment of milestones with certification criteria and proactive coordination can mitigate delays. The results suggest that projects that embed green requirements into the schedule from the outset experience fewer disruptions and smoother execution.

Third, quality management systems should be expanded to include environmental performance criteria alongside traditional construction standards. The strong statistical relationship between green building certification and quality outcomes indicates that sustainability-focused quality controls enhance durability, functionality and long-term asset value. Project teams should therefore align inspection, testing and commissioning processes with certification benchmarks.

Table 4.23 Collinearity diagnostics for regression predictors

Dimension	Eigen value	Condition index	(Constant)	Cost management	Time management	Quality management	Stakeholder engagement
1	4.500	1.000	0.01	0.02	0.00	0.01	0.00
2	0.400	3.350	0.25	0.30	0.10	0.15	0.05
3	0.050	15.120	0.05	0.20	0.25	0.05	0.20
4	0.030	20.430	0.05	0.10	0.20	0.30	0.30
5	0.020	25.670	0.64	0.38	0.45	0.49	0.45

Note: dependent variable: impact of GBC.

Fourth, stakeholder engagement emerges as a critical driver of project impact. Green building projects involve a broader stakeholder ecosystem, including regulators, certification bodies, consultants and end-users. Clear communication strategies, early involvement and collaborative decision-making improve trust, reduce conflict and strengthen overall project outcomes.

Finally, industry practitioners and policy makers should invest in training, digital tools and standardised guidelines to support the delivery of green projects. Enhancing competencies in sustainable project management, BIM and performance monitoring will enable more consistent and scalable implementation of green building certifications across the Malaysian construction sector.

4.5. Conclusion and recommendations

This chapter highlights that green building certification (GBC) has significant implications for project management in Malaysia. Although green-certified projects require higher upfront costs due to sustainable materials and technologies, long-term savings from reduced operational and maintenance expenses make them cost-effective.

4.5.1 Implications and strategic considerations

Time management is initially more demanding because certification requires extensive documentation and coordination, but early problem identification ultimately reduces design changes and improves construction efficiency. GBC also enhances project quality, as certification standards require superior materials and eco-friendly practices that exceed typical industry requirements. Furthermore, stakeholder engagement increases with the involvement of multiple parties, including architects, engineers, contractors and regulatory bodies, thereby improving communication, collaboration and overall project success.

4.5.2 Limitations

Despite its contributions, the study has several limitations. The sample lacked diversity across industries and geographic regions, potentially limiting generalisability. Reliance on self-reported survey data introduces the risk of respondent bias, particularly for subjective measures such as engagement and satisfaction. The cross-sectional design restricts insights into long-term outcomes or changes over a project's lifecycle. Additionally, rapid developments in green technologies and evolving regulations may affect the relevance of the findings over time.

4.5.3 Recommendations and future research

Future research should expand the sample to include more industries and regions, especially rural areas and other construction sectors. Longitudinal studies would provide deeper insights into the lasting effects of GBC across different project stages. Combining quantitative surveys with qualitative methods, such as interviews or case studies, could yield richer insights into challenges and best practices. Further exploration into emerging technologies such as AI, Big Data and IoT could reveal how these tools enhance GBC implementation. Cost–benefit analyses, especially for small and medium-sized enterprises (SMEs), would also help organisations better understand the financial implications of adopting GBC.

In conclusion, GBC significantly influences project management practices in Malaysia, improving long-term cost efficiency, quality and stakeholder participation despite initial complexities and longer preparation times. The findings emphasise the need for an integrated approach to sustainability throughout all project phases. As the construction sector shifts toward greener practices, GBC is poised to become a key driver of sustainable development.

REFERENCES

Abolghasemi M, Al-Mhdawi MKS, Rahimian F and Qazi A (2026) Exploring the nexus of stakeholder management, project performance and stakeholder satisfaction in Malaysian residential building projects: a PLS-SEM approach. *Engineering, Construction and Architectural Management* **33(3)**: 2089–2117.

Cao Y, Kamaruzzaman SN and Aziz NM (2022) Green building construction: a systematic review of BIM utilization. *Buildings* **12(8)**: 1205.

Frost J (2019) *Introduction to Statistics: An Intuitive Guide for Analyzing Data and Unlocking Discoveries*. Statistics By Jim Publishing, State College, PA, USA.

Hayes J (2025) *The Theory and Practice of Change Management*. Bloomsbury, London, UK.

Liu T, Chen L, Yang M *et al.* (2022) Sustainability considerations of green buildings: a detailed overview on current advancements and future considerations. *Sustainability* **14(21)**: 14393.

Lubis NW (2022) Resource-based view (RBV) in improving company strategic capacity. *Research Horizon* **2(6)**: 587–596.

Mahajan R, Lim WM, Sareen M, Kumar S and Panwar R (2023) Stakeholder theory. *Journal of Business Research* **166**: 114104.

Mio C, Costantini A and Panfilo S (2022) Performance measurement tools for sustainable business: a systematic literature review on the sustainability balanced scorecard use. *Corporate Social Responsibility and Environmental Management* **29(2)**: 367–384.

Mustaffa NK, Mat Isa CM, Ekundayo D and Joseph VRA (2022) Barriers and strategies for improving carbon emissions management approaches in Malaysian construction. *Construction Economics and Building* **22(3)**: 99–123.

Muthusamy K and Chew L (2020) Critical success factors for the implementation of building information modeling (BIM) among Construction Industry Development Board (CIDB) G7 contractors in the Klang Valley, Malaysia. In *2020 IEEE European Technology and Engineering Management Summit (E-TEMS)*. Institute of Electrical and Electronics Engineers, New York, NY, USA, pp. 1–6.

Risi D, Vigneau L, Bohn S and Wickert C (2023) Institutional theory-based research on corporate social responsibility: bringing values back in. *International Journal of Management Reviews* **25(1)**: 3–23.

Solla M, Elmesh A, Memon ZA *et al.* (2022) Analysis of BIM-based digitising of Green Building Index (GBI): assessment method. *Buildings* **12(4)**: 429.

Zakaria Z, Ariff NRM, Noorhani NMA and Hussain MM (2024) Sustainable green building initiatives in Malaysia: issues in the implementation practices. *Planning Malaysia* **22(3)**: 530–542.

Zhang X, Nutakor F, Minlah MK and Li J (2023) Can digital transformation drive green transformation in manufacturing companies? Based on a socio-technical systems theory perspective. *Sustainability* **15(3)**: 2840.

Advancing Project Management: Frameworks, Digital Transformation, Sustainability and Governance

Mohammed Reyasudin Basir Khan
ISBN 978-1-83742-799-4
https://doi.org/10.1108/978-1-83742-798-720261006

Chapter 5
Managing projects amid geopolitical challenges: lessons from engineering firms

Nur Dayana Fatin Zulkhurnain and Mohammed Reyasudin Basir Khan

Tun Razak Graduate School, Universiti Tun Abdul Razak (UNIRAZAK), Kuala Lumpur, Malaysia

5.1. Introduction

Penang, located in Northwestern Malaysia, has evolved into one of Southeast Asia's key industrial and engineering hubs. Its strong manufacturing base, supportive infrastructure and skilled workforce have attracted multinational firms across electronics, machinery, construction and civil engineering (Athukorala, 2017). Over recent decades, Penang has transitioned from a regional manufacturing base to an international engineering and technology hub, playing a critical role in global supply chains, particularly in the semiconductor, electronics and automotive industries (Raj-Reichert, 2020).

Engineering firms in Penang, both large and small, operate within a highly interconnected ecosystem that depends heavily on the smooth cross-border movement of materials, components, services and skilled labour. While this interconnectedness is a competitive strength, it also increases vulnerability to geopolitical tensions, including trade wars, sanctions, regulatory shifts and regional political instability. Such tensions can disrupt trade flows, alter market dynamics and create uncertainty in international relations, with direct implications for project-based industries.

Geopolitical tensions often lead to tariffs, trade barriers, regulatory changes and currency volatility, which affect project timelines and resource allocation. Sudden changes in trade policy, export controls or sanctions can delay the procurement of raw materials, specialised equipment and technological components (Asadollah *et al.*, 2024). For Penang's engineering firms that rely on global suppliers, these disruptions can lead to project delays, cost overruns and difficulties in meeting contractual deadlines. Delayed projects not only undermine profitability but also harm client relationships, reputational standing and future business opportunities.

Resource allocation is also severely affected. Engineering projects require continuous access to raw materials (e.g. steel, copper, aluminium), machinery, skilled labour and financial capital. When geopolitical tensions affect supplier countries or trade routes, firms may face shortages, price spikes or longer lead times. Political unrest and stricter immigration rules can limit access to foreign skilled labour, thereby delaying projects and reducing quality. Frequent regulatory changes such as new trade rules, environmental standards or labour regulations further complicate compliance and planning, especially in a trade-dependent region like Penang.

In response, engineering companies are increasingly adopting risk management practices and technology-enabled solutions such as project management software, digital supply chain tools and real-time monitoring to anticipate disruptions and proactively adjust plans (Tortorella *et al.*, 2025). However, despite these efforts, geopolitical risks continue to pose complex challenges to project performance and resource allocation, underscoring the need for systematic study in the Penang context.

Engineering companies in Penang are experiencing rising project delays and escalating costs due to intensifying geopolitical tensions. Global rivalries, particularly between the USA and China, along with trade policy fluctuations, sanctions and regional instability, disrupt supply chains and labour flows essential to project execution. These disruptions complicate material procurement, access to specialised equipment, availability of skilled labour and compliance with evolving regulations, thereby undermining operational effectiveness and financial performance.

Trade wars and sanctions lead to tariffs and export restrictions, which increase costs and reduce the availability of critical inputs. Political instability in key supplier regions further amplifies uncertainty, causing shortages, delivery delays and currency volatility. For Penang firms that depend on global sourcing and foreign talent, even small disruptions can cascade into significant project delays and cost overruns.

Technological dimensions add another layer of complexity. Geopolitical conflicts can trigger export controls on advanced technologies, restrict access to high-tech components and force rapid compliance with new standards. This may delay technology adoption, increase development and testing time, and discourage investment in innovation due to heightened uncertainty. In parallel, tightening immigration policies and labour-movement restrictions can worsen talent shortages, affecting project quality and schedules.

Although global studies widely acknowledge that geopolitical tensions disrupt supply chains and project delivery, there is a limited body of empirical research focusing specifically on Penang's engineering sector. Existing studies often examine global or multinational perspectives, with less attention to

- the localised impact of geopolitical tensions on project performance indicators (delays, cost overruns, scope changes) in Penang-based engineering firms
- how such tensions affect resource allocation and procurement processes in high-speed, capital-intensive engineering projects
- the experience of small and medium-sized engineering firms, which form the backbone of Penang's industrial landscape

- the role of technology adoption in mitigating or amplifying geopolitical risks
- the specific impact of human resource constraints (e.g. foreign worker restrictions, skill shortages) on project timelines and quality in Penang
- how reputational risks from geopolitically induced delays translate into customer dissatisfaction and long-term loss of competitiveness.

The study presented in this chapter analysed how geopolitical challenges affect project performance and resource allocation in engineering companies in Penang. The specific objectives were

- to analyse how geopolitical challenges influence project performance and resource allocation in Penang engineering companies
- to investigate the effects of supply chain disruption on project timelines and resource management in Penang engineering companies
- to examine the impact of human resource challenges, such as talent shortages and workforce management issues, on project performance and resource allocation in engineering companies in Penang
- to assess the role of technology adoption in improving project performance and optimising resource allocation in Penang engineering companies.

In line with the above objectives, the study addresses the following research questions.

- What are the impacts of geopolitical challenges on project performance and resource allocation in Penang engineering companies, and what strategies can be implemented to mitigate these effects?
- What are the impacts of supply chain disruption on project timelines and resource management in Penang engineering companies, and what solutions can enhance supply chain resilience?
- What influence do human resource challenges such as talent shortages and workforce management issues have on project performance and resource allocation in Penang engineering companies, and what strategies can be developed to overcome these challenges?
- What role does technology adoption play in improving project performance and optimising resource allocation in Penang engineering companies, and what are the best practices for effective technology integration?

Several theories provide the foundation for understanding how geopolitical tensions affect project performance and resource allocation.

- *Institutional theory* explains how external institutions, government policies, regulations and sociopolitical dynamics shape organisational behaviour. Under geopolitical tensions, regulatory changes, tariffs and international agreements create uncertainty, affecting engineering project timelines and resource deployment. Firms in Penang must adapt flexibly to evolving institutional environments to remain resilient (Scott, 2005).
- *Risk management theory* emphasises the identification, assessment and mitigation of risks that threaten project objectives. In this context, geopolitical risks such as trade

disputes, sanctions or sudden policy changes must be systematically analysed so that firms can plan contingencies, diversify sources and adjust schedules to minimise project delays and cost overruns (Cagliano *et al.*, 2015).

- *Supply chain management (SCM) theory* focuses on the efficient flow of materials, information and services across interconnected entities. Under geopolitical stress, disruptions in logistics, suppliers and transport infrastructure can create cascading delays. Effective SCM strategies, including multi-sourcing, buffer stocks and digital monitoring, are essential to sustain project performance (Van der Vorst, 2004).
- *Human capital theory* views skills, knowledge and competencies as critical assets for organisational productivity. In volatile geopolitics, a highly skilled and adaptable workforce can anticipate risks, innovate and sustain performance despite external pressures. Conversely, skill shortages and high turnover amplify the negative effects of disruptions (Wuttaphan, 2017).
- *The technology acceptance model (TAM)* posits that technology adoption is driven by perceived usefulness and perceived ease of use. In engineering projects, implementing project management software, building information modelling (BIM) and digital collaboration tools can improve scheduling, communication and resource allocation, provided that project personnel perceive these tools as useful and easy to use (Al-Momani and Ramayah, 2025).

Empirical studies show that geopolitical risk is associated with increased project delays, higher costs and scope changes, especially in sectors reliant on global logistics and international coordination (Ji, 2025). In Malaysia, studies highlight the importance of a clear project mission, top management support, skilled personnel and effective communication for project success (Haron *et al.*, 2017; Ong and Bahar, 2019). However, many engineering projects still experience delays due to regulatory and technical challenges.

Research on supply chain disruption identifies logistics, collaboration and financial risks as major drivers of performance issues. Factors such as transport delays, weak supplier collaboration and currency volatility lead to cost overruns and schedule slippages. Studies recommend supplier diversification, contingency planning, integrated information systems and robust financial risk management (Bolomope *et al.*, 2025; Katsaliaki *et al.*, 2022).

Empirical work on technology adoption demonstrates that automation, integration tools, digital project management systems and advanced communication technologies improve cost control, schedule adherence, quality and risk management (Chourasiya and Malviya, 2025; Mishra *et al.*, 2025). Yet, many Malaysian firms face barriers such as high costs, resistance to change and limited technical expertise.

Drawing on this literature, the proposed conceptual framework positions four key independent variables – geopolitical challenges, supply chain disruption, human resource challenges and technology adoption – as determinants of project performance, measured primarily by project timelines and resource allocation efficiency. The framework assumes that effective risk management, human capital development and technology integration can moderate the adverse effects of geopolitical tensions on project outcomes. These proposed relationships are illustrated in Figure 5.1.

Figure 5.1 Conceptual framework

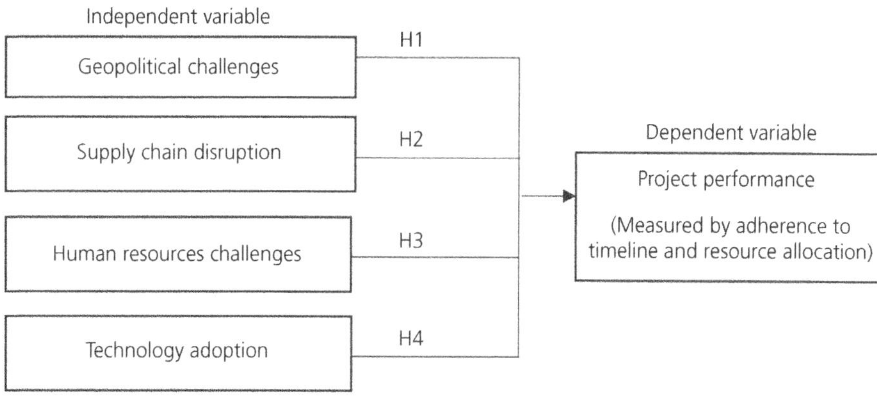

Based on the theoretical and empirical review, the study proposes the following hypotheses.

- H1: increasing geopolitical challenges negatively affect project performance.
- H2: supply chain disruption reduces the efficiency of resource allocation for engineering projects in Penang.
- H3: human resource challenges, such as skills shortages and high turnover, result in delayed project performance in engineering companies in Penang.
- H4: higher levels of technology adoption improve the efficiency of resource allocation within engineering projects in Penang.

5.2. Methodology

5.2.1 Research design

The study adopted a quantitative research design to gain an in-depth understanding of how geopolitical challenges, supply chain disruption, human resource challenges and technology adoption affect project performance (Slater and Hasson, 2025).

The research focused on identifying factors influencing project performance in engineering companies in Penang, Malaysia. Project performance, the dependent variable, was assessed based on the timeliness and efficiency of resource allocation. Overall, this research design supports the study's aim of developing a rich, context-specific understanding of how multiple factors interact to influence project performance in Penang's engineering sector.

5.2.2 Population and sampling procedures

The study population comprised key informants from engineering firms in Penang who have direct involvement or significant knowledge of the research variables. These include project managers, supply chain professionals, engineers and members of technology adoption teams. Their roles and expertise make them well-suited to provide relevant insights into how geopolitical challenges, supply chain disruption, human resource (HR) challenges and technology adoption affect project performance. Selecting participants from these groups ensured that multiple perspectives were represented. Penang's Bayan Lepas Free Industrial Zone, known as the 'Silicon Valley of the East', hosts over 300 multinational corporations, including Intel, AMD, Bosch, HP and Motorola, with a labour force of 939 400 and an employment rate of 71.3% in 2023. The state offers a robust pool of experienced professionals suitable for this study.

A purposive sampling method was employed to intentionally select individuals with specialised experience relevant to the research topics. Project managers provide insights on timelines, resources and the effects of geopolitical and HR challenges. Supply chain professionals provide information on logistics, procurement and disruptions that affect project performance. Technology team members offer perspectives on technology adoption and its challenges. This sampling approach ensures the collection of rich and meaningful qualitative data.

The sample size was determined using Raosoft's sample size calculator, which accounts for the confidence level, margin of error and population size (Renthlei and Lallawmkima, 2025). Using a 95% confidence level, a 10% margin of error and a 50% response distribution for Penang's 939 400 labour force, the recommended sample size is 98 respondents. This provides a statistically valid and manageable sample that enhances the reliability and validity of the study's findings.

5.2.3 Data collection method

The study used structured questionnaires to collect data on how geopolitical challenges, supply chain disruption, human resource challenges and technology adoption affect project performance. The questionnaires were designed to gather detailed, relevant information from key personnel at engineering companies in Penang, including project managers, engineers, supply chain professionals and other stakeholders. This method is appropriate for capturing diverse perspectives from individuals directly involved in project-related activities.

The questionnaire was distributed electronically through Google Forms and shared through email, WhatsApp and other communication channels to ensure broad participation and efficient data collection. Strict ethical procedures were followed, including obtaining informed consent, maintaining confidentiality, ensuring anonymity and allowing voluntary participation. The questionnaire includes demographic items followed by sections on geopolitical challenges, supply chain disruption, human resource challenges, technology adoption and project performance. Likert-scale items measure perceptions quantitatively, while open-ended questions provide qualitative insights. Responses from open-ended items were analysed using thematic analysis to identify patterns and themes.

5.2.4 Operationalisation and measurement

Operationalisation and measurement help translate abstract concepts into clear, measurable variables, ensuring consistency, accuracy and replicability in research. In this study, the independent and dependent variables identified in the conceptual framework were operationalised to support effective data collection and analysis. A five-point Likert scale (ranging from strongly disagree to strongly agree) was used to measure respondents' attitudes and perceptions, particularly regarding the effects of geopolitical tensions on project performance and resource allocation.

The independent variables (geopolitical challenges, supply chain disruption, human resource challenges and technology adoption) were operationalised to assess their effects on resource allocation and project performance. Geopolitical challenges were examined through institutional theory and risk management theory, which highlight how political instability, regulatory changes and international relations influence organisational decisions and risk mitigation. Their impact was measured using Likert-scale items that assess how these factors affect project outcomes. Supply chain disruption was operationalised using supply chain management theory, which emphasises integration, collaboration and resilience in managing material and information flows. Disruptions such as supplier failures, transportation delays or unexpected events were measured using Likert-scale questions evaluating their influence on resource allocation and performance.

Human resource challenges were grounded in human capital theory, which links employee skills, training and retention to organisational productivity. Issues such as skill shortages, turnover and motivation were assessed using Likert-scale items measuring their impact on project timelines and resource use. Technology adoption was operationalised through TAM, which focuses on perceived usefulness and ease of use as drivers of technology acceptance. Likert-scale questions assessed how adopting new technologies affects resource allocation efficiency and project performance.

The dependent variable (project performance) was operationalised using two key criteria: adherence to timelines and the efficiency of resource allocation. Timeline adherence assesses how well projects adhere to planned schedules and the extent to which geopolitical tensions contribute to delays or missed milestones. Resource allocation efficiency examines how effectively resources are utilised and reallocated in response to disruptions, ensuring minimal waste and sustained project progress. Both criteria were measured using Likert-scale items that capture the perceptions of project managers and team members. This approach aligns with established frameworks for assessing the effects of geopolitical challenges on project outcomes and provides a comprehensive understanding of how such tensions influence project execution. The measurement items, adapted from the existing literature and refined for the study's context, support the development of strategies to mitigate geopolitical risks, enhance resource utilisation and improve project performance.

5.2.5 Data analysis techniques

The collected survey data were imported into the statistical package for the social sciences (SPSS) for comprehensive analysis using both descriptive and inferential statistics. Access to the dataset was restricted to authorised personnel to ensure confidentiality. SPSS enables a detailed examination of the relationships between the dependent variable (project performance) and the independent variables (geopolitical challenges, supply chain disruption, human resource challenges and technology adoption). Statistical techniques such as multiple regression, analysis of variance (ANOVA) and correlation analysis were used to test the research hypotheses and generate meaningful conclusions.

Descriptive analysis was used to summarise and interpret the data, providing insights into trends, patterns and the overall impact of geopolitical tensions on project performance and resource allocation in engineering companies in Penang. This includes analysing respondents' demographics, such as industry sector, role, experience and company size, to contextualise the findings. Key descriptive techniques such as measures of central tendency (mean, median, mode) and measures of dispersion (standard deviation, variance, range) help identify typical delays, common cost impacts and the variability of disruptions caused by geopolitical factors. Frequency distributions and percentage analysis further illustrate how often specific delays or resource allocation issues occur, revealing patterns that can guide risk mitigation strategies. Together, these descriptive methods provide a comprehensive understanding of how geopolitical challenges influence project outcomes, enabling companies to make informed, proactive decisions to improve efficiency, strengthen resource planning and maintain project performance despite external uncertainties.

Inferential analysis in this study used reliability testing, normality assessment, multiple regression and Pearson correlation to examine relationships between the independent variables (geopolitical challenges, supply chain disruption, human resource challenges and technology adoption) and the dependent variable (project performance). Reliability was assessed using Cronbach's alpha to ensure questionnaire item internal consistency. At the same time, normality checks involved skewness, kurtosis, histograms and Q-Q plots to confirm that data meet statistical assumptions. Multiple regression analysis, conducted in SPSS, evaluated how each independent variable influences project performance, measured by timeline adherence and resource allocation efficiency, while controlling for other factors. This yielded ANOVA results, regression coefficients and standardised beta values. Pearson correlation coefficients further determined the strength and direction of relationships between variables and helped identify multicollinearity issues. Model validation included checking assumptions of linearity, independence, homoscedasticity and normality of residuals to ensure accuracy and reliability. Together, these inferential techniques provide a comprehensive understanding of how geopolitical tensions and related factors affect project performance and support data-driven decision-making for improved project outcomes.

5.3. Results and analysis

5.3.1 Demographic profile

The demographic analysis of the 98 valid respondents covers four key dimensions: industry sector, role in the company, years of professional experience and company size. Respondents represent a range of engineering-related sectors, including construction, manufacturing, energy and technology, enabling sector-specific comparisons. Their roles vary across managerial, technical and operational positions, providing a balanced view of organisational perspectives. The distribution of industry experience highlights both seasoned professionals and newer entrants, offering diverse insights into project management practices. Company size, categorised by number of employees, helps contextualise differences in resources, structure and operational challenges between small, medium and large firms. The distribution of respondents by industry sector is presented in Figure 5.2.

The industry distribution of respondents is diverse but highly concentrated in the manufacturing electronic and electrical sector, which accounts for 70.4% of the sample. This strong representation indicates that the study's findings are especially relevant to high-precision and technology-driven engineering environments common in this sector. Smaller proportions of

Figure 5.2 Respondent career field

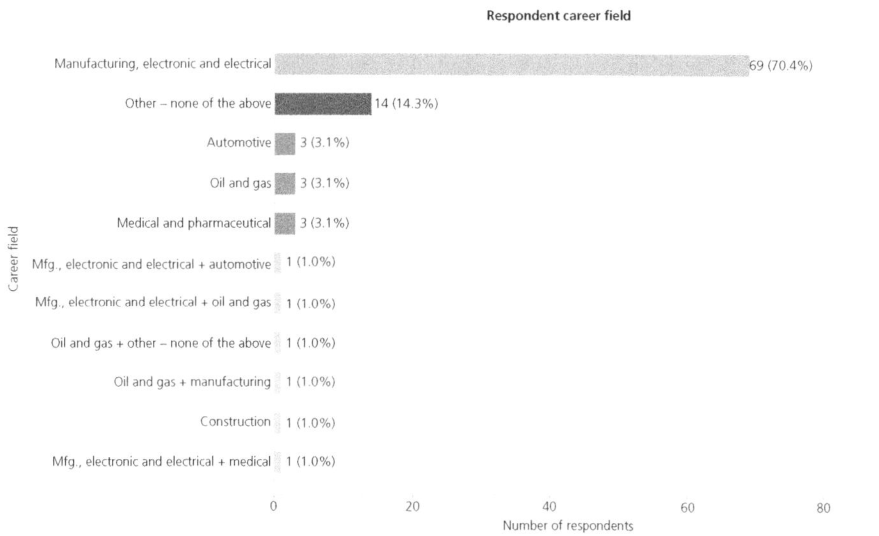

respondents come from the automotive, medical and pharmaceutical, and oil and gas industries, each contributing 3.1%, while another 1.0% are in hybrid sectors combining manufacturing and electronic and electrical with automotive or medical fields. Additionally, 14.3% fall under the 'other – none of the above' category, reflecting niche or unspecified engineering sectors. Overall, the distribution highlights the dominance of the electronic and electrical sector while still capturing the diversity of engineering activities present in Penang.

The distribution of respondents' professional roles shows that engineers account for 50.0% of the sample, indicating that technical practitioners form the core of the surveyed workforce. Other roles are represented in smaller proportions, including project managers (8.2%), designers/drafters (7.1%), quality professionals (6.1%), planners (4.1%) and project engineers (3.1%). A few respondents (1.0% each) hold hybrid or specialised roles such as engineering manager or supply chain. Additionally, 17.3% fall under the 'other – none of the above' category, reflecting diverse or multidisciplinary positions not listed in the predefined options. Overall, the sample captures a broad mix of technical, managerial, design and quality functions, ensuring that the study's findings reflect a wide range of professional viewpoints within engineering.

The respondents' professional experience shows a well-balanced mix of industry tenure. The largest group, 39.8%, has 5–10 years of experience, indicating a strong representation of mid-career professionals. Another 29.6% have more than 10 years of experience, reflecting seasoned practitioners with extensive industry knowledge. Additionally, 28.6% of respondents have 1–5 years of experience, contributing perspectives from early-career engineers familiar with newer practices. A small minority (2.0%) selected overlapping categories, likely due to transitional career stages or reporting errors. Overall, the distribution is 70.4% with up to 10 years of experience and 29.6% with more than 10 years, ensuring a diverse range of

Figure 5.3 Respondents by company size

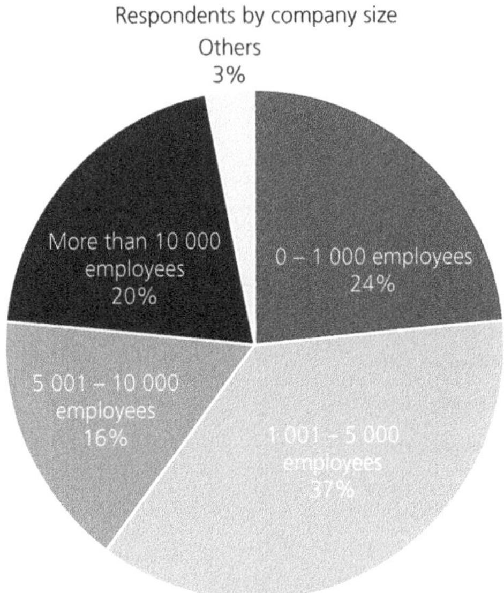

viewpoints and strengthening the study by incorporating insights from both newer and highly experienced engineering professionals. The distribution of respondents by company size is illustrated in Figure 5.3.

The largest group (36.7%) works in companies with 1001–5000 employees, indicating that mid-sized to large organisations form a significant portion of the sample. Smaller firms with 0–1000 employees account for 23.5%, while very large corporations make up a substantial share: 20.4% of respondents are from companies with over 10 000 employees and 16.3% from firms with 5001–10 000 employees. A small number of respondents (3.0%) selected overlapping size categories, possibly due to transitional growth stages or complex structures. Overall, 62.2% of participants work in organisations with up to 5000 employees and 37.8% in larger firms, ensuring a balanced representation of small, medium and large companies. This diversity enriches the study by capturing different operational dynamics, resource capacities and management approaches across various organisational sizes.

5.3.2 Reliability analysis

The reliability statistics for all constructs are presented in Table 5.1. The internal consistency results in Table 5.1 show strong Cronbach's alpha (α) values ranging from 0.833 to 0.923, confirming the high reliability of all measured constructs. Project performance demonstrates the highest reliability ($\alpha = 0.923$), indicating exceptional consistency across its five items. Supply chain disruption also shows near-ideal reliability ($\alpha = 0.912$), while human resource challenges ($\alpha = 0.841$) and technology adoption ($\alpha = 0.843$) exhibit very good internal consistency. Geopolitical challenges ($\alpha = 0.833$) remains well above the acceptable threshold of 0.70, confirming that its items reliably measure the construct. Because reliability values

Table 5.1 Reliability test

Variable	Number of items	Cronbach's alpha, α
Geopolitical challenges	4	0.833
Supply chain disruption	5	0.912
Human resource challenges	5	0.841
Technology adoption	5	0.843
Project performance	5	0.923

above 0.7 and ideally above 0.9 are considered strong, these coefficients indicate that respondents consistently interpreted and responded to the survey items. Overall, the high reliability across constructs supports the instrument's quality and strengthens the validity of conclusions about how operational challenges relate to project outcomes in engineering contexts.

5.3.3 Normality analysis

A normality test is essential for determining whether the dataset follows a normal distribution, a key assumption for parametric analyses such as correlation and regression. Assessing normality ensures the validity and accuracy of subsequent statistical tests. In this study, skewness, kurtosis, histograms and Q-Q plots were examined to assess the data distribution and confirm its suitability for further analysis. The histogram illustrating data distribution is presented in Figure 5.4. The normal Q-Q plot, which assesses linearity and distributional alignment, is shown in Figure 5.5.

Figure 5.4 Histogram of normality

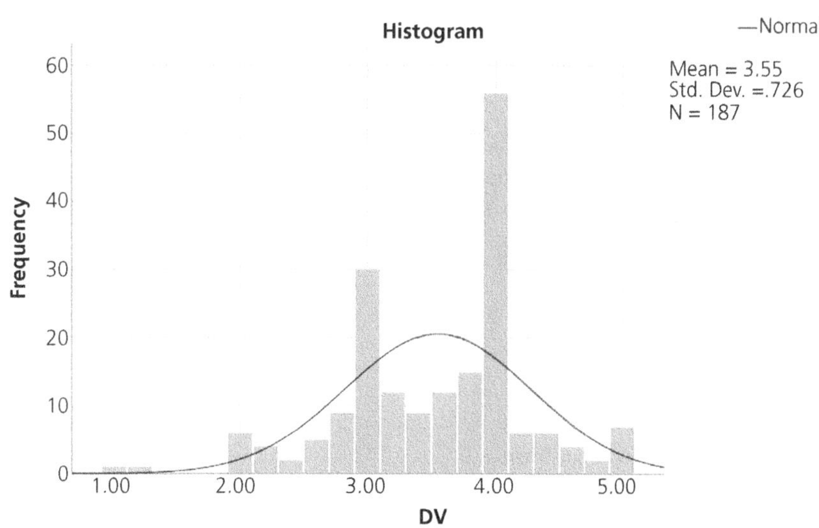

Figure 5.5 Q-Q plot

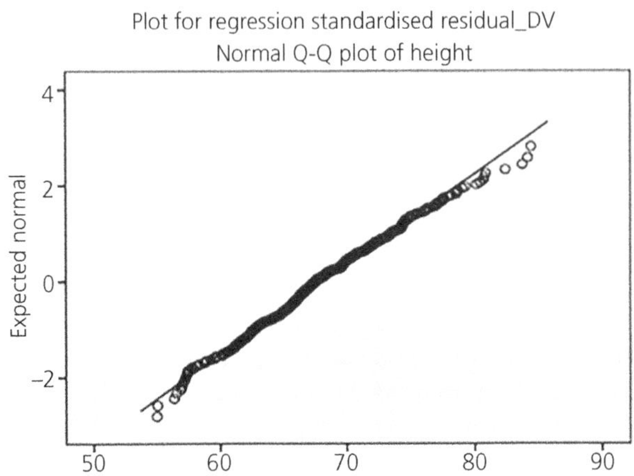

5.3.4 Descriptive analysis

Descriptive statistics for project performance indicators are summarised in Table 5.2. The table shows that engineering project performance scores range from 3.59 to 3.82 on a five-point scale. Effective resource allocation received the highest rating (mean = 3.82, standard deviation (SD) = 0.878), indicating generally strong alignment between planned and actual

Table 5.2 Descriptive analysis of project performance

	n	Minimum	Maximum	Mean	SD
Most of the engineering project was completed within the originally scheduled timeframe	98	1	5	3.61	0.970
All key project milestones were met on schedule throughout the project's duration	98	1	5	3.70	0.976
Resources were allocated effectively in alignment with the initial project plan to the end of the project	98	2	5	3.82	0.878
The engineering project used fewer resources than initially anticipated while maintaining quality standards	98	1	5	3.59	1.092
The project always remained within the approved budget, even after adjustments to resource allocation	98	1	5	3.74	1.008
Valid *n* (listwise)	98				

resource use. Budget adherence (mean = 3.74) and milestone achievement (mean = 3.70) also reflect satisfactory performance, although not at an exceptional level. Timeliness scored lower (mean = 3.61), suggesting challenges in meeting original deadlines. Resource efficiency had the lowest mean (3.59, SD = 1.092), highlighting variability in maintaining quality with minimal resource consumption. Overall, the results indicate that project performance is acceptable but leaves room for improvement, particularly in timeliness and productivity. These performance gaps may be influenced by previously identified operational challenges such as geopolitical tensions, supply chain disruption, human resource challenges and technology adoption, which collectively constrain optimal project outcomes.

5.3.5 Skewness and kurtosis

The skewness and kurtosis values for all variables are reported in Table 5.3. The skewness and kurtosis values in the descriptive statistics table indicate the overall shape and distribution of the variables geopolitical risk (GR), supply chain disruption (SCD), human resource challenges (HRC), technology adoption (TA) and project performance (PP). Skewness values for GR (–0.688), SCD (–1.358), HRC (–0.271) and TA (–0.337) show slight left-skewness, while PP is nearly symmetrical with a skewness of –0.097. Among them, SCD is the most skewed. Since all skewness and kurtosis values fall within the acceptable range of –2 to + 2, the data can be considered approximately normal, with no extreme asymmetry or unusual peak patterns.

5.3.6 Correlation analysis

5.3.6.1 Pearson correlation coefficient

Table 5.4 presents the Pearson correlation analysis, revealing significant interrelationships between the key operational challenge variables: geopolitical risk (GR), supply chain disruption (SCD), human resource challenges (HRC) and technology adoption (TA). Their strong positive correlations ($r = 0.594$–0.669) indicate that these challenges tend to co-occur in engineering projects. The strongest association is between HRC and TA ($r = 0.669$), suggesting that workforce capabilities are directly tied to the effectiveness of technology implementation.

Project performance (PP) shows a differentiated pattern of relationships: it is moderately correlated with SCD ($r = 0.277$) and HRC ($r = 0.326$) and strongly correlated with TA

Table 5.3 Skewness and kurtosis

Variable	*n*	Min	Max	Mean	SD	Skewness	SE skew	Kurtosis	SE kurtosis
GR	98	2.00	5.00	4.1939	0.65658	–0.688	0.244	0.589	0.483
SCD	98	2.00	5.00	4.4143	0.66317	–1.358	0.244	1.923	0.483
HRC	98	2.60	5.00	4.1878	0.61246	–0.271	0.244	–0.654	0.483
TA	98	2.00	5.00	4.1204	0.61844	–0.337	0.244	0.262	0.483
PP	98	2.00	5.00	3.6939	0.86308	–0.097	0.244	–0.872	0.483

Note: SE, standard error

Table 5.4 Pearson correlation coefficient

		GR	SCD	HRC	TA	PP
GR	Pearson correlation	1	0.661[a]	0.596[a]	0.660[a]	0.196
	Sig. (2-tailed)		< 0.001	< 0.001	< 0.001	0.053
	n	98	98	98	98	98
SCD	Pearson correlation	0.661[a]	1	0.659[a]	0.594[a]	0.277[a]
	Sig. (2-tailed)	< 0.001		< 0.001	< 0.001	0.006
	n	98	98	98	98	98
HRC	Pearson correlation	0.596[a]	0.659[a]	1	0.669[a]	0.326[a]
	Sig. (2-tailed)	< 0.001	< 0.001		< 0.001	0.001
	n	98	98	98	98	98
TA	Pearson correlation	0.660[a]	0.594[a]	0.669[a]	1	0.488[a]
	Sig. (2-tailed)	< 0.001	< 0.001	< 0.001		< 0.001
	n	98	98	98	98	98
PP	Pearson correlation	0.196	0.277[a]	0.326[a]	0.488[a]	1
	Sig. (2-tailed)	0.053	0.006	0.001	< 0.001	
	n	98	98	98	98	98

[a]Correlation is significant at the 0.01 level (2-tailed).

($r = 0.488$, $p < 0.01$). Its weaker, non-significant association with GR ($r = 0.196$, $p = 0.053$) further underscores the varying influence of these factors. Overall, the matrix supports discriminant validity, with all coefficients below the threshold. Notably, the strong TA–PP link ($r = 0.488$) emphasises technology adoption as a critical success factor. Although the operational variables are interrelated, their impacts on project outcomes differ, with technological capabilities emerging as the most influential determinant.

5.3.7 Multiple regression

Table 5.5 presents the multiple regression results, indicating that operational challenges collectively explain a significant proportion of variance in project performance. The model

Table 5.5 Model summary

R	R^2	Adj. R^2	Std. error of the estimate	Change statistics				
				R^2 change	F change	*df1*	*df2*	Sig. *F* change
0.521	0.271	0.240	0.75251	0.271	8.650	4	93	< 0.001

Note: predictors: (constant), TA, SCD, GR, HRC.

shows a moderate fit ($R = 0.521$; $R^2 = 0.271$), meaning 27.1% of the variability in performance is accounted for by technology adoption (TA), supply chain disruption (SCD), geopolitical challenges (GR) and human resource challenges (HRC). The adjusted R^2 of 0.240 reinforces the model's robustness after adjusting for sample size and number of predictors.

The overall model is statistically significant (F-change < 0.001), confirming a meaningful improvement over the null model. The standard error of 0.75251 suggests that estimates of project performance are reasonably precise with these predictors. While the model explains a notable portion of variance, the remaining 72.9% highlights the influence of other unexamined factors. The results also align with earlier evidence emphasising the strong effect of technology adoption, underscoring the need for further coefficient analysis to clarify the differential impacts of each predictor. Overall, the four operational challenges are important although not exhaustive determinants of engineering project performance.

As shown in Table 5.6, the analysis of variance (ANOVA) results confirm that the regression model significantly explains variation in project performance using the four operational challenge variables. The model is statistically significant ($F(4, 93) = 8.650$, $p < 0.001$), indicating that technology adoption, supply chain disruption, geopolitical challenges and human resource challenges jointly explain project performance.

The model accounts for 19.593 units of explained variance and 52.664 units of unexplained variance, totalling 72.256. The very low p-value ($p < 0.001$) and the significant F-ratio demonstrate that the explained variance is far greater than would be expected by chance. This confirms that the set of operational challenges is meaningfully related to project performance.

Although the model is significant, the large residual sum of squares indicates considerable unexplained variance, suggesting that additional factors beyond these four challenges also influence project outcomes. Nonetheless, the findings validate these challenges as statistically sound predictors of engineering project performance.

The regression coefficients for each predictor variable are presented in Table 5.7.

5.3.8 Hypothesis testing

A summary of hypothesis testing results is provided in Table 5.8. The hypothesis on geopolitical challenges was supported ($\beta = -0.271$, $p = 0.043$), indicating that rising geopolitical risks

Table 5.6 ANOVA

Model	Sum of squares	*df*	Mean square	*F*	Sig.
Regression	19.593	4	4.898	8.650	< 0.001
Residual	52.664	93	0.566		
Total	72.256	97			

Note: dependent variable: PP; predictors: (constant), TA, SCD, GR, HRC.

Table 5.7 Coefficient regression

Coefficient							
Model	Unstandardised coefficient,		Standardised coefficient, β	t	Sig.	95.0% confidence interval for B	
	B	Std. error				Lower bound	Upper bound
(Constant)	1.095	0.599		1.827	0.071	−0.095	2.286
GR	−0.357	0.174	−0.271	−2.051	0.043	−0.702	−0.011
SCD	0.102	0.172	0.078	0.592	0.555	−0.240	0.443
HRC	0.053	0.188	0.038	0.283	0.778	−0.320	0.426
TA	0.831	0.186	0.595	4.470	<0.001	0.462	1.200

Note: dependent variable: PP.

Table 5.8 Summary of hypotheses results

Hypothesis	Results
Hypothesis 1: increasing geopolitical challenges hurt adherence to project performance in engineering companies in Penang	Supported
Hypothesis 2: supply chain disruption reduces the efficiency of resource allocation for engineering projects in Penang	Partially supported
Hypothesis 3: human resource challenges, such as skills shortages and high turnover, result in delayed project performance in engineering companies in Penang	Not supported
Hypothesis 4: higher levels of technology adoption improve the efficiency of resource allocation within engineering projects in Penang	Supported

negatively affect project performance through delays, cost overruns and scope changes. This aligns with high descriptive ratings (mean = 4.11–4.41), indicating strong agreement among respondents about their disruptive impact.

Hypothesis 2 (supply chain disruption) received partial support. Although its regression coefficient was non-significant ($\beta = 0.078$, $p = 0.555$), the moderate correlation ($r = 0.277$, $p = 0.006$) and high descriptive means (4.27–4.48) suggest that supply chain issues do affect operations, potentially acting indirectly through variables such as technology adoption.

The third hypothesis on human resource challenges was not supported, with a non-significant regression effect ($\beta = 0.038$, $p = 0.778$). However, HR challenges still show a significant correlation with performance ($r = 0.326$, $p = 0.001$) and high mean ratings (4.12–4.24), implying they remain important barriers whose effects may only become measurable with more targeted interventions.

The strongest support was for hypothesis 4 (technology adoption), which showed a substantial and highly significant effect ($\beta = 0.595$, $p < 0.001$) and a strong correlation ($r = 0.488$, $p < 0.001$). High descriptive means (4.05–4.21) further highlight technology adoption as a critical driver of efficiency, quality and timely project delivery.

5.4. Practical insights

The findings of this study offer several practical insights for engineering firms in Penang and similar industrial regions operating under geopolitical uncertainty, supply chain volatility, workforce constraints and rapid technological change. By translating empirical results into actionable lessons, this section shows how firms can enhance project performance and improve resource-allocation efficiency in real-world settings.

First, the significant negative effect of geopolitical challenges on project performance underscores the need for firms to move beyond reactive risk management toward proactive geopolitical awareness. Engineering companies should systematically monitor geopolitical developments such as trade disputes, sanctions and regulatory shifts and embed these considerations into project planning and procurement decisions. The results suggest that firms which anticipate geopolitical risks early are better positioned to adjust schedules, diversify sourcing strategies and allocate buffers for time and cost, thereby reducing the likelihood of severe project delays.

Second, although supply chain disruption did not show a strong direct effect in the regression analysis, its significant correlation with project performance indicates that it remains an important operational concern. In practice, this suggests that supply chain risks may influence project outcomes indirectly, particularly through their interactions with technology adoption and managerial decision-making. Firms should therefore focus on strengthening supply chain visibility and coordination rather than relying solely on cost-based procurement strategies. Practices such as maintaining alternative suppliers, increasing collaboration with local vendors and using digital tracking systems can help mitigate the cascading effects of disruptions on project timelines and resource availability.

Third, the non-significant regression result for human resource challenges does not imply that workforce issues are unimportant. On the contrary, the moderate correlation between human resource challenges and project performance indicates that skill shortages, turnover and workforce adaptability still affect project execution. In practice, these challenges may be absorbed or masked by organisational routines, overtime practices or technological support. Engineering firms should therefore invest in continuous training, cross-skilling and knowledge-sharing mechanisms to ensure that workforce limitations do not become critical bottlenecks during periods of external disruption.

Most importantly, technology adoption emerged as the strongest and most consistent determinant of project performance. The significant positive relationship between technology adoption and project performance highlights technology's role as a key enabler of resilience.

In practical terms, this finding suggests that tools such as project management software, digital collaboration platforms and data-driven resource planning systems can help firms compensate for uncertainty in the external environment. Technology enables faster decision-making, real-time resource monitoring and more flexible responses to unexpected disruptions. However, to fully realise these benefits, firms must ensure that technologies are user-friendly, well-integrated into existing workflows and supported by adequate training, in line with the technology acceptance model.

Collectively, these insights indicate that improving project performance in geopolitically sensitive environments requires an integrated approach rather than isolated interventions. Firms that combine geopolitical risk awareness, resilient supply chain practices, workforce capability development and effective technology adoption are better equipped to maintain project timelines and optimise resource allocation. For practitioners, the key lesson is that technology acts not merely as an operational tool but as a strategic buffer that strengthens organisational resilience against external shocks.

5.5. Conclusion and recommendations

5.5.1 Limitations

Although this study offers useful insights into the operational challenges affecting engineering project performance in Penang, several limitations should be acknowledged. First, the study relies heavily on quantitative, self-reported survey data, which may not fully capture the breadth of factors shaping project outcomes and may be subject to response bias. The cross-sectional design also limits the ability to observe changes over time or establish causal relationships between variables such as geopolitical risks, supply chain disruption, human resource challenges and technology adoption.

The sample composition presents another limitation. Most respondents were from the manufacturing electronics and electrical sector, reflecting Penang's industrial profile but limiting the generalisability of the findings to other sectors, such as construction, automotive, and oil and gas, which operate under different risk conditions. Similarly, the geographical focus on Penang limits applicability to other Malaysian states or Southeast Asian regions with differing regulatory environments, market structures and workforce characteristics.

Methodologically, the regression model explains 27.1% of the variance in project performance, suggesting that many other influential factors, such as project complexity, environmental uncertainty, payment delays, contractual clarity and stakeholder management, were not examined in depth. Existing literature suggests these issues play a substantial role in project underperformance in Malaysia and warrant deeper investigation.

5.5.2 Recommendations and future research

To enhance engineering project performance in Penang's rapidly evolving industrial landscape, firms must adopt targeted strategies that address the most critical operational challenges identified in the study: geopolitical risks, supply chain disruption, workforce readiness and technological integration. The following recommendations translate the empirical findings into actionable policy and industry measures designed to strengthen resilience, improve efficiency and support sustainable long-term growth.

- *Strengthen geopolitical challenge mitigation*: only 41% of engineering firms in Penang have dedicated geopolitical risk teams (Liang and Lean, 2024), indicating the need for wider adoption. Scenario-based modelling can help anticipate export controls and trade barriers, reducing delays by up to 34% as shown in case studies (Connolly, 2022). Firms should collaborate with INVEST Penang to access political risk insurance and diversify suppliers across the Association of Southeast Asian Nations (ASEAN) and South Asia to lessen reliance on US–China trade routes. For megaprojects, contracts should include contingency allocations of 15–20% to cushion against permit and regulatory delays, consistent with ASEAN resilience practices.
- *Accelerate digital supply chain adoption*: close the small and medium-sized enterprise–multinational corporation (SME–MNC) technology gap. Penang's science, technology, engineering and mathematics (STEM) grant schemes should subsidise artificial intelligence (AI)-enabled procurement tools and Internet of Things (IoT)-based inventory tracking, which have reduced excess material costs by 24% among adopting firms (Neumann, 2024). Public tenders should require BIM and radio-frequency identification (RFID) integration to increase material planning accuracy by 41% (Radzi and Yaacob, 2022). Developing regional supplier clusters within free industrial zones can further reduce lead times by 41% (Hutchinson, 2025) by enabling localised sourcing for semiconductor and precision-machinery components.
- *Reform workforce development systems*: to address the 20–25% gap in engineering graduate employability, curricula should be redesigned in collaboration with industry partners such as ViTrox, which plans to hire 4000 workers by 2034 and train 2500 technicians annually in automation and robotics (Tan *et al.*, 2023). Using virtual reality (VR)-based modules can reduce skill-related delays by 19%. Reducing turnover, responsible for 34% of project lead-time extensions, requires retention incentives and cross-training initiatives (Dhivandram and Arumugam, 2026). Because SME reskilling costs are 47% lower than those of MNCs, tax incentives should offset 50% of SMEs' upskilling expenses (Kee *et al.*, 2025).
- *Institutionalise circular tech integration*: allocate 30% of Penang's Digital Transformation Fund to circular technologies such as digital material passports, which yield 22–25% cost savings through reuse tracking (Barsa and Rakesh, 2025). Mandating industrialised building systems (IBS) in public projects can reduce onsite waste by 29% and boost resource efficiency (Ahmed *et al.*, 2024). Expanding 5G smart infrastructure pilots has been shown to improve decision-making speed by 31% (Lu and Moorthy, 2026), and establishing a centralised 5G project monitoring hub will support real-time resource redirection during disruptions.

Future research should therefore consider longitudinal designs to track how operational challenges evolve with technological and geopolitical shifts. Qualitative methods, including interviews and case studies, can add contextual richness and uncover mechanisms that are not observable through surveys. Expanding the study to other regions and industry sectors would enable broader comparisons and generalisations. Further work should also assess the effectiveness of mitigation strategies such as flexible supply chains, targeted workforce development and dual-purpose resource allocation. Finally, comprehensive research into integrated digital technologies and circular economy practices could illuminate pathways for achieving world-class engineering project performance in Malaysia and beyond.

REFERENCES

Ahmed S, Abd Aziz N, Haque R *et al.* (2024) Digital transformation in Malaysian manufacturing: a study of change sensing and seizing capabilities. *Cogent Business Management* **11(1)**: 2392046.

Al-Momani AAM and Ramayah T (2025) Analyzing EHR technology adoption: a comparative review of the technology acceptance model in different economic contexts. In *Intelligence-Driven Circular Economy* (Hannoon A and Mahmood A (eds)). Springer, Cham, Switzerland, pp. 327–344.

Asadollah O, Carmy LS, Hoque MR and Yilmazkuday H (2024) Geopolitical risk, supply chains, and global inflation. *The World Economy* **47(8)**: 3450–3486.

Athukorala PC (2017) Global production sharing and local entrepreneurship in developing countries: evidence from Penang export hub, Malaysia. *Asia & the Pacific Policy Studies* **4(2)**: 180–194.

Barsa JZ and Rakesh HA (2025) Digital transformation and operational efficiency in medium-sized enterprises in Penang, Malaysia. *JBMI Insight* **2(12)**: 16–28.

Bolomope MT, Nwadike AN and Aigwi IE (2025) Managing supply chain disruptions in the construction industry: an institutional approach. *Journal of Financial Management of Property and Construction* **30(2)**: 212–231.

Cagliano AC, Grimaldi S and Rafele C (2015) Choosing project risk management techniques: a theoretical framework. *Journal of Risk Research* **18(2)**: 232–248.

Chourasiya R and Malviya RK (2025) Factors influencing technology adoption for sustainability in manufacturing firms: an empirical analysis and literature review. *Sustainable Development* **33(5)**: 6734–6757.

Connolly C (2022) Introduction: governing urban transformations in Penang. In *Political Ecologies of Landscape*. Bristol University Press, Bristol, UK, pp. 1–19.

Dhivandram PD and Arumugam P (2026) Rebuilding microenterprises' brand identity through technology adaptation: a case study of Penang. *International Journal of Academic Research in Economics and Management Sciences* **15(1)**: 616–628.

Haron NA, Devi P, Hassim S, Alias AH, Tahir MM and Harun AN (2017) Project management practice and its effects on project success in Malaysian construction industry. *IOP Conference Series: Materials Science and Engineering* **291(1)**: 012008.

Hutchinson FE (2025) A look back at how Penang learned to shine through its electrical and electronics sector. *Economic History Malaysia*. https://www.ehm.my/publications/articles/a-look-back-at-how-penang-learned-to-shine-through-its-electrical-and-electronics-sector (accessed 18/05/2026).

Ji J (2025) Geopolitical risk and its impact on capital allocation and stock market performance. *Applied Economics Letters*: 1–6.

Katsaliaki K, Galetsi P and Kumar S (2022) Supply chain disruptions and resilience: a major review and future research agenda. *Annals of Operations Research* **319(1)**: 965–1002.

Kee DMH, Cordova M and Khin S (2025) The key enablers of SMEs readiness in Industry 4.0: a case of Malaysia. *International Journal of Emerging Markets* **20(3)**: 1042–1062.

Liang H and Lean HH (2024) Fifty years of excellence and beyond: the electrical and electronic industry of Penang, Malaysia. In *Economic Growth and Development in the Tropics*. Routledge, Abingdon, UK, pp. 30–48.

Lu Q and Moorthy R (2026) Strategic responses and regional pressures: Malaysia in the US-China semiconductor competition. *The Pacific Review* **39(1)**: 177–207.

Mishra NK, Sahoo S, Agarwal S, Sharma PP and Ilahi F (2025) Impact of institutional pressures and security on blockchain technology adoption and organization performance: an empirical study. *The Journal of Technology Transfer* **50(1)**: 245–270.

Neumann AL (2024) Malaysia's hot hand in tech investment. *Global Asia* **19(2)**: 50–55.

Ong CH and Bahar T (2019) Factors influencing project management effectiveness in the Malaysian local councils. International Journal of Managing Projects in Business 12(4): 1146–1164.

Radzi RM and Yaacob Z (2022) Identifying business potential for small and medium-sized enterprises (SMEs) based on developmental plans in Penang, Malaysia. *Journal of Technology Management and Business* **9(2)**: 34–50.

Raj-Reichert G (2020) Global value chains, contract manufacturers, and the middle-income trap: the electronics industry in Malaysia. *The Journal of Development Studies* **56(4)**: 698–716.

Renthlei Z and Lallawmkima C (2025) Choosing the right sample size in social science studies: a methodological review. *Mizoram Educational Journal* **1**.

Scott WR (2005) Institutional theory: Contributing to a theoretical research program. In *Great Minds in Management: The Process of Theory Development* (Smith KG and Hitt MA (eds)). Oxford University Press, Oxford, UK, pp. 460–484.

Slater P and Hasson F (2025) Quantitative research designs, hierarchy of evidence and validity. *Journal of Psychiatric and Mental Health Nursing* **32(3)**: 656–660.

Tan LP, Choe KL, Ng YK, Lai KF and Khor HC (2023) The environmental, social, and governance (ESG) in Malaysia SME: a case study of Vitrox. In *Proceedings of the 11th International Conference on Business, Accounting, Finance and Economics (BAFE 2023)*, vol. 272. Springer, Cham, Switzerland, p. 344.

Tortorella G, Gloet M, Samson D *et al.* (2025) Food supply chain resilience through digital transformation: a mixed-method approach. *The International Journal of Logistics Management* **36(2)**: 381–412.

Van der Vorst JGAJ (2004) Supply chain management: theory and practices. *Bridging Theory and Practice* **1(1)**: 105–128.

Wuttaphan N (2017) Human capital theory: the theory of human resource development, implications, and future. *Life Sciences and Environment Journal* **18(2)**: 240–253.

emerald PUBLISHING ice Publishing

Part II

Digital transformation

Advancing Project Management: Frameworks, Digital Transformation, Sustainability and Governance

Mohammed Reyasudin Basir Khan
ISBN 978-1-83742-799-4
https://doi.org/10.1108/978-1-83742-798-720261008

Chapter 6
Digital project management systems: adoption, integration and project performance

Suganya Sukhumaran and Mohammed Reyasudin Basir Khan

Tun Razak Graduate School, Universiti Tun Abdul Razak (UNIRAZAK), Kuala Lumpur, Malaysia

6.1. Introduction

The increasing complexity of modern projects, particularly in engineering, construction and software development, has heightened the need for robust, integrated project management tools. Project management software (PMS) has emerged as a critical enabler of enhanced efficiency, productivity and overall project performance. Empirical evidence across industries suggests that PMS strengthens control over scheduling, budgeting, communication and resource allocation, thereby reducing common project risks such as delays, cost overruns and coordination failures (Gilal *et al.*, 2023).

Over time, PMS has evolved from basic task-tracking applications into comprehensive digital platforms that integrate real-time collaboration, workflow automation and predictive analytics capabilities (Nenni *et al.*, 2025). In the construction sector, PMS has contributed significantly to workflow optimisation, improved stakeholder coordination and enhanced cost management. For instance, Hamada (2023) demonstrated that PMS improves communication efficiency and resource allocation in construction projects, while Parikh *et al.* (2021) highlighted the value of mobile-based PMS tools in supporting onsite coordination and operational flexibility.

Beyond traditional functionalities, contemporary PMS increasingly incorporates machine learning (ML) and artificial intelligence (AI) to enhance forecasting and decision-making capabilities. Mahdi *et al.* (2021) showed that AI-enabled PMS tools can improve risk prediction accuracy, optimise resource utilisation and support strategic planning. Similarly, Waseem *et al.* (2024) demonstrated how AI-driven tools such as ChatGPT can streamline documentation processes and automate routine project tasks, thereby improving operational efficiency.

The importance of PMS is particularly evident in geographically dispersed and globally distributed project environments. Alqahtani *et al.* (2024) emphasised the role of PMS in facilitating coordination and performance management in international software development teams. Furthermore, integrating big data analytics into PMS platforms enhances data-driven decision-making and performance monitoring (Obaid and Abu-Naser, 2023). However, despite these technological advancements, organisations continue to face inconsistent implementation outcomes due to factors such as limited user acceptance, insufficient organisational readiness, high project complexity and inadequate integration with legacy systems.

Although PMS has become integral to contemporary project management practices, many organisations still encounter persistent challenges, including schedule delays, budget overruns and ineffective communication, even when adopting complementary methodologies such as Agile, Scrum, building information modelling (BIM) and enterprise systems (Gilal *et al.*, 2023). In software development contexts, global teams frequently struggle with communication barriers, cultural differences and coordination inefficiencies despite structured Agile implementation (Alqahtani *et al.*, 2024). While AI- and ML-based PMS tools offer promising mechanisms for managing complexity (Mahdi *et al.*, 2021), their widespread and effective application remains limited. Moreover, Azizi *et al.* (2024) observed that although PMS adoption aims to enhance operational efficiency and financial performance, consistent empirical evidence demonstrating sustained cross-industry success is still emerging.

Given these gaps, the study presented in this chapter investigated how PMS influences project outcomes across sectors, identifies contextual and organisational factors affecting its effectiveness, and proposes a structured framework to optimise PMS implementation. By examining the interrelationships between planning quality, stakeholder management, risk management and user satisfaction, the study provides empirical insights to improve time, cost and quality performance. These findings are particularly relevant for industrial environments such as Senai Industrial Park, where technology-enabled coordination and operational efficiency are essential for sustaining competitiveness.

6.1.1 Theoretical contributions

This study contributes to the theoretical development of project management and technology adoption literature by conceptualising project management software (PMS) as a strategic capability rather than merely an operational tool. Building on empirical evidence that PMS enhances coordination, monitoring and workflow integration (Hamada, 2023), the study positions PMS within a broader theoretical framework linking planning quality, stakeholder management, risk management and user satisfaction to project outcomes.

Second, the research extends the technology acceptance model (TAM) by demonstrating that system usability and perceived usefulness influence not only behavioural intention but also measurable performance indicators such as time, cost and quality outcomes. The integration of AI-driven analytics and machine learning into PMS platforms (Mahdi *et al.*, 2021) further strengthens the theoretical argument that digital intelligence enhances decision-making quality and predictive control.

Third, the findings reinforce governance and stakeholder theory perspectives by highlighting the role of structured communication and risk alignment in achieving project success

(Lesum *et al.*, 2024). The study also supports Agile theory by recognising adaptability and iterative coordination as critical success mechanisms in dynamic environments (Rasool *et al.*, 2023).

Collectively, these contributions advance theoretical understanding of how digitally enabled project systems reshape traditional management paradigms across industries.

6.1.2 Academic contributions

This study contributes to the expanding literature on project management, digital transformation and technology adoption in several ways.

- *Closing literature gaps*: it provides a comprehensive cross-sector analysis of PMS impacts, particularly within industrial settings such as Malaysian industrial parks.
- *Technological relevance*: it advances theoretical understanding of AI-, ML- and data-driven PMS applications in project management (Mahdi *et al.*, 2021; Waseem *et al.*, 2024).
- *Cross-sector evaluation*: by comparing construction, IT and software contexts, it broadens insights into the applicability of PMS across industries (Alqahtani *et al.*, 2024; Hamada, 2023).
- *Emerging technologies*: it integrates discussion on big data analytics, collaboration platforms and remote work systems (Obaid and Abu-Naser, 2023).
- *Foundation for future research*: it establishes empirical grounding for further investigation into PMS integration within digital ecosystems and global virtual teams.

6.1.3 Theoretical foundation

The adoption of PMS has transformed how organisations plan, execute and monitor projects. In competitive, technologically diverse environments such as industrial parks, projects are increasingly complex and require integrated coordination mechanisms. PMS platforms enable automation, real-time collaboration and data-driven decision-making, thereby mitigating schedule delays, cost overruns and workflow inefficiencies.

However, the effectiveness of PMS depends on organisational readiness, project complexity, software usability and user acceptance. In industrial zones characterised by varying levels of digital maturity, understanding contextual influences on PMS performance is critical.

TAM provides a suitable theoretical framework for analysing PMS adoption. TAM explains technology usage through two core constructs: perceived usefulness (PU) and perceived ease of use (PEOU). PU reflects the extent to which users believe PMS enhances job performance. In project environments, this includes improved planning, communication, resource control and efficiency (Hamada, 2023; Mahdi *et al.*, 2021). PEOU refers to the degree to which PMS is perceived as user-friendly and intuitive. Systems with simple interfaces reduce resistance, increase user satisfaction and promote productivity.

These perceptions influence behavioural intention and actual system usage. Regular PMS utilisation supports accurate planning, real-time monitoring and enhanced collaboration, all of which improve time, cost and quality performance.

6.1.4 Digital transformation and industry applications

Digital transformation has significantly reshaped project execution across construction, IT, manufacturing and education sectors. As project complexity increases, organisations increasingly rely on advanced PMSs that integrate AI, machine learning and analytics.

In construction, PMS enhances scheduling accuracy, resource tracking and stakeholder coordination (Hamada, 2023). Mobile-based PMS tools further strengthen onsite transparency and communication (Parikh *et al.*, 2021). In software development, Agile PMS tools such as Scrum and Kanban improve collaboration, adaptability and iterative delivery cycles (Rasool *et al.*, 2023). Nevertheless, resistance to change and integration challenges remain significant barriers (Kennedyd *et al.*, 2026).

Emerging technologies have expanded PMS capabilities. Machine learning supports automated risk prediction and performance forecasting (Mahdi *et al.*, 2021), while AI tools such as ChatGPT streamline documentation and routine tasks (Waseem *et al.*, 2024). Big data analytics enhances forecasting, resource optimisation and stakeholder communication (Obaid and Abu-Naser, 2023). However, concerns regarding data security, skill dependency and over-reliance on automation persist.

Strong governance frameworks further improve alignment between stakeholders and project goals (Lesum *et al.*, 2024). Effective stakeholder engagement is particularly important for complex, sustainable projects (Köksal and Ersoy Maraş, 2026). In global software development contexts, PMS mitigates cultural, language and time-zone barriers by enabling real-time collaboration and structured communication (Alqahtani *et al.*, 2024). The rise of remote work has further amplified the importance of digital coordination systems.

6.1.5 Challenges in PMS adoption

Despite its advantages, PMS implementation faces several challenges:

- high initial implementation costs
- digital security and data privacy risks
- continuous maintenance and system upgrades
- training and competency gaps
- organisational resistance to change.

Addressing these barriers is essential to realise the benefits of PMS fully.

6.1.6 Conceptual framework

The proposed relationships between planning quality, stakeholder management, risk management, user satisfaction and project outcomes are illustrated in Figure 6.1.

- *Quality of planning process*: effective planning is fundamental to successful project execution as it enhances scheduling accuracy, optimises resource utilisation and strengthens task coordination. PMS facilitates structured planning by integrating timelines, milestones and performance indicators into a unified platform. Empirical evidence demonstrates that PMS improves planning precision and monitoring

Figure 6.1 Conceptual framework

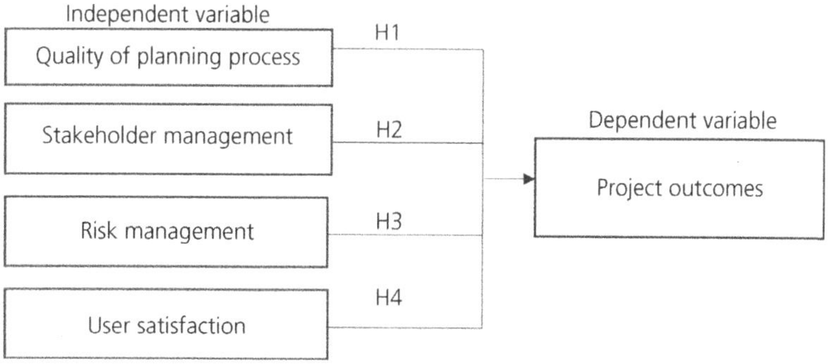

capability, thereby reducing schedule delays and cost overruns (Hamada, 2023). By supporting systematic forecasting and real-time tracking, PMS enhances control and predictability in project environments.

- *Stakeholder management*: stakeholder engagement plays a vital role in improving teamwork, communication and collective decision-making. Effective stakeholder management aligns project objectives with stakeholder expectations, reducing misunderstandings and conflict. In digitally supported environments, PMS enhances transparency and collaboration, thereby strengthening coordination and long-term project sustainability.
- *Risk management*: risk identification, assessment and mitigation are essential for maintaining project stability and continuity. Structured risk management processes reduce uncertainty and protect projects from operational disruptions. Evidence suggests that proactive governance and systematic risk controls significantly enhance performance outcomes (Ongesa *et al.*, 2025). PMS tools support this process by enabling early risk detection, monitoring and documentation.
- *User satisfaction*: user satisfaction determines the extent to which PMS is effectively adopted and utilised. Intuitive, user-friendly systems encourage frequent use, improve collaboration and minimise operational stress. Research indicates that well-designed digital interfaces enhance team productivity and engagement (Arvikar *et al.*, 2025). High levels of user satisfaction, therefore, strengthen system effectiveness and project performance.
- *Project outcomes*: are typically measured in terms of time, cost, scope and quality performance. Prior studies demonstrate that enhanced planning, structured communication and proactive risk control facilitated by PMS collectively contribute to improved overall project performance (Hamada, 2023; Ongesa *et al.*, 2025).

6.1.7 Hypotheses development

- H1: quality of planning positively affects project outcomes; structured planning aligns objectives, resources and schedules, reducing uncertainty and enhancing performance.

- H2: stakeholder management positively affects project outcomes; effective stakeholder engagement improves communication, strengthens decision-making and minimises conflict.
- H3: risk management positively affects project outcomes; proactive identification and mitigation of risks minimises disruptions and enhances project stability (Ongesa *et al.*, 2025).
- H4: user satisfaction with PMS usability positively affects project outcomes; intuitive, user-friendly PMS tools encourage consistent use, enhance collaboration and improve monitoring effectiveness (Arvikar *et al.*, 2025).

6.2. Methodology

6.2.1 Research design

This study adopted a quantitative, positivist research design that focuses on measurable data, enabling the examination of statistically significant relationships between PMS use and project outcomes. Structured data were collected from a representative sample of companies to test the proposed hypotheses. Statistical techniques were applied to analyse relationships between the independent variables (planning process quality, stakeholder management, risk management and satisfaction with PMS usability) and the dependent variable (project outcomes). This design enabled an objective assessment of the extent to which PMS influences project performance across time, cost and quality.

6.2.2 Population and sampling procedures

Using G*Power, the required sample size for this study was estimated at approximately 393 respondents. The target population comprises 87 business units operating within i-Park at Senai Airport City, which serves as the basis for participant recruitment. This sample size ensures adequate statistical power for analysing the relationships among the study variables.

The study population comprises individuals and organisations within Senai Industrial Park that actively use PMS. This includes project managers responsible for planning, executing and monitoring projects; team leaders and key project personnel, such as engineers, IT specialists and supervisors, who engage with PMS in their daily tasks; and companies across the manufacturing, construction and service sectors that have integrated PMS into their operations. Additionally, IT personnel and system administrators involved in configuring, maintaining and supporting PMS were included as they provide essential technical perspectives on software integration and functionality within organisational workflows.

6.2.3 Data collection method

Quantitative data for this research were collected through structured questionnaires distributed to team leaders and project managers who use PMS. The survey captured their perceptions of project outcomes, the frequency of PMS use and their experiences with planning, stakeholder engagement and risk management. On completion of data collection, the responses from all 393 participants were analysed using the statistical package for the social sciences (SPSS).

6.2.4 Operationalisation and measurement

The operationalisation and measurement of variables in this study ensured that abstract concepts related to PMS and project outcomes were translated into quantifiable indicators. The independent variables (quality of the planning process, stakeholder management, risk management and user satisfaction) were assessed using structured survey items rated on a five-point Likert scale. Each variable was operationalised according to its role within PMS use: planning quality reflects the software's ability to support task definition, scheduling and resource allocation; stakeholder management captures the effectiveness of communication and engagement throughout the project lifecycle; risk management evaluates the identification, assessment and mitigation of project risks; and user satisfaction measures the perceived usability, functionality and performance of PMS. Sample items for each construct allowed respondents to indicate their level of agreement, with higher scores indicating greater perceived effectiveness in that dimension.

The dependent variable (project outcomes) was similarly operationalised using both subjective and objective measures, including timeliness, budget adherence, deliverable quality and customer satisfaction. A five-point Likert scale captures user perceptions, while objective organisational data, such as actual project timelines and financial records, supplement the analysis. Together, these operational definitions enabled a systematic evaluation of how PMS usage influences project performance. By employing standardised measurement tools, the study can reliably assess the relationship between software utilisation and project success within companies in the Senai Industrial Park, thereby providing empirical insights into the effectiveness of PMS in real-world project environments. The operational definitions, measurement indicators and scales for each construct are summarised in Table 6.1.

Table 6.1 Summary of operationalisation and measurement

Variable	Operationalisation	Measurement	Scale
Quality of planning process	Efficiency of task planning with PMS for scheduling and deadline management	Likert scale (1–5)	Level of planning effectiveness
Stakeholder management	Stakeholder communication and engagement effectiveness during the project	Likert scale (1–5)	Level of stakeholder management effectiveness
Risk management	Efficiency in recognising, evaluating and reducing project risks	Likert scale (1–5)	Level of risk management
User satisfaction	Satisfaction with the software's functionality and effectiveness	Likert scale (1–5)	Level of satisfaction with PMS usage
Project outcomes (DV)	Achievement in terms of user satisfaction, quality, timeliness and budget	Likert scale (1–5)	Project success on multiple dimensions

Note: DV, dependent variable.

6.2.5 Data analysis techniques

Descriptive and inferential analysis techniques are essential for interpreting the data collected in this study and for understanding how PMS influences project outcomes. Descriptive analysis summarises the dataset using measures such as mean, standard deviation and frequency distributions to provide an overview of planning quality, stakeholder management, risk management, PMS usability and project outcomes. These techniques highlight trends, patterns and variations in the data, offering an initial understanding of how organisations in the Senai Industrial Park use PMS and how these practices relate to project performance. Tools such as charts, cross-tabulations and descriptive statistics generated in SPSS support this foundational analysis and help identify preliminary associations in the dataset.

Inferential analysis extends beyond description by testing hypotheses and determining the statistical significance of relationships between variables. Techniques such as correlation and regression analysis were used to examine the strength and direction of relationships – for example, whether effective planning, stakeholder management, risk control and PMS usability significantly predict project outcomes. These methods allow the findings from the sample of respondents to be generalised to the broader population of companies in the Senai Industrial Park. Through hypothesis testing and model interpretation, inferential analysis provides evidence-based insights into how PMS contributes to improved timeliness, cost adherence, quality performance and overall project success, supporting informed decision-making and strategic enhancement of project management practices.

6.3. Results and analysis

6.3.1 Demographic profile

The demographic findings indicate a balanced gender distribution among respondents – 54.5% female and 45.5% male – suggesting equitable representation in PMS-related roles within the Senai Industrial Park. This balance reflects broader trends of increasing female participation in management, technical and project-oriented positions. Such representation strengthens the study's validity by capturing gender-diverse experiences and perceptions of PMS use.

The age distribution shows that most respondents were between 21 and 40 years old, accounting for over 60% of the sample. This indicates a predominantly young to mid-career workforce that is likely technologically proficient and actively involved in operational project management tasks. Meanwhile, respondents aged 41–50 (22.6%) and those aged over 50 (7.6%) represent experienced professionals who may hold strategic or supervisory responsibilities. This mix ensures that perspectives from varying career stages are incorporated, enhancing the generalisability of the findings across age cohorts.

In terms of ethnicity, the sample reflects Malaysia's multicultural workforce, consisting of Chinese (35.1%), Indian (30.8%), Malay (27.2%) and other ethnic groups (6.9%). This diversity ensures that the study captures culturally varied perspectives on PMS adoption, communication practices and collaboration, all of which may influence project management outcomes.

The respondents' organisational positions indicate broad involvement from different hierarchical levels. Executives (30.0%) and team leaders (27.7%) represent the majority, followed by key personnel (23.7%) and managers (18.6%). This distribution demonstrates that PMS is used across operational, supervisory and managerial roles, providing a comprehensive understanding of how various job functions interact with PMS.

Regarding experience with PMS, most respondents had used such systems for 2–4 years (26.0%) or 5–7 years (24.9%), showing a well-established adoption of PMS within organisations. Additionally, 19.8% had less than 1 year of experience, while 15.5% had used PMS for over 8 years, indicating both novice users and highly experienced practitioners. This range of experience supports robust analysis of how familiarity with PMS impacts perceptions of efficiency and project outcomes.

Finally, respondents represent diverse sectors, with substantial participation from manufacturing (23.9%), logistics and warehousing (20.4%), health and food (16.3%) and construction (14.2%). The remaining 25.2% fall under 'other' sectors, highlighting the broad applicability of PMS across various industries. The strong participation in manufacturing and logistics reflects the importance of PMS in coordinating complex operations. At the same time, the presence of respondents from service-based and specialised sectors underscores the widespread relevance of project management systems in modern organisational environments.

6.3.2 Reliability analysis

The internal consistency of the measurement constructs is reported in Table 6.2. Cronbach's alpha (α) was used to assess internal consistency across all four sections, and the results indicate strong internal consistency, with coefficients ranging from 0.849 to 0.873. Quality of the planning process demonstrated a reliability score of 0.871, with item means ranging from 3.84 to 3.94. Stakeholder management recorded a Cronbach's alpha of 0.849, with item means ranging from 3.85 to 3.93. Risk management had the highest reliability (0.873), with item means ranging from 3.90 to 4.01. Lastly, user satisfaction reported a Cronbach's alpha of 0.862, accompanied by item means ranging from 3.91 to 4.09. These results confirm that the measurement items used in the study are highly reliable and suitable for further statistical analysis.

Table 6.2 Reliability analysis

Variable	Total item	Reliability, α
Quality of the planning process	5	0.87
Stakeholder management	5	0.85
Risk management	5	0.87
User satisfaction	5	0.86
Project outcomes	5	0.86

6.3.3 Assumption testing and data screening

Prior to conducting correlation and multiple regression analyses, several statistical assumptions were examined to ensure the validity and robustness of the findings. Data screening procedures were performed to assess normality, multicollinearity, linearity, homoscedasticity and error independence.

Normality was assessed using skewness and kurtosis, as well as visual inspection of Q-Q plots. The values fell within the acceptable threshold range (±2), indicating no serious deviations from normal distribution. Visual inspection of the Q-Q plots further confirmed that the data points closely followed the diagonal reference line, suggesting approximate normality.

Multicollinearity was assessed using the variance inflation factor (VIF) and tolerance values. All VIF values were below the conservative threshold of 5.0, and tolerance levels exceeded 0.10, indicating that multicollinearity was not a concern and that the independent variables were sufficiently distinct.

The linearity and homoscedasticity assumptions were examined using a residual scatterplot. The residual plots demonstrated no clear patterns or funnel-shaped distributions, supporting the assumption of linear relationships and constant variance across predicted values.

Finally, independence of errors was assessed using the Durbin–Watson statistic, which fell within the acceptable range of 1.5–2.5, indicating no significant autocorrelation.

Overall, the results confirm that all key statistical assumptions were met, allowing reliable interpretation of the regression and hypothesis-testing results.

6.3.4 Pearson's correlation analysis

The correlation matrix of the study variables is shown in Table 6.3. Pearson's correlation analysis shows that all independent variables are positively and significantly associated with project outcomes at the $p < 0.01$ level, confirming the hypothesised relationships and underscoring their importance to project success. Risk management shows the strongest correlation with project outcomes ($r = 0.75$), indicating that effective risk identification and mitigation greatly enhance performance. Quality of planning process ($r = 0.72$) also shows a strong relationship, demonstrating that clear objectives, accurate resource allocation and realistic timelines contribute substantially to successful project execution. User satisfaction ($r = 0.71$) further highlights the value of adopting user-friendly systems that support collaboration and decision-making. Stakeholder management, although slightly lower ($r = 0.68$), remains

Table 6.3 Pearson's correlation analysis

Variable	Quality of planning process	Stakeholder management	Risk management	User satisfaction
Project outcomes	$r = 0.72$ $p < 0.01$	$r = 0.68$ $p < 0.01$	$r = 0.75$ $p < 0.01$	$r = 0.71$ $p < 0.01$

significantly correlated, emphasising the importance of communication, expectation alignment and stakeholder engagement. Collectively, these findings indicate that strengthening planning, risk management, stakeholder communication and PMS usability can meaningfully improve project outcomes.

6.3.5 Multiple regression analysis

The regression coefficients for predicting project outcomes are presented in Table 6.4. The multiple regression analysis shows that all four independent variables (quality of planning process, stakeholder management, risk management and user satisfaction) significantly predict project outcomes ($p < 0.001$), highlighting their collective importance. Risk management is the strongest predictor ($\beta = 0.46$), indicating that effective risk identification, assessment and mitigation are essential for managing uncertainty and ensuring project continuity. Quality of planning process follows closely ($\beta = 0.42$), demonstrating that well-defined goals, timelines and resource allocations contribute substantially to timely and efficient project delivery. User satisfaction ($\beta = 0.39$) also plays a key role, suggesting that user-friendly, reliable project management systems enhance collaboration, decision-making and overall performance. Stakeholder management, although the weakest predictor, remains significant ($\beta = 0.35$) and underscores the importance of clear communication, alignment of expectations and stakeholder engagement. Overall, the regression results confirm that strengthening risk management, planning processes, PMS usability and stakeholder relationships can meaningfully improve project success.

6.4. Practical insights

The findings of this chapter offer several practical insights for organisations seeking to improve project outcomes through effective adoption of PMS, particularly in industrial environments such as Senai Industrial Park in Malaysia. First, the strong influence of planning quality on project outcomes highlights the need for organisations to fully utilise PMS capabilities for task definition, scheduling and resource allocation. Managers should ensure that PMS is embedded into early project planning stages rather than used solely as a monitoring tool.

Second, risk management emerged as the strongest predictor of project success, indicating that organisations should leverage PMS features to systematically identify, track and mitigate risks. Integrating risk registers, alerts and analytics within PMS platforms can significantly reduce disruptions and enhance project stability.

Table 6.4 Regression coefficients for predicting project outcomes

Predictor variable	*B*	SE *B*	β	*t*	*p*
Quality of planning process	0.45	0.05	0.42	9.00	< 0.001
Stakeholder management	0.38	0.04	0.35	8.50	< 0.001
Risk management	0.50	0.06	0.46	10.00	< 0.001
User satisfaction	0.41	0.05	0.39	8.20	< 0.001

Third, stakeholder management plays a critical role in project performance. PMS tools that support real-time communication, shared dashboards and transparent reporting can improve coordination, align expectations and reduce conflicts between internal and external stakeholders.

Fourth, user satisfaction with PMS usability significantly affects project outcomes. This underscores the importance of selecting intuitive, user-friendly systems and investing in adequate training and change-management initiatives to encourage consistent use across teams.

Finally, organisations should align PMS adoption with their broader digital transformation strategies. By integrating PMS with enterprise systems and embracing emerging technologies such as analytics and automation, firms can enhance decision-making, improve efficiency and achieve sustainable project performance across industries.

6.5. Conclusion and recommendations

6.5.1 Implications

The study's findings support all four hypotheses, demonstrating that planning quality, stakeholder management, risk management and PMS usability satisfaction each have a significant positive impact on project outcomes. Effective planning increases the likelihood of project success by ensuring clear objectives, realistic timelines and efficient resource allocation, consistent with prior research linking strategic planning to improved operational performance. Similarly, strong stakeholder management enhances collaboration, reduces conflict and ensures project outputs meet expectations, consistent with studies highlighting the value of continuous communication and stakeholder engagement.

Risk management is also a critical driver of project success, as proactive risk identification and mitigation reduce delays and cost overruns, consistent with evidence that adaptive risk strategies increase project resilience. Finally, the results confirm that user satisfaction with PMS usability significantly improves project performance, with intuitive systems enhancing productivity, coordination and decision-making. These findings align with earlier research emphasising the role of advanced PMS tools in promoting transparency and accountability.

Overall, the study underscores the importance of an integrated project management approach that combines strong planning, stakeholder engagement, risk control and effective utilisation of PMS. When applied together, these elements significantly enhance project outcomes and help organisations achieve strategic and competitive advantages.

6.5.2 Limitations

This study highlights the importance of integrating key project management elements, namely stakeholder management, risk management, planning quality and user satisfaction, with PMS to improve project outcomes. While the findings underscore the value of adopting a holistic approach aligned with contemporary project management practices, several limitations should be acknowledged. Although the research demonstrates the benefits of proactive risk management in minimising delays, cost overruns and operational disruptions, the study relies on self-reported data, which may be subject to respondent bias. Additionally, the cross-sectional nature of the study restricts the ability to infer long-term effects or causal relationships between PMS usage and project outcomes.

The study also indicates that satisfaction with PMS enhances decision-making, collaboration and productivity; however, it does not examine differences in PMS functionalities across various software platforms, which may influence user perceptions. Organisational factors such as culture, leadership style and digital maturity were not explored in depth and may further moderate the effectiveness of PMS. Future studies should consider comparative analyses across industries that employ different project management methodologies, particularly those operating in technology-intensive environments. Research on integrating AI, machine learning and advanced analytics into PMS could also provide deeper insights into enhancing usability and project success. Furthermore, examining the role of cultural diversity, Agile frameworks and stakeholder dynamics may strengthen the understanding of how project management practices can be tailored to diverse organisational contexts.

6.5.3 Recommendations and future research

Future research should examine how different industries, particularly technologically advanced and highly dynamic ones, vary in planning quality, stakeholder management, risk management and PMS satisfaction. Longitudinal studies would be particularly valuable for assessing the long-term effects of integrated project management practices on project success and wider organisational performance. Given the rapid evolution of digital technologies, further investigation into the use of AI, machine learning and advanced analytics to enhance PMS usability and influence project outcomes is also warranted.

In addition, future studies could examine how stakeholder dynamics, including cross-functional collaboration and cultural diversity, shape project performance. Investigating the adoption and effectiveness of PMS tools in small and medium-sized enterprises may reveal unique challenges and advantages compared to larger organisations. Research into the adaptability of project management systems to varying organisational and project-specific needs, as well as the influence of training and capacity-building initiatives on user adoption and satisfaction, would also offer valuable insights.

Moreover, examining the integration of PMS tools with Agile project management methodologies and assessing their combined impact on project success represents an important area for further inquiry. Comparative research across sectors or countries may help clarify how cultural, economic and regulatory factors influence project management effectiveness. Finally, future work should consider the potential role of PMS in supporting sustainability-oriented project management, particularly in relation to environmental, social and governance (ESG) objectives.

REFERENCES

Alqahtani J, Siddique A, Aseere AM, Alasiry A and Naveed QN (2024) Evaluating success factors of software project management in global software development. *IEEE Access* **12**: 22345–22358.

Arvikar S, Aung PPW, Cha G and Park S (2025) Augmented reality-based construction site management using optimized BIM and project schedule integration. *Automation in Construction* **175**: 106204.

Azizi J, Nikravesh R and Golshani T (2024) Investigating the impact of an integrated project management system on financial performance. *International Journal of Management Studies* **18(2)**: 55–70.

Gilal R, Omar M and Rejab MM (2023) Key factors contributing to time pressure in software development projects: a review. *International Journal of Advanced and Applied Sciences* **10(10)**: 155–165.

Hamada MA (2023) Investigating the efficiency of project management software in construction projects. In *Proceedings of the International Conference on Basic Sciences, Engineering and Technology (ICBASET 2023)* 22: 247–257.

Kennedyd SI, Zadeh AA, Alborz S, Ramezani R and Choi J (2026) Agile implementation and its dual impact on employee satisfaction and project outcomes in IT. *Information Systems Management* 1–16.

Köksal A and Ersoy Maraş S (2026) Bayesian network model for stakeholder management in large scale housing projects: prediction of project success. *Journal of Asian Architecture and Building Engineering* **25(2)**: 1406–1423.

Lesum SA, Akthar SR, Islam MR, Sadia F and Hasan M (2024) Project governance to improve the performance of software projects by mitigating the software risk factors: the moderating role of project leadership. *Procedia Computer Science* **239**: 1863–1870.

Mahdi MN, Zabil MHM, Ahmad AR *et al.* (2021) Software project management using machine learning techniques: a review. *Applied Sciences* **11(11)**: 51–83.

Nenni ME, De Felice F, De Luca C and Forcina A (2025) How artificial intelligence will transform project management in the age of digitization: a systematic literature review. *Management Review Quarterly* **75(2)**: 1669–1716.

Obaid T and Abu-Naser SS (2023) Big data analytics in project management: a key to success. *International Journal of Academic Engineering Research* **7(7)**: 1–8.

Ongesa TN, Ugwu OPC, Ugwu CN *et al.* (2025) Optimizing emergency response systems in urban health crises: a project management approach to public health preparedness and response. *Medicine* **104(3)**: e41279.

Parikh JH, Mody J and Pitroda JR (2021) Effective project management using mobile applications for construction projects: a review. *International Research Journal of Modernization in Engineering Technology and Science* **3(3)**: 1197–1198.

Rasool N, Khan SY, Haseeb U *et al.* (2023) SCRUM and the agile procedure's impact on software project management. *Journal of Jilin University (Engineering and Technology Edition)* **43(2)**: 380–382.

Waseem M, Das T, Ahmad A *et al.* (2024) ChatGPT as a software development bot: a project-based study. *Journal of Software Engineering Research and Development* **12(1)**: 1–15.

Advancing Project Management: Frameworks, Digital Transformation, Sustainability and Governance
Mohammed Reyasudin Basir Khan
ISBN 978-1-83742-799-4
https://doi.org/10.1108/978-1-83742-798-720261009

Chapter 7
Risk management in infrastructure projects: examining schedule, cost and quality impacts on project success

Rozlano Muhamad Tobat and Mohammed Reyasudin Basir Khan

Tun Razak Graduate School, Universiti Tun Abdul Razak (UNIRAZAK), Kuala Lumpur, Malaysia

7.1. Introduction

Malaysia's construction industry has expanded rapidly due to economic growth and urbanisation, with the market valued at US$30.1 billion in 2023 and projected to grow at an average annual growth rate (AAGR) of over 5% from 2025 to 2028. Despite this growth, project success is often hindered by multiple risk factors that lead to delays, cost overruns and quality issues (Alsaadi and Norhayatizakuan, 2021; Shehu *et al.*, 2014). A significant portion of Malaysian construction projects, particularly government-funded ones, experience delays of 10–20%, often due to material shortages, labour constraints, supply disruptions and weak project management.

Economic instability, including inflation and currency fluctuations, has contributed to rising construction costs. Between 2020 and 2024, steel prices increased by 56% and cement prices by 30%, adding 8–15% to total project costs. Labour shortages and reliance on foreign workers, exacerbated by political uncertainties, have further strained project timelines and productivity. External shocks such as the COVID-19 pandemic disrupted supply chains, delayed project execution and increased material costs. Safety-related risks also remain significant; from 2018 to 2023, construction accidents steadily increased, with 159 cases reported in 2023, often linked to inadequate safety supervision and poor site management. Global geopolitical tensions and trade disruptions have driven up material prices and heightened procurement uncertainty. Combined with domestic challenges, these factors create a complex environment in which project risks significantly affect schedule, cost and quality. Effective identification and mitigation of these risks are therefore essential for ensuring project success.

This chapter explores how risk factors influence construction outcomes in Malaysia by examining the interconnected effects of economic conditions, political instability, labour availability and environmental events on project performance. Understanding these dynamics will provide insights to improve risk management and strengthen the resilience of infrastructure development.

The construction sector faces persistent challenges that threaten project timelines, budgets and quality standards. The study presented in this chapter focuses on three core problem areas: schedule delays, cost overruns and quality deficiencies (Štrok, 2025). Project schedule delays often stem from regulatory complications, labour shortages, poor planning, supply chain disruptions and unforeseen events. Government projects, in particular, frequently encounter these issues, resulting in extended timelines, financial penalties and damage to organisational reputation. Cost overruns pose another major challenge, often driven by inaccurate cost estimates, material price fluctuations, unexpected site conditions and changes in project scope. Between 2020 and 2024, substantial increases in material prices placed financial strain on projects, with many exceeding budgets by 5–10%. Quality deficiencies further complicate project success, as inadequate quality control, poor materials, limited skilled labour and accelerated schedules lead to defects, rework, safety risks and reduced sustainability.

Although the study focuses on internal risk factors, external risks such as economic instability, supply chain disruptions, regulatory changes, environmental hazards, political and policy shifts and labour market constraints also significantly contribute to setbacks in construction projects. These risks collectively jeopardise project performance: delays arise from poor planning, limited resources, regulatory obstacles, labour strikes and environmental disruptions; project cost is affected by inflation, material price volatility, inaccurate estimation and regulatory modifications; quality issues emerge from insufficient supervision, substandard materials, unskilled labour and inadequate testing. Together, these risks threaten compliance with standards, increase the need for rework and reduce overall stakeholder satisfaction.

In light of these challenges, the study analysed the impact of key risk factors on project success by examining how delays, cost overruns and quality issues influence final project outcomes. The primary research objectives were

- to investigate how schedule-related risks influence timely project completion
- to explore how cost-related risks contribute to project cost overruns and budget deviation
- to examine how quality-related risks affect the final quality of project deliverables.

Accordingly, the study addresses three research questions.

- How do work-schedule risks contribute to project delays?
- How do cost-related risks affect budget overruns?
- How do quality-related risks lead to defects in project outcomes?

The study offers substantial contributions to both academia and industry. From an academic perspective, it deepens insights into how different risk factors influence project outcomes, thereby enhancing theoretical understanding of risk management in construction. It also provides empirical evidence to validate and refine existing project risk models and encourages multidisciplinary research by linking engineering, finance and management perspectives. In addition, the study offers methodological improvements for assessing and mitigating project risks, which can be adopted in future research and academic work.

From an industry perspective, the findings of this study can help project managers to better prioritise risks and implement more effective mitigation strategies in construction projects. The insights gained can also support the development and improvement of standard operating procedures (SOPs), organisational policies and risk management training programmes. Ultimately, by improving risk planning and control, the study helps reduce delays, cost overruns and quality defects, thereby supporting more successful and sustainable construction project delivery.

The construction industry is a key driver of Malaysia's economic development, underpinning infrastructure growth and social advancement. However, construction projects frequently face risks that affect schedules, costs, quality and, ultimately, project success. These risks arise from technological, environmental, financial, political, social and organisational factors, making risk management a complex but essential domain in construction project management.

The following review of the literature provides an overview of theories and empirical studies related to risk factors influencing infrastructure development and the successful completion of construction projects in Malaysia. It synthesises scholarly work, industry reports and empirical evidence to identify the main categories of risks, their interactions and their impact on project outcomes. It establishes the theoretical foundations of risk management and project success, summarises empirical research on construction risk factors in Malaysia, develops a conceptual framework linking schedule, cost and quality risks to project outcomes, and supports the formulation of hypotheses for the study's empirical phase. The study draws on key theoretical perspectives from construction management, project management and organisational studies, particularly risk management theory and project success theory.

Risk management theory provides a structured approach for identifying, assessing, responding to and monitoring risks that may affect project objectives (Sotamaa *et al.*, 2025). It emphasises a proactive, systematic handling of uncertainty through processes such as risk identification (using tools like expert judgement, checklists, brainstorming and strengths, weaknesses, opportunities and threats (SWOT) analyses), risk analysis (using qualitative methods such as matrices and probability–impact grids, and quantitative methods such as decision tree analysis and Monte Carlo simulation), risk response planning (through avoidance, mitigation, transfer or acceptance) and continuous risk monitoring. Widely recognised frameworks, including the project management body of knowledge (PMBOK: PMI, 2021), ISO 31000 (ISO, 2018) and the COSO ERM Framework (COSO, 2023), advocate embedding risk management into organisational strategy, governance and operations to strengthen resilience and improve decision-making (Varajao *et al.*, 2025).

Project success theory examines the factors contributing to effective project completion. While the 'iron triangle' of time, cost and quality has traditionally been used, modern perspectives broaden these criteria to include stakeholder satisfaction, alignment with strategic and business objectives, and long-term sustainability outcomes (Mirzaei *et al.*, 2025). These expanded dimensions reflect the increasing complexity of construction projects and the need to evaluate success beyond short-term performance metrics. Together, these theories provide the conceptual foundation for analysing how risk factors influence the success of construction projects in Malaysia.

Related theories, such as agency theory, highlight the roles of contracts, incentives and information asymmetry between principals (owners) and agents (contractors or consultants). Better alignment of interests and clearer outcome-based arrangements can improve project performance and reduce the risk of failure. Project success theory thus integrates traditional performance metrics with broader considerations of stakeholder value, strategic alignment and sustainability. In this study, construction project success is conceptualised as a multidimensional outcome influenced by schedule, cost and quality risk factors.

A substantial body of empirical research has examined risk management and project success in construction, including in Malaysia. Bahamid and Doh (2017) describe risk management as the classification, analysis/assessment and response to unpredictable risks during project implementation, emphasising its continuous and dynamic nature. Many studies agree that uncertainty is central to risk and that risk management increases the likelihood of project success but does not guarantee it. Alsaadi and Norhayatizakuan (2021) note that unmanaged uncertainty can significantly affect project outcomes, underscoring the need for systematic risk management processes.

Risk management typically involves identifying risks that may affect project objectives, analysing their likelihood and impact, and planning appropriate responses such as reducing, transferring, accepting or eliminating risks. Although practices in this area have improved, greater robustness and maturity are still needed to meet industry expectations. Malaysian studies further highlight several limitations: human error and the absence of systematic methodologies weaken risk-handling efforts, whereas adopting knowledge-based frameworks and lessons learned from past projects can significantly enhance risk management effectiveness.

Empirical findings in Malaysia reveal that time and cost overruns remain major challenges. Shehu *et al.* (2014) reported an average cost deviation of 2.08% and a significantly higher time deviation of 49.71%, indicating that schedule overruns are a more pressing issue than cost variation. Some studies found that top management support, a clear project mission, effective communication and skilled personnel are strong predictors of project success, whereas scheduling and planning alone do not guarantee successful outcomes without strong organisational and human factors (Sotamaa *et al.*, 2025). Hoffmann (2017) further underscores that weaknesses in risk management practices, particularly human-related and methodological gaps, can compromise project sustainability and long-term performance.

Collectively, these studies show that construction projects in Malaysia frequently face schedule and cost risks, risk management practices are present but insufficiently systematic, and human and organisational factors play a critical role in shaping project outcomes. Building on these insights, the conceptual framework for this research centres on three independent variables: schedule risk factors (such as delays, resource issues and project complexity), cost risk factors (including cost overruns, inaccurate estimates and price volatility) and quality risk factors (such as poor workmanship, weak quality control and non-compliance with standards). The dependent variable is construction project success, measured through adherence to schedule and budget, meeting quality standards, client satisfaction, durability and the achievement of intended project outcomes. The conceptual relationships between these variables and construction project success are illustrated in Figure 7.1.

Based on the conceptual framework and existing literature, three hypotheses are proposed to examine the relationship between key risk factors and construction project success. The first hypothesis (H1) posits that schedule-related risk factors significantly influence project success. These risks may arise from delays in material delivery, adverse weather conditions, labour shortages, equipment failures, regulatory changes and unforeseen site issues, all of which can disrupt timelines, negatively affect client satisfaction and compromise cost control. The second hypothesis (H2) asserts that cost-related risk factors have a major effect on project success. Cost risks stem from inaccurate estimates, fluctuations in material and labour prices, unexpected site conditions and scope changes, leading to cost overruns and financial strain that undermine project outcomes.

The third hypothesis (H3) suggests that quality-related risk factors substantially affect project success. Quality risks, including poor workmanship, substandard materials, design flaws, inadequate supervision and weak quality-control processes, can lead to defects, rework, safety hazards and reduced stakeholder satisfaction. Developing these hypotheses clarifies the variables of interest and the expected relationships between them, guiding data collection and analysis. Testing these hypotheses allows the study to determine the extent to which schedule, cost and quality risks affect construction project success in Malaysia and to provide evidence-based recommendations to enhance risk management practices.

Figure 7.1 Conceptual framework

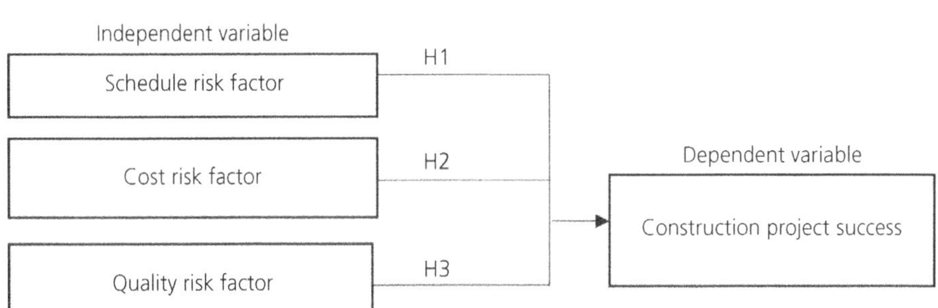

7.2. Methodology

7.2.1 Research design

The research strategy for this study adopted a quantitative approach to provide a comprehensive understanding of the risk factors influencing the success of construction projects. This method validates and measures the prevalence and impact of these risk factors across different subsectors of the construction industry. This method's design began with identifying the research problem, followed by developing research questions, a conceptual model and a theoretical framework. It then selected suitable data collection techniques, chose analysis methods and collected, interpreted and reported data.

7.2.2 Population and sampling procedures

For the sampling design, a stratified random sampling technique was employed to ensure that the study captured diverse perspectives from across the construction industry. The overall population was divided into relevant strata by project type (infrastructure, residential, commercial and industrial) and geographical location within the Klang Valley. From each stratum, respondents from the construction sector and the general public were randomly selected to participate. This approach enhances representativeness and enables the quantitative phase to effectively validate and measure the severity and interrelationships of the identified risk factors.

7.2.3 Data collection method

A review of previous studies on risk management, mainly by Malaysian and other relevant construction researchers, was conducted to identify common risk factors within the local context. Integrating findings from these empirical works allowed the study to focus on risks most prevalent in Malaysia's construction industry. Stratified random sampling is used to select participants with diverse roles and expertise across the construction industry, ensuring broad representation of perspectives. Questionnaires were distributed until the target sample was reached, with 126 completed responses collected over a one-month period. A structured questionnaire was developed to collect quantitative data on schedule, cost and quality risk factors. The questions were designed to capture both perceived and actual risks, as well as their impact on project success indicators such as time, cost and quality.

7.3. Results and analysis

7.3.1 Response rate

A total of about 200 surveys were distributed through WhatsApp, Telegram, LinkedIn and Facebook, with participants given one month to respond. Of these, 126 completed surveys were returned, representing a 63% response rate, significantly higher than the average online survey response rate of 44.1% reported by Wu *et al.* (2022).

7.3.2 Demographic profile

The demographic distribution of respondents by age group is presented in Figure 7.2. Based on the demographic analysis, the largest age group among respondents is 45–54 years (38.1%), followed by 35–44 years (33.3%), indicating that the sample is predominantly middle-aged. Younger respondents (under 25) and older individuals (over 55) together account

Figure 7.2 Percentage of respondents by age group

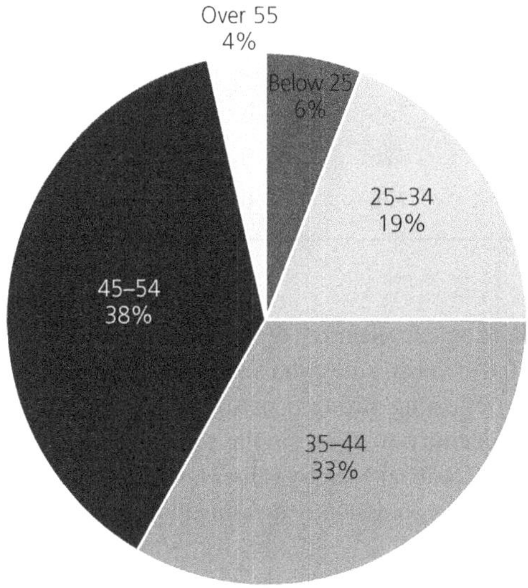

for only 9.5% of the total, indicating limited participation from these age groups. In terms of education, nearly half of the respondents hold a Bachelor's degree (46.0%), followed by a Master's degree (36.5%). Another 15.9% possess a diploma, while none hold a doctorate. This suggests a highly educated sample, with more than 80% having tertiary qualifications.

Respondents also demonstrate substantial professional experience: 41.3% have more than 20 years of experience and another 30.2% have 11–20 years. This means that over 70% of the sample consists of seasoned professionals, many of whom are likely in senior or leadership roles. Employment status data show that 81% are employed full-time, 9.5% work on a contract basis and 6.3% fall under other categories, reflecting a workforce largely engaged in stable, continuous employment. By job level, middle management accounts for the largest share (37.1%), followed by senior management (27.4%), indicating that participants are predominantly in mid- to high-level positions.

Regarding industry involvement, most respondents (60.3%) do not work in the construction sector, suggesting a diverse sample beyond construction-related professions. Among those in the industry, the infrastructure and civil engineering sub-sector is the most represented (15.9%), followed by the commercial/industrial sector (12.7%) and other miscellaneous sub-sectors (6.3%). This distribution shows moderate representation from across the construction industry, while still including perspectives from outside the sector.

7.3.3 Reliability analysis

The reliability results for all constructs are summarised in Table 7.1. The reliability analysis shows strong internal consistency across all variables. Project success has a Cronbach's

Table 7.1 Summary of reliability analysis findings

Variable	n	Number of items	Cronbach's alpha, α
Project success (DV)	126	6	0.913
Schedule risk factor (IV1)	126	6	0.815
Cost risk factor (IV2)	126	6	0.872
Quality risk factor (IV3)	126	6	0.904

alpha (α) of 0.913, indicating excellent reliability among its six items. Schedule risk (IV1) shows good reliability (0.815), well above the acceptable threshold. Cost risk (IV2) has very good reliability (0.872), reflecting strong consistency across its items. Quality risk (IV3) demonstrates excellent reliability (0.904), similar to project success. Overall, all variables show high reliability, with Cronbach's alpha values ranging from 0.815 to 0.913, confirming that the items for each construct consistently measure their intended concepts and are suitable for further analysis.

7.3.4 Descriptive analysis

The descriptive statistics for all study variables are presented in Table 7.2. The descriptive statistics show that project success has a mean of 3.66, with a median and mode of 4.00, indicating that most respondents rated it at the higher end of the scale. Its standard deviation (SD) of 0.883 reflects relatively low variability. Schedule risk has a slightly higher mean of 3.84, with a median and mode of 4.00 and a standard deviation of 0.963, suggesting slightly greater variability in perceptions. Cost risk has a mean of 3.72, with a median and mode of 4.00, but the highest standard deviation (0.991), indicating the widest variation in responses among the variables. Quality risk has a mean of 3.73, a median and mode of 4.00 and a standard deviation of 0.886, reflecting moderate variability. Overall, all variables have the same median and mode of 4.00, indicating strong agreement among respondents. Standard deviations below 1.0 indicate consistent responses, with slightly more variability observed for the three risk factors than for project success. The distributions of the dependent and independent variables are shown in Figure 7.3.

Table 7.2 Descriptive statistics of variables

Variable	Mean	Median	Mode	SD
Project success (DV)	3.66	4.00	4.00	0.883
Schedule risk factor (IV1)	3.84	4.00	4.00	0.963
Cost risk factor (IV2)	3.72	4.00	4.00	0.991
Quality risk factor (IV3)	3.73	4.00	4.00	0.886

Figure 7.3 Distributions of the dependent and independent variables

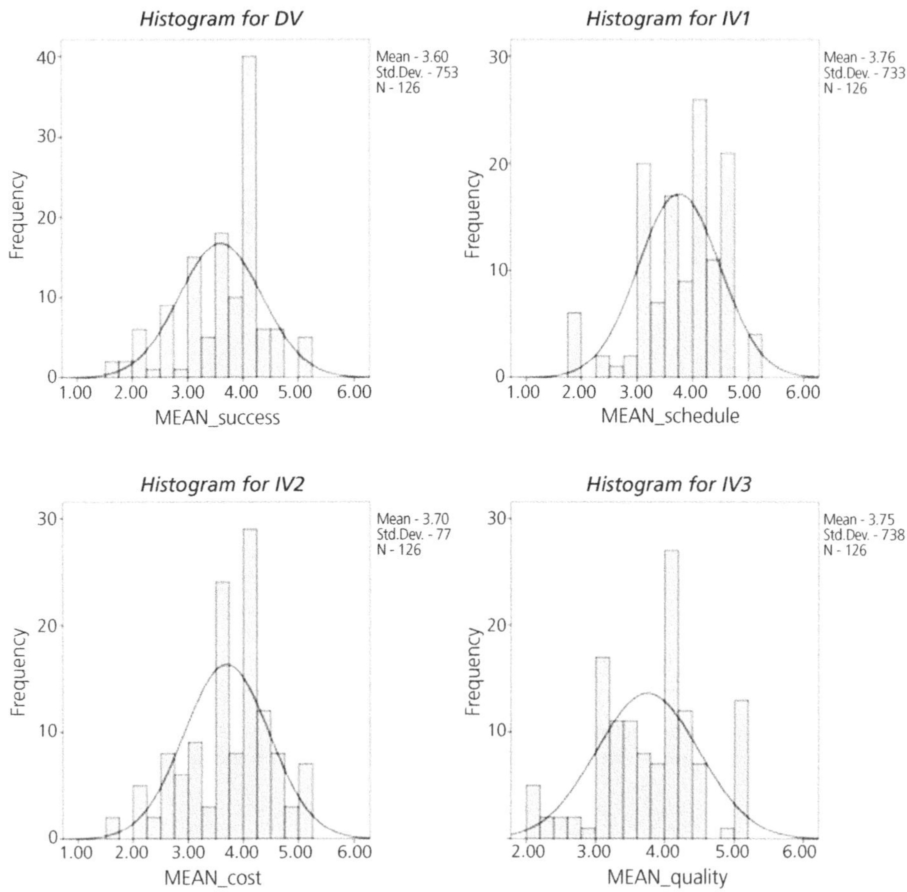

To evaluate whether the data satisfy the assumption of linearity, scatter plots were used as an effective visual assessment tool. Based on the scatter plots generated in the statistical package for the social sciences (SPSS), the observed patterns closely followed a straight-line trend, indicating that the linearity assumption was met (Figure 7.4).

Figure 7.5 presents the Q-Q plots for all independent variables (schedule, cost and quality risk) and the dependent variable (project success). The plots show that the data points for all variables closely align with the 45° reference line, with only minor deviations, indicating that the distributions are approximately normal. This confirmation of normality supports the use of parametric statistical methods, thereby increasing the reliability and validity of the study's analyses.

Figure 7.4 Scatter plots for the dependent and independent variables

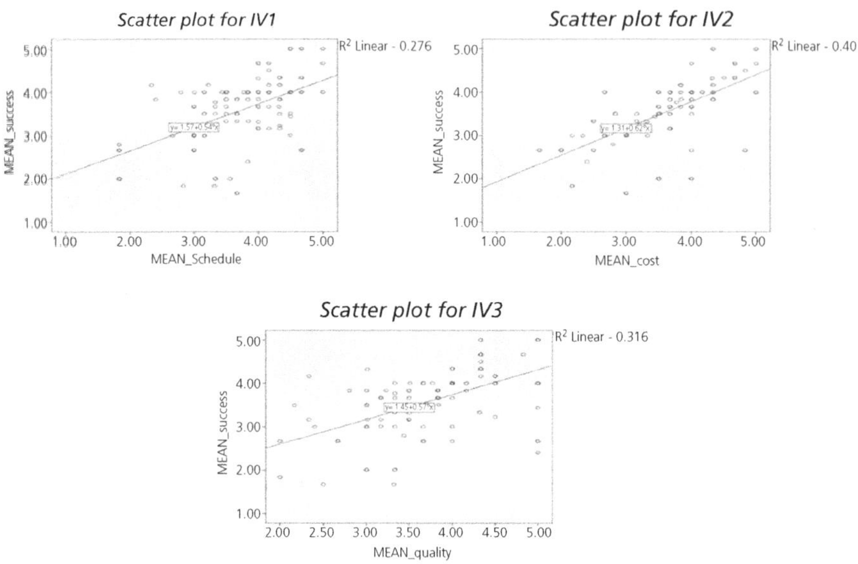

Figure 7.5 Q-Q plots for DV and IVs

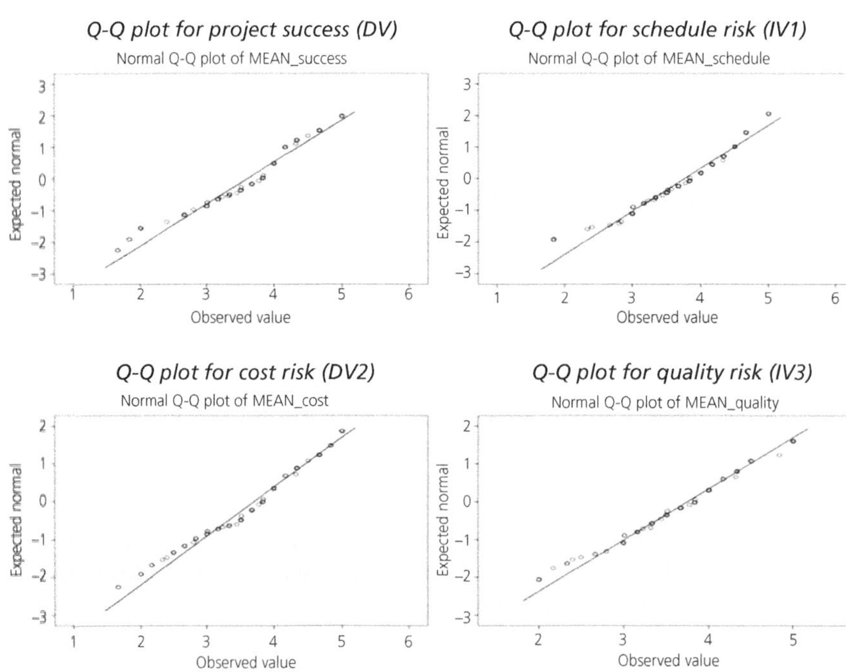

Table 7.3 Findings of Pearson's correlation

		Project success	Schedule risk factor	Cost risk factor	Quality risk factor
Project success	Pearson correlation	1	0.608[a]	0.649[a]	0.685[a]
	Sig. (2-tailed)		0.000	0.000	0.000
	n	126	126	126	126
Schedule risk factor	Pearson correlation	0.608[a]	1	0.674[a]	0.579[a]
	Sig. (2-tailed)	0.000		0.000	0.000
	n	126	126	126	126
Cost risk factor	Pearson correlation	0.649[a]	0.674[a]	1	0.737[a]
	Sig. (2-tailed)	0.000	0.000		0.000
	n	126	126	126	126
Quality risk factor	Pearson correlation	0.685[a]	0.579[a]	0.737[a]	1
	Sig. (2-tailed)	0.000	0.000	0.000	
	n	126	126	126	126

[a]Correlation is significant at the 0.01 level (2-tailed).

7.3.5 Correlation analysis

The Pearson correlation coefficients among the study variables are presented in Table 7.3. The results indicate that all three hypotheses are supported. The significance values reported as 0.000 correspond to $p < 0.001$, indicating highly significant relationships rather than an exact probability of 0. Schedule risk shows a strong positive correlation with project success ($r = 0.608$, $p < 0.001$), suggesting that effective scheduling risk management is associated with improved project outcomes. Similarly, cost risk shows a strong positive relationship with project success ($r = 0.649$, $p < 0.001$), indicating that proactive cost risk management contributes to improved project performance. Among the three dimensions, quality risk shows the strongest correlation with project success ($r = 0.685$, $p < 0.001$), underscoring its critical importance for successful infrastructure delivery. All correlations are statistically significant at the 0.01 level, confirming robust relationships between the identified risk factors and overall project success.

The correlations between independent variables are also notable: schedule and cost risks are strongly related ($r = 0.674$); schedule and quality risks are moderately positively related ($r = 0.579$); cost and quality risks are highly correlated ($r = 0.737$). These significant correlations suggest potential multicollinearity considerations for further regression analysis.

7.3.6 Multiple regression analysis

Regression analysis was conducted to examine the relationship between the three key risk factors and project success. This analysis quantifies the influence of each factor on project outcomes and identifies which risks significantly affect success. The regression results provide insight into how these variables interact and highlight areas for improvement to strengthen project management strategies. A summary of the findings from the multiple regression model is shown in Table 7.4.

Based on the model summary, the three independent variables (schedule risk, cost risk and quality risk factors) collectively explain 54.9% of the variance in project success. This indicates that these factors contribute substantially to construction project performance in the Klang Valley. The remaining 45.1% of the variance, however, is attributable to other external or unexamined factors not included within the scope of this study.

Table 7.5 presents the unstandardised and standardised coefficients, showing the influence of schedule, cost and quality risk factors on construction project success in the Klang Valley. The constant term 0.450 is not statistically significant ($t = 1.135$, $p = 0.261$), indicating that the baseline prediction provides no meaningful explanatory value. Schedule risk demonstrates a significant positive effect on project success, with an unstandardised coefficient of 0.271 and a moderate standardised beta (β) of 0.251 ($t = 2.091$, $p = 0.041$), suggesting that higher schedule risk is associated with increased project success in this model. Cost risk shows a positive but statistically insignificant relationship, with $B = 0.170$, $\beta = 0.179$, $t = 1.236$ and $p = 0.221$, indicating that its influence is weak and not meaningful. Quality risk exhibits the strongest effect, with $B = 0.413$ and the highest standardised β at 0.408, supported by a strong

Table 7.4 Multiple regression model summary

R	R^2	Adjusted R^2	Std. error of the estimate
0.741[a]	0.549	0.526	0.50856

Note: predictors: (constant), schedule risk (IV1), cost risk (IV2), quality risk (IV3); dependent variable: project success (DV).

Table 7.5 Coefficients

Model	Unstandardised coefficients		Standardised coefficient	*t*	Sig.
	B	Std. error	β		
(Constant)	0.450	0.396		1.135	0.261
Schedule risk	0.271	0.130	0.251	2.091	0.041
Cost risk	0.170	0.137	0.179	1.236	0.221
Quality risk	0.413	0.133	0.408	3.111	0.003

Note: dependent variable: project success (DV).

Table 7.6 Analysis of variance (ANOVA)

Model	Sum of squares	*df*	Mean square	*F*	Sig.
Regression	18.542	3	6.181	23.898	0.000
Residual	15.259	59	0.259		
Total	33.802	62			

Note: dependent variable: project success (DV); predictors: (constant), schedule risk (IV1), cost risk (IV2), quality risk (IV3).

t-value of 3.111 and a highly significant *p*-value of 0.003. This confirms that quality risk is the most influential predictor of project success. Overall, the regression results indicate that schedule and quality risks significantly affect project success, while cost risk does not show a statistically significant impact in this model.

The *F*-statistic reported in Table 7.6 evaluates the overall significance and fit of the regression model. The *F*-statistic of 23.898 indicates that the model provides a strong explanation of the variation in project success. The result is statistically significant ($p < 0.001$), demonstrating that the combined effects of schedule, cost and quality risk factors meaningfully predict project outcomes. This finding confirms that at least one independent variable significantly contributes to explaining the variance in the dependent variable, supporting the regression model's robustness and explanatory power.

7.3.7 Hypothesis testing

A summary of the hypothesis testing results from both correlation and regression analyses is presented in Table 7.7. In conclusion, the results provide strong support for hypotheses 1 (H1) and 3 (H3), as both schedule risk and quality risk demonstrate statistically significant and positive effects on construction project success in the multiple regression and correlation analyses. Although hypothesis 2 (H2) shows a significant correlation with project success, it does not exhibit a significant influence within the regression model. This suggests that its

Table 7.7 Multiple regression and correlation analysis findings

Hypothesis	Multiple regression	Correlation
H1: project schedule-related risk factors have a significant effect on construction project success	Supported	Supported
H2: project cost-related risk factors have a major effect on construction project success	Not supported	Supported
H3: project quality-related risk factors have a substantial effect on construction project success	Supported	Supported

effect diminishes when schedule and quality risks are considered simultaneously, indicating a comparatively weaker contribution to project success.

7.4. Practical insights

The findings of this study provide several important practical insights for construction practitioners, policy makers and project stakeholders involved in infrastructure development in Malaysia. First, the strong influence of schedule risk on project success underscores the need for more rigorous planning and monitoring. Project managers should prioritise realistic scheduling, early identification of potential delays and proactive coordination with suppliers, regulators and subcontractors to minimise disruptions to project timelines.

Second, quality risk emerged as the most influential predictor of project success. This underscores the importance of strengthening quality management systems through regular site inspections, stricter material control, competent supervision and continuous skills development for site personnel. Emphasising quality at every stage of the project lifecycle can significantly reduce rework, safety incidents and long-term maintenance costs.

Third, although cost risk showed a positive correlation with project success, it was not a significant predictor when schedule and quality risks were considered simultaneously. This suggests that cost overruns may often result from schedule delays and quality failures rather than being an independent driver of project performance. Practitioners should therefore focus on controlling upstream risks related to time and quality to indirectly manage cost outcomes.

Finally, the results support adopting structured risk management frameworks, such as ISO 31000 and PMBOK, to systematically identify, assess and mitigate risks. Embedding risk management into organisational policies and decision-making processes can enhance resilience, improve project predictability and support the successful delivery of infrastructure projects in the Klang Valley and similar construction environments.

7.5. Conclusion and recommendations

This study examined public and industry perspectives on risk factors affecting construction project success in Klang Valley using survey data analysed through correlation and multiple regression. The results show that schedule and quality risks are the strongest predictors of project success, while cost risk has no significant impact. These findings indicate that, although several risks influence construction outcomes, schedule and quality risks play a more critical role in the Klang Valley context. The analysis confirms that cost risk does not significantly contribute to project success, underscoring the need to prioritise key risk factors to enhance project performance.

7.5.1 Limitations

This study has several limitations. The small sample size and the short one-month data-collection period may have reduced the accuracy and reliability of the findings. A larger and more diverse sample would increase statistical power and improve the precision of the results. Future research should seek broader participation through industry collaboration or the use of historical datasets, and meta-analytic approaches could further strengthen the evidence. Participant selection bias is another concern; transparent recruitment procedures and the use of stratified or random sampling would help minimise this issue. Rigorous data collection protocols, such as standardised instruments, inter-rater reliability checks and robust data validation practices, are essential for improving data quality. Using multiple data sources or external validation could also enhance confidence in the results. While this study provides useful insights into the effects of schedule, cost and quality risks on project success in Klang Valley, addressing these limitations and examining additional variables in future research would support a more comprehensive understanding of construction project performance.

7.5.2 Recommendations and future research

Future research should incorporate additional independent variables to better capture the complex, interconnected risks that influence construction projects in the Klang Valley. Factors such as project size, complexity, location, market conditions, organisational culture, leadership, stakeholder engagement and regulatory compliance could enhance the model's predictive accuracy. Subgroup analyses based on project type, size or geographic region would also provide deeper insights into how risk profiles differ across Malaysia's construction sector. Finally, validating the model with external datasets such as industry reports or historical project data would strengthen its generalisability and ensure its reliability across diverse project environments.

REFERENCES

Alsaadi N and Norhayatizakuan N (2021) The impact of risk management practices on the performance of construction projects. *Studies of Applied Economics* **39(4)**.

Bahamid RA and Doh SI (2017) A review of risk management process in construction projects of developing countries. *IOP Conference Series: Materials Science and Engineering* **271**: 012042.

COSO (2023) *COSO ERM Framework*. https://www.coso.org/erm-framework (accessed 30/04/2026).

Hoffmann CH (2017) Strengths and weaknesses of scenario planning as a risk management tool. In *Assessing Risk Assessment: Towards Alternative Risk Measures for Complex Financial Systems*. Springer Fachmedien Wiesbaden, Wiesbaden, Germany, pp. 213–218.

ISO (International Organization for Standardization) (2018) ISO 31000: *Risk Management – Guidelines*. ISO, Geneva, Switzerland.

Mirzaei M, Mabin VJ and Zwikael O (2025) Customising hybrid project management methodologies. *Production Planning & Control* **36(9)**: 1188–1205.

PMI (Project Management Institute) (2021) *A Guide to the Project Management Body of Knowledge (PMBOK® Guide)*, 7th edn. Project Management Institute, Newtown Square, PA, USA.

Shehu Z, Endut IR, Akintoye A and Holt GD (2014) Cost overrun in the Malaysian construction industry projects: a deeper insight. *International Journal of Project Management* **32(8)**: 1471–1480.

Sotamaa T, Reiman A and Kauppila O (2025) Manufacturing SME risk management in the era of digitalisation and artificial intelligence: a systematic literature review. *Continuity and Resilience Review* **7(1)**: 1–28.

Štrok KT (2025) An overview of the relations within the iron triangle on a sample of construction projects in Croatia. *Journal of Applied Engineering Science* **23(1)**: 131–140.

Varajao J, Lopes L and Tenera A (2025) Framework of standards, guides and methodologies for project, program, portfolio, and PMO management. *Computer Standards and Interfaces* **92**: 103888.

Wu MJ, Zhao K and Fils-Aime F (2022) Response rates of online surveys in published research: a meta-analysis. *Computers in Human Behavior Reports* **7**: 100206.

Advancing Project Management: Frameworks, Digital Transformation, Sustainability and Governance
Mohammed Reyasudin Basir Khan
ISBN 978-1-83742-799-4
https://doi.org/10.1108/978-1-83742-798-720261010

Chapter 8
Impact of Lean Six Sigma on cost reduction in electronics manufacturing service companies

Khugan Kanagasabapathy and Mohammed Reyasudin Basir Khan

Tun Razak Graduate School, Universiti Tun Abdul Razak (UNIRAZAK), Kuala Lumpur, Malaysia

8.1. Introduction

In the global manufacturing sector, organisations continually strive to enhance efficiency, reduce operational costs and improve overall performance. Lean Six Sigma (LSS) has emerged as one of the most effective methodologies for accomplishing these goals (Ibrahim and Kumar, 2025). It integrates the principles of lean manufacturing, which focuses on eliminating non-value-added activities, with Six Sigma, which emphasises reducing process variations and defects. Together, these approaches form a comprehensive framework that helps organisations achieve streamlined operations, improved quality and sustainable cost efficiency.

The relevance of LSS is particularly significant within the electronics manufacturing service (EMS) sector. EMS companies operate in a highly competitive, cost-sensitive environment marked by rising material costs, intense global competition, rapid technological advancement and increasing customer expectations for reliability and timely delivery. Because these companies work on narrow profit margins, they must consistently optimise their processes, reduce waste and improve product quality to remain competitive. Over time, many organisations in this sector have adopted LSS as a strategic approach to address these challenges.

Lean principles originated from the Toyota Production System, which emphasised waste elimination, continuous flow and a culture of ongoing improvement. Six Sigma, first developed by Motorola and later championed by General Electric, introduced a data-driven methodology that aims to minimise defects through structured techniques such as the define, measure, analyse, improve and control (DMAIC) cycle and statistical process control (Coccolini and Bragadin, 2025). The integration of these methodologies has demonstrated significant performance improvements in a wide range of industries, including manufacturing, healthcare, logistics and finance. Within the EMS environment, LSS helps reduce defects, optimise inventory, improve production flow and strengthen quality control. These improvements are particularly crucial because defects in electronic components often lead to rework, material waste, production delays and compromised customer satisfaction.

Numerous studies have highlighted the substantial benefits of LSS implementation. Previous research has shown that its application can reduce operational costs, improve equipment effectiveness and yield long-term financial gains for organisations (Aryan *et al.*, 2025; Gomaa, 2026; Rathi *et al.*, 2022). Furthermore, LSS requires not only technical tools but also active employee engagement. Successful implementation depends on workers' commitment to identifying inefficiencies, participating in improvement activities and sustaining changes over time. Strong leadership support is also essential, as organisations with clear strategic alignment, adequate resource allocation and a long-term commitment to continuous improvement tend to achieve more sustainable outcomes. For EMS companies, LSS represents a valuable approach for enhancing process efficiency, improving quality and maintaining competitiveness in an increasingly demanding industry.

Managing operational costs is a critical concern for EMS companies because it directly affects profitability and long-term sustainability. Although LSS is widely promoted as a powerful methodology for improving process efficiency, eliminating defects and reducing waste, there is limited empirical evidence of its specific contribution to cost reduction in the EMS sector. Existing studies confirm the general benefits of LSS across various industries (Antony *et al.*, 2023; Elmarzouki and Jiuhe, 2026). Yet, the extent to which it directly influences cost reduction in EMS organisations has not been sufficiently explored.

Past research also indicates inconsistent success rates in LSS implementation (Khourshed, 2023; Sohal *et al.*, 2022). While some EMS companies have achieved significant financial savings, others struggle due to challenges such as high upfront investment, a shortage of skilled personnel and employee resistance. These variations suggest that the impact of LSS on cost reduction may depend on several moderating factors, including the choice of specific tools, the level of employee engagement and the degree of leadership support. The aim of the study presented in this chapter was to clarify the actual contribution of LSS to cost reduction in the EMS context and to identify the factors that influence its successful implementation.

The research examined how the methodology affects operational performance, identified the tools that contribute most effectively to cost savings, evaluated the role of employee involvement and determined how leadership support and training influence LSS outcomes. By addressing these goals, the study provides deeper insights into the extent to which LSS enhances financial performance and operational efficiency within the EMS environment.

This research was guided by several key questions: the relationship between LSS implementation and cost reduction; the effectiveness of specific LSS tools; the role of employee engagement; and the influence of leadership support and training. These questions aimed to clarify how different aspects of LSS contribute to improved performance in EMS organisations and to identify the conditions that facilitate successful implementation.

The findings of this study will be valuable for multiple stakeholders within the manufacturing industry. For EMS companies, the research provides empirical evidence demonstrating how

LSS can enhance process efficiency, reduce operational costs and strengthen competitiveness. It also offers practical insights into the challenges of LSS implementation and potential strategies to address them. Industry practitioners, including engineers and process improvement specialists, can benefit from the study's recommendations on effective LSS tools and approaches for achieving cost savings. Policy makers and academic institutions may also find the research useful for promoting advanced manufacturing practices and supporting LSS training programmes. Overall, the study helps bridge the gap between academic research and industry application, encouraging broader adoption of LSS in the EMS sector.

This research examined the impact of LSS implementation in EMS companies, focusing on tools such as 5S, DMAIC, value stream mapping, control charts and Pareto analysis. The study involved collecting data from organisations operating across various regions. It emphasises functional areas where LSS is typically applied, including production, quality control, procurement, logistics and inventory management. Statistical techniques such as descriptive analysis, correlation and regression modelling were used to assess the relationship between LSS practices and cost reduction. However, the study was limited to organisations that have already implemented LSS, which may reduce the general applicability of the findings to companies at earlier stages of adoption.

LSS is widely recognised as a powerful approach for improving efficiency, reducing costs and enhancing quality across many industries, including EMS companies. It combines lean manufacturing, which focuses on eliminating waste and streamlining workflows, with Six Sigma, which emphasises defect reduction and statistical process control. In an increasingly competitive global manufacturing environment, EMS providers face rising raw material prices, intense market pressure and evolving customer expectations regarding quality and delivery. These conditions compel organisations to adopt systematic, data-driven methods for process optimisation. LSS offers a structured means of reducing non-value-added activities, improving process stability and achieving sustainable operational performance, making it highly relevant for EMS firms operating with narrow margins and high-quality requirements.

This section establishes the theoretical foundation for understanding the role of LSS in EMS organisations. It explains the core principles of lean manufacturing and Six Sigma, discusses their integration into a unified LSS framework and reviews empirical studies on the impact of LSS on cost reduction and operational performance. It also examines the key tools used in LSS initiatives, the implementation challenges and the development of a conceptual framework and hypotheses for this study. In doing so, it provides a basis for evaluating how LSS can support cost reduction and process improvement in EMS companies.

Lean manufacturing is a systematic approach to identifying and eliminating non-value-added activities, or waste, from production processes. Originating from the Toyota Production System, Lean manufacturing emphasises value identification, value stream mapping, smooth flow, pull-based production and continuous improvement. It typically addresses seven classic forms of waste: overproduction, waiting, excessive transport, excess inventory, unnecessary motion, overprocessing and defects. By targeting these inefficiencies, organisations can improve productivity, reduce costs and deliver products more quickly and reliably.

In EMS environments, Lean principles are used to minimise material waste, reduce cycle times, remove bottlenecks and allocate resources more efficiently along the production line.

Six Sigma, developed by Motorola and later popularised by companies such as General Electric, is a data-driven methodology focused on reducing process variation and defects. It is commonly structured around the DMAIC cycle: define, measure, analyse, improve and control, which provides a disciplined framework for problem-solving and process optimisation. Six Sigma aims to achieve very low defect levels by using statistical tools such as control charts, Pareto analysis, regression analysis and failure mode and effects analysis. For EMS companies, Six Sigma methods help identify the root causes of defects, reduce rework and improve process reliability, thereby supporting higher customer satisfaction and lower quality-related costs.

Integrating Lean and Six Sigma into LSS combines the speed and waste-reduction capabilities of Lean with the analytical rigour and defect-reduction focus of Six Sigma (Lúcio *et al.*, 2025). This integration provides a comprehensive methodology for improving processes and reducing costs. Lean value stream mapping can simplify and streamline workflows, while Six Sigma techniques stabilise and control the resulting processes. For EMS organisations, this combination supports shorter lead times, reduced waste and improved product quality. However, the integration also demands cultural change, commitment of resources and sustained effort, as organisations must align both philosophies within their operational and strategic contexts.

Empirical research has consistently shown that LSS can contribute to significant cost reductions in manufacturing organisations. Studies report that companies that implement LSS often achieve substantial savings by eliminating inefficient processes, reducing defect rates and improving asset utilisation (Wankhede *et al.*, 2025). In EMS contexts, LSS implementation has been associated with measurable improvements in operational performance, including reductions in operational costs and enhancements in overall equipment effectiveness over several years of sustained application.

The successful implementation of LSS depends on the effective use of its tools. Commonly used tools in EMS companies include value stream mapping to visualise and eliminate non-value-added activities, the DMAIC framework to systematically address process problems and Kaizen to support continuous, incremental improvement. Workplace organisation practices such as 5S help maintain clean, orderly and efficient workspaces. At the same time, statistical process control tools, including control charts and Pareto analysis, enable organisations to monitor process stability and prioritise high-impact problems. Together, these tools support waste reduction, improved process flow and more reliable product quality, all of which contribute to cost reduction.

Despite these benefits, researchers also highlight several challenges in implementing LSS (Vashishth *et al.*, 2024). Employee resistance to change, particularly when LSS is perceived as an additional burden or a threat to existing work routines, can hinder adoption. The financial costs of training, software and monitoring systems present another barrier, especially for small and medium-sized EMS firms. A lack of internal expertise in statistical analysis and process improvement can limit the effective use of LSS tools. In contrast, difficulties in sustaining improvements over time may arise when leadership priorities shift or ongoing monitoring declines. These challenges underscore the importance of strong communication, adequate training, leadership commitment and a culture of continuous improvement in achieving lasting benefits from LSS.

The study presented in this chapter proposed a conceptual framework to examine the relationship between LSS and cost reduction in EMS companies. The framework considers several independent variables related to LSS implementation, including the extent of LSS adoption, the duration of its use within the organisation, the application of specific LSS tools, the level of employee engagement and the degree of training and leadership support. These elements were expected to influence the cost-reduction dependent variable, defined as the extent to which LSS initiatives reduce operational expenses.

The model also introduces process efficiency as a mediating variable between LSS and cost reduction (Sahoo, 2022). Process efficiency captures improvements such as smoother workflows, fewer defects and the elimination of waste. The framework posits that LSS may reduce costs both directly and indirectly by improving process efficiency. By incorporating human factors such as employee engagement and leadership support, the framework acknowledges that LSS outcomes are shaped not only by technical tools but also by organisational culture and management practices. This conceptual structure helps to explain how LSS influences financial performance in EMS companies and provides a basis for empirical testing. The proposed relationships between the variables are illustrated in Figure 8.1.

Figure 8.1 Conceptual framework

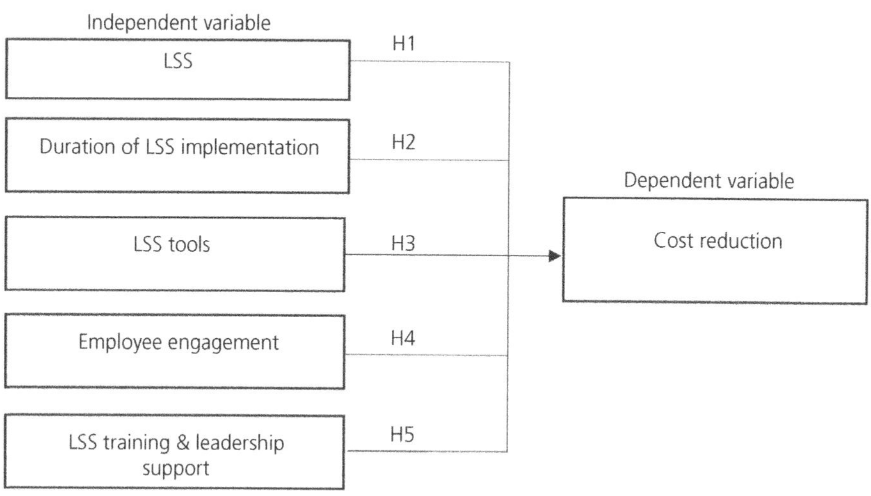

The hypotheses in this study were derived from the reviewed literature and the proposed conceptual framework. Prior research suggests that LSS implementation can reduce waste, improve quality and lower costs (Alzoubi *et al.*, 2022; Thakur *et al.*, 2023). It is therefore expected that higher levels of LSS implementation will be associated with greater cost reduction in EMS companies. Studies also indicate that organisations with more mature or longer-standing LSS programmes tend to experience greater long-term financial benefits than those in the early stages of adoption, suggesting that the duration of implementation may strengthen cost-saving effects over time.

The literature further suggests that specific LSS tools, such as DMAIC, value stream mapping and 5S, are instrumental in improving process efficiency, thereby leading to cost savings. Evaluating their impact helps to identify which tools contribute most significantly to cost reduction in EMS environments. Research on continuous improvement emphasises that employee engagement is critical to the success of LSS initiatives; when employees actively participate in identifying problems and implementing solutions, organisations tend to achieve lower defect rates, higher productivity and better cost outcomes. Finally, numerous studies underline the importance of leadership commitment and training in supporting LSS projects (Alnadi *et al.*, 2024; Tan *et al.*, 2024). Adequate training ensures employees can apply LSS techniques effectively, while leadership support provides clear objectives, resources and a culture that values continuous improvement. Together, these insights support the development of hypotheses linking LSS implementation, duration, tools, employee engagement and leadership support to cost reduction, with process efficiency acting as a key mechanism through which these factors translate into improved financial performance.

8.2. Methodology

8.2.1 Research design

This study adopted a quantitative research design to examine the relationship between LSS implementation and cost reduction in EMS organisations. A quantitative approach is appropriate because it enables objective measurement and statistical analysis, allowing the findings to be generalised (Xue *et al.*, 2025). Data were collected through a structured survey administered to industry experts, using a cross-sectional design that captures information at a single point in time. This approach is suitable for assessing the effects of LSS practices on cost savings without requiring long-term tracking of organisations. The survey method also facilitates efficient data collection from a large sample within a relatively short period, ensuring systematic and comprehensive exploration of the research variables.

8.2.2 Population and sampling procedures

The target population for this study comprises professionals with direct experience of implementing LSS within EMS companies. These include engineers, production managers, quality assurance personnel and process improvement staff, whose diverse roles provide comprehensive insights into the impact of LSS across organisational levels. A purposive sampling technique was used to select respondents with relevant expertise. A total of 110 EMS personnel participated in the study, with this sample size guided by previous research and established recommendations for reliable data collection.

8.2.3 Data collection method

Primary data were collected using a structured questionnaire distributed through Google Forms. The instrument was designed to gather respondents' views on LSS implementation, the extent of its use and its impact on cost reduction. It included multiple-choice items, Likert-scale statements and categorical questions to facilitate easy responses and statistical analysis. A pilot test was conducted to assess reliability and refine the questionnaire for clarity and relevance. The final instrument consisted of four sections covering the respondents' background, LSS implementation, its cost-reduction impact and the challenges associated with adoption. The questionnaire was distributed electronically through email and professional networks such as LinkedIn, allowing respondents to complete it conveniently and reducing potential interviewer bias.

8.2.4 Operationalisation and measurement

The study examined the relationships between LSS implementation, employee engagement and leadership support as independent variables and cost reduction as the dependent variable. To measure these constructs, several dimensions were assessed. LSS implementation was evaluated based on the number of tools applied, the duration of implementation and the level of employee involvement. Cost reduction was measured by actual cost savings, improved efficiency and reduced waste. Employee engagement and leadership support were measured using Likert-scale items assessing participation in LSS activities and managerial support. All responses were recorded using a five-point Likert scale ranging from 'strongly disagree' to 'strongly agree', enabling meaningful statistical analysis of trends and relationships among the variables.

8.2.5 Data analysis techniques

Data analysis was conducted using the statistical package for the social sciences (SPSS) to ensure accuracy and reliability. Several statistical techniques were applied to examine the relationship between LSS implementation and cost reduction. Descriptive analysis was first used to summarise the dataset through means, standard deviation and percentage distribution, providing an overview of respondent demographics and the extent of LSS adoption in EMS companies. Reliability analysis, using Cronbach's alpha with an acceptable threshold of 0.70, assessed the internal consistency of the survey items. Pearson's correlation analysis measured the strength and direction of the relationships between LSS practices and cost reduction. Multiple regression analysis was then applied to evaluate the combined effects of LSS tools, employee engagement and leadership support on cost reduction and to identify significant predictors of cost efficiency. The hypotheses were tested using regression results at the $p < 0.05$ significance level. The findings were presented using tables, charts and graphs to support clear interpretation.

8.3. Results and analysis

8.3.1 Descriptive analysis

The survey findings indicate that Kaizen and 5S were the most widely used tools, each reported by 98.2% of respondents, followed closely by DMAIC at 97.3%. Control charts were utilised by 92.7% of participants, while value stream mapping and Pareto analysis were each applied by 87.3%. These results demonstrate that EMS companies commonly employ a broad range of Lean and Six Sigma tools to enhance operational performance, suggesting a strong organisational emphasis on continuous improvement and process optimisation. A visual representation of this distribution is shown in Figure 8.2.

The survey results show strong support for the positive effects of LSS in EMS companies. A large majority of respondents agreed that LSS reduces waste (80.9%), improves process efficiency (80.9%), enhances product quality (78.2%) and shortens cycle time (80%). Additionally, 73.6% agreed that LSS increases employee engagement in improvement activities, indicating its role in promoting a culture of continuous improvement.

A total of 36.4% of respondents reported achieving a cost reduction of 10–25%, while 62.7% reported reductions of 25–50%. These findings suggest that most EMS companies achieve substantial cost savings by implementing LSS. The distribution of cost-reduction levels is shown in Figure 8.3.

Figure 8.2 LSS tool usage in EMS companies

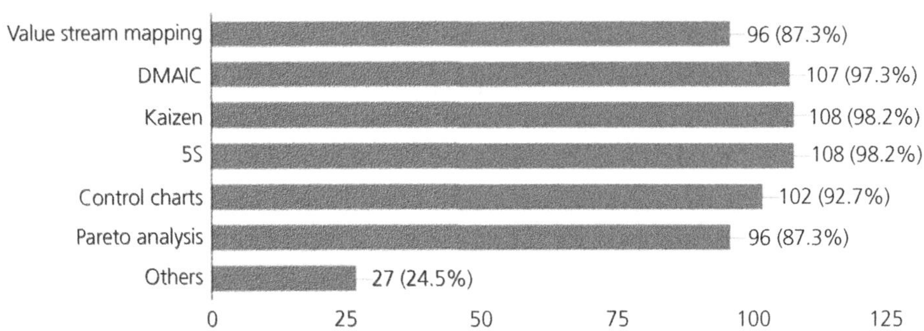

Figure 8.3 Cost reduction observed due to LSS

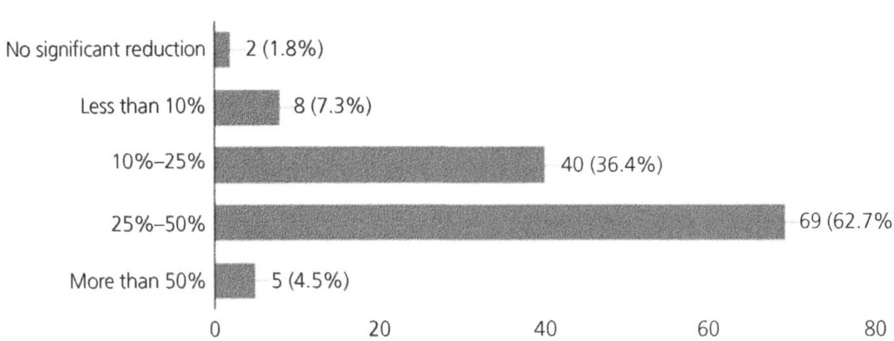

Figure 8.4 shows the functional areas that benefit most from LSS implementation. The survey identified production (94.5%), quality control (76.4%) and maintenance (68.2%) as the primary areas where cost reductions were achieved, indicating that LSS has its greatest impact within these operational functions.

A visual overview of these implementation challenges is presented in Figure 8.5. The key challenges reported were insufficient training (78.2%), employee resistance (65.5%) and difficulties in sustaining improvements (60.9%). These findings indicate that although LSS delivers clear benefits, organisations face significant obstacles in maintaining consistent, effective implementation.

Figure 8.6 illustrates the key measures identified for strengthening LSS effectiveness. The figure shows that 90% of respondents indicated that expanding training programmes

Figure 8.4 Areas of cost reduction

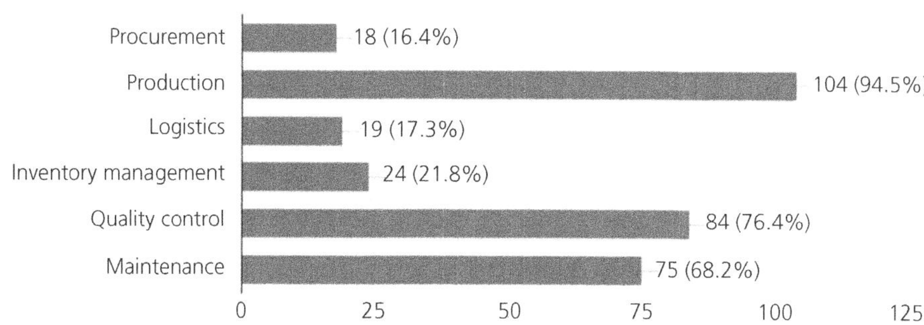

Figure 8.5 Challenges in LSS implementation

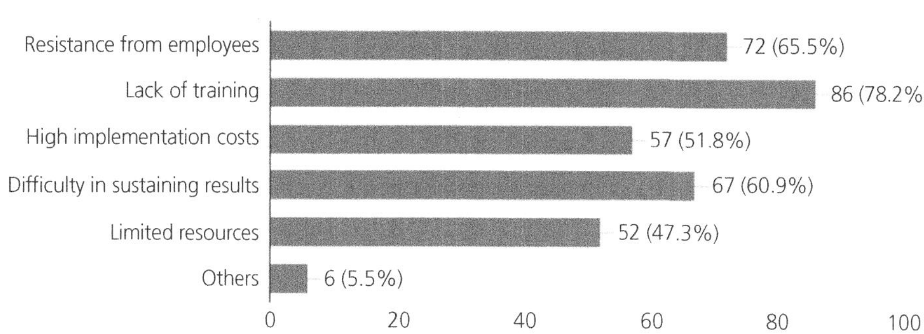

Figure 8.6 Measures to improve LSS effectiveness

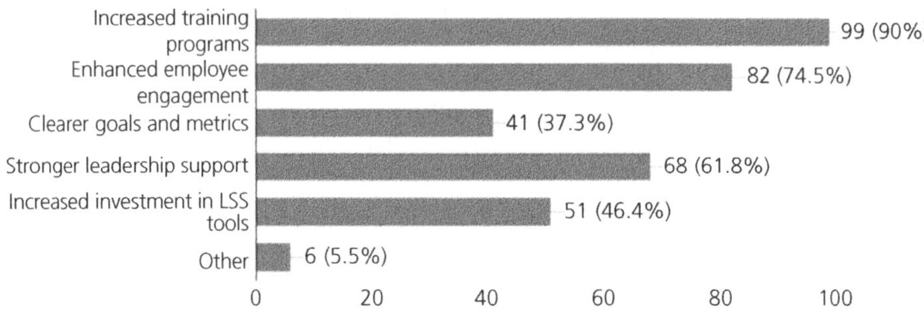

would enhance the effectiveness of LSS, underscoring the importance of continuous employee development.

8.3.2 Pearson's correlation analysis

As shown in Table 8.1, the study found strong, statistically significant positive associations between LSS practices and key performance indicators. LSS implementation showed a strong positive correlation with cost reduction ($r = 0.72, p < 0.001$), validating that the adoption of a systematic methodology directly translates into financial savings. LSS tools usage showed a strong positive correlation with process efficiency ($r = 0.78$, $p < 0.001$), and employee engagement was positively associated with reduced cycle time ($r = 0.65, p < 0.01$).

8.3.3 Multiple regression analysis

Multiple regression analysis was conducted to evaluate the predictors of cost reduction. As shown in Table 8.2, the model explained 68% of the variance in cost reduction (adjusted $R^2 = 0.68$). Duration ($\beta = 0.58$) emerged as the strongest predictor, followed by employee engagement ($\beta = 0.43$) and LSS tools ($\beta = 0.39$), indicating that the length of LSS implementation, human involvement and toolset application are the primary drivers of cost reduction.

8.4. Practical insights

The study provides important practical and theoretical implications for EMS companies, managers and policy makers. The findings confirm that sustained LSS implementation

Table 8.1 Pearson's correlation analysis

Variable	Correlation coefficient, r	Significance, p
LSS implementation → cost reduction	0.72	< 0.001
Use of LSS tools → process efficiency	0.78	< 0.001
Employee engagement → reduced cycle time	0.65	< 0.01

Table 8.2 Multiple regression analysis

Independent variable	Beta coefficient, β	Significance, p
LSS implementation duration	0.58	< 0.001
Employee engagement in LSS	0.43	< 0.01
Use of LSS tools	0.39	< 0.05

produces meaningful cost savings and operational improvements. Organisations that apply LSS consistently over time report greater reductions in defects, improved process stability and enhanced product quality. This reinforces Snee's (2010) argument that LSS delivers cumulative and compounding performance benefits when embedded as a long-term organisational strategy rather than a short-term improvement initiative.

The results further emphasise the critical role of employee involvement in achieving sustainable outcomes. Firms that actively engage employees in structured improvement activities experience fewer defects and reduced operational downtime, aligning with the leadership and engagement perspectives highlighted by Laureani and Antony (2017). Continuous training and visible leadership commitment are also essential to sustaining performance gains. As noted by Jeyaraman and Teo (2010), strong leadership support is a decisive factor in successful LSS implementation, particularly in complex manufacturing environments.

From an operational perspective, waste reduction and process efficiency are the primary drivers of LSS's cost performance. EMS managers should therefore prioritise structured tools such as DMAIC, value stream mapping, 5S and Kaizen to eliminate non-value-added activities and stabilise workflows. Improvement efforts should focus particularly on production, quality control and maintenance functions, where the strongest financial impact was observed. To optimise LSS outcomes, managers should prioritise structured training programmes and visible leadership commitment. Because the strongest financial impact was observed in production (94.5%) and quality control (76.4%), improvement efforts should be concentrated in these functional areas. Sustaining benefits requires embedding LSS within the organisational culture rather than treating it as a series of isolated projects.

Finally, the findings highlight that leadership support, resource allocation and digital integration are necessary to overcome implementation barriers and sustain long-term gains. Embedding LSS within organisational culture, supported by performance metrics and data-driven decision-making systems, enhances both efficiency and competitiveness in demanding manufacturing environments.

Theoretically, this study contributes empirical evidence demonstrating how LSS tools translate into measurable cost efficiency within EMS firms. It strengthens the body of knowledge linking continuous improvement practices to financial performance outcomes. It provides a foundation for policy makers and industry leaders seeking to develop structured guidelines that maximise returns on LSS adoption.

8.5. Conclusion and recommendations

This section summarises the study's overall findings and highlights their practical implications for LSS implementation in electronics manufacturing service companies. It begins by revisiting the research objectives and synthesising the key outcomes from the literature review, research methodology and empirical analysis. It emphasises the study's central themes, particularly the influence of LSS on cost reduction, process efficiency, waste minimisation and employee engagement. It demonstrates its value in enhancing operational performance within EMS firms. It also addresses the research's limitations, including sample size, industry-specific factors and potential data reliability concerns, noting that these constraints provide direction for future studies. Suggestions include expanding research across multiple industries, conducting longitudinal assessments of LSS adoption and integrating digital technologies into LSS frameworks. The chapter concludes by offering practical insights for EMS organisations. Together, these elements reinforce the study's contribution to ongoing discussions on effective LSS implementation in the EMS sector.

8.5.1 Limitations

Despite the clear benefits, the data reveal significant barriers to successful adoption. The primary obstacle is a lack of training (78.2%), followed by employee resistance (65.5%) and difficulties in sustaining improvements (60.9%). These findings suggest that while EMS companies are proficient in using tools like Kaizen, 5S and DMAIC, they struggle with the cultural and educational components required for long-term LSS maturity.

8.5.2 Future research and recommendations

Based on the study's findings and limitations, several directions for future research are recommended. Expanding the sample size and including organisations from a wider range of manufacturing sectors would enhance the generalisability of the results. Incorporating qualitative methods, such as interviews or focus groups, could yield richer insights into the lived experiences of employees and managers involved in LSS initiatives. Longitudinal studies would also be valuable, as they could track the long-term effects of LSS on cost reduction, defect rates and employee engagement. Future research may further explore how digital technologies, such as artificial intelligence, automation and data analytics, can support and enhance LSS practices within Industry 4.0 environments.

Addressing challenges commonly faced during LSS adoption, such as employee resistance, insufficient training and implementation costs, could help organisations refine their strategies and improve outcomes. Practical insights for EMS organisations include strengthening training initiatives, fostering leadership-driven improvement cultures, utilising data analytics to support decision-making and addressing employee resistance through clear communication.

Finally, comparative studies between companies that successfully sustain LSS and those that struggle could offer important insights into best practices and effective change management approaches.

REFERENCES

Alnadi M, McLaughlin P and Abushaikha I (2024) Leadership behaviours for Lean Six Sigma: Jordan as a case study. *International Journal of Lean Six Sigma* **15(7)**: 1530–1554.

Alzoubi HM, In'airat M and Ahmed G (2022) Investigating the impact of total quality management practices and Six Sigma processes to enhance the quality and reduce the cost of quality: the case of Dubai. *International Journal of Business Excellence* **27(1)**: 94–109.

Antony J, McDermott O, Powell D and Sony M (2023) The evolution and future of Lean Six Sigma 4.0. *The TQM Journal* **35(4)**: 1030–1047.

Aryan JK, Albaadani SAMM, Gebre DG, Boateng BS and Olalekan AD (2025) Application of Lean Six Sigma in construction project management: reducing delays and cost overruns. *Journal of Engineering Research and Reports* **27(11)**: 245–261.

Coccolini D and Bragadin MA (2025) History and overview of lean construction fundamentals. In *Proceedings of the International Conference of Architectural Engineering (Scientific Society of Architectural Engineering)*. Springer, Cham, Switzerland, pp. 197–207.

Elmarzouki M and Jiuhe W (2026) Hybrid innovation models for productivity growth: the role of Lean, Six Sigma and Industry 4.0 integration. *International Journal of Lean Six Sigma* **17(1)**: 71–117.

Gomaa AH (2026) Achieving operational excellence in manufacturing supply chains using Lean Six Sigma: a case study approach. *International Journal of Lean Six Sigma* **17(2)**: 614–648.

Ibrahim A and Kumar G (2025) A framework for integrating Lean Six Sigma and Industry 4.0 for sustainable manufacturing. *International Journal of Production Research*: 1–23.

Jeyaraman K and Teo LK (2010) A conceptual framework for critical success factors of Lean Six Sigma implementation on the performance of the electronic manufacturing service industry. *International Journal of Lean Six Sigma* **1(3)**: 191–215.

Khourshed NF (2023) Investigating the critical success factors for integrating Lean Six Sigma and Industry 4.0. *Quality Management Journal* **30(4)**: 259–278.

Laureani A and Antony J (2017) Leadership characteristics for Lean Six Sigma. *Total Quality Management and Business Excellence* **28(3–4)**: 405–426.

Lúcio JS, Miranda RDC, dos Santos CH and Lúcio TAV (2025) Bringing together Lean Six Sigma and discrete event simulation: a systematic literature review. *International Journal of Lean Six Sigma* **16(3)**: 635–659.

Rathi R, Kaswan MS, Garza-Reyes JA, Antony J and Cross J (2022) Green Lean Six Sigma for improving manufacturing sustainability: framework development and validation. *Journal of Cleaner Production* **345**: 131130.

Sahoo S (2022) Lean practices and operational performance: the role of organizational culture. *International Journal of Quality & Reliability Management* **39(2)**: 428–467.

Sohal A, De Vass T, Vasquez T *et al.* (2022) Success factors for Lean Six Sigma projects in healthcare. *Journal of Management Control* **33(2)**: 215–240.

Snee RD (2010) Lean Six Sigma – getting better all the time. *International Journal of Lean Six Sigma* **1(1)**: 9–29.

Tan AB, Van Dun DH and Wilderom CP (2024) Lean innovation training and transformational leadership for employee creative role identity and innovative work behavior in a public service organization. *International Journal of Lean Six Sigma* **15(8)**: 1–31.

Thakur V, Akerele OA, Brake N *et al.* (2023) Use of a Lean Six Sigma approach to investigate excessive quality control (QC) material use and resulting costs. *Clinical Biochemistry* **112**: 53–60.

Vashishth A, Lameijer BA, Chakraborty A, Antony J and Moormann J (2024) Implementing Lean Six Sigma in financial services: the effect of motivations, selected methods and challenges on LSS program and organizational performance. *International Journal of Quality & Reliability Management* **41(2)**: 509–531.

Wankhede VA, Vinodh S and Antony J (2025) Empirical analysis of key operational characteristics of Lean Six Sigma and Industry 4.0 integration. *The TQM Journal* **37(2)**: 611–634.

Xue F, Tan Y and Anwar S (2025) Innovation strategy, digital transformation and competitive advantage of manufacturing enterprises: evidence from China. *European Journal of Innovation Management* **28(8)**: 3395–3417.

emerald PUBLISHING ice Publishing

Part III

Sustainability and strategy

Advancing Project Management: Frameworks, Digital Transformation, Sustainability and Governance

Mohammed Reyasudin Basir Khan
ISBN 978-1-83742-799-4
https://doi.org/10.1108/978-1-83742-798-720261012

Chapter 9
Engineering management practices for optimising EPC project performance

Robeatul Adawiah Osman and Mohammed Reyasudin Basir Khan

Tun Razak Graduate School, Universiti Tun Abdul Razak (UNIRAZAK), Kuala Lumpur, Malaysia

9.1. Introduction

The construction industry has experienced rapid shifts driven by economic growth, intensified competition and increasing project complexity. Despite these developments, concerns persist about the continued underperformance of construction projects, particularly regarding schedule delays and cost overruns (Sanni-Anibire *et al.*, 2022). Although these issues are well documented in literature, past research has predominantly focused on construction activities, often overlooking the engineering phase, which forms the backbone of engineering, procurement and construction (EPC) projects. Studies have emphasised that although engineering accounts for a relatively small share of total project expenditure, errors or inefficiencies during this phase can have significant consequences, affecting cash flow, procurement readiness, construction progress and, ultimately, the project's overall performance (Peansupap and Ly, 2015).

Quality is another critical dimension of project success and is closely linked to client satisfaction, adherence to functional requirements and compliance with established standards. Poor engineering procedures, inadequate planning and ineffective coordination during early project stages frequently lead to design inconsistencies, defects and rework, all of which increase costs and extend timelines. Consequently, the ability of EPC contractors to deliver projects within the allocated budget, timeline and quality expectations is strongly influenced by the effectiveness of engineering management practices.

EPC project implementation typically follows a structured lifecycle comprising engineering, procurement and construction phases. Before EPC execution, clients undertake feasibility studies, conceptual design and front-end engineering design (FEED), which form the basis for cost estimation and tender documentation. While traditional project delivery approaches proceed with engineering and construction sequentially, the modern demand for shorter delivery times has encouraged parallel execution, requiring stronger collaboration and integration across disciplines (Huang *et al.*, 2023). The procurement phase links engineering and construction, and inadequate coordination between the two often leads to late material deliveries and construction delays, particularly when engineering teams fail to deliver design outputs on time.

In major industrial sectors, especially oil and gas, petrochemicals, and energy, EPC lump-sum contracts are widely adopted because they centralise responsibility under a single contractor. This approach offers benefits such as fixed pricing, clearer accountability and streamlined management (Wan *et al.*, 2023). However, these advantages are accompanied by significant risks, including high contractor premiums and potential ambiguities in contractual requirements. Large-scale EPC projects also involve multidisciplinary engineering teams, making technical integration and collaboration essential for effective execution. Poorly defined scopes, slow decision-making, inadequate design reviews and miscommunication between stakeholders frequently lead to disputes, rework, claims and delays. Standardised contract frameworks, such as FIDIC (International Federation of Consulting Engineers), are therefore crucial for mitigating ambiguity, protecting stakeholder interests and reducing the likelihood of conflict.

In Malaysia, EPC contracting has become a preferred approach for major oil and gas infrastructure projects, which play a substantial role in national economic development. With the increasing need for midstream and downstream facilities, Malaysia has emerged as an active market for both local and international EPC contractors. Despite the sector's maturity and adoption of globally recognised project management standards, challenges such as cost overruns, inadequate planning, inexperienced contractors, material shortages, communication breakdowns and slow regulatory approvals remain common. These issues highlight the need for a deeper understanding of engineering management practices and their impact on schedule performance within the Malaysian EPC context.

Although global research acknowledges the significance of engineering performance in shaping overall project outcomes, limited scholarly attention has been paid to engineering management during the early project phases in Malaysia. Moreover, while digital tools such as building information modelling (BIM), cloud platforms and integrated modelling software have demonstrated their potential to support collaboration and reduce miscommunication, their adoption within Malaysian EPC organisations remains limited (Ahmed *et al.*, 2024). Understanding these challenges and the opportunities digitalisation offers is essential to improving coordination, transparency and project efficiency.

This chapter uses project management theory as a framework to examine how engineering management practices affect EPC project schedule performance. Anchored in principles from the project management body of knowledge (PMBOK: PMI, 2021), it evaluates key elements such as stakeholder engagement, resource allocation, schedule planning, interdisciplinary integration, quality assurance, risk management and digitalisation (Faraji *et al.*, 2022). Due to practical constraints and the need for deeper contextual understanding, the research focuses on a single Malaysian EPC contractor with extensive experience executing major local industrial projects. Although this limits the generalisability of the findings, it enables a focused and realistic assessment of engineering management practices within a real organisational environment.

The research addresses the gap between theoretical knowledge and industry practice by examining how engineering management decisions affect project timelines. Through this investigation, the study presented in this chapter identifies weaknesses in current practices, evaluates the effectiveness of existing management approaches and proposes strategies that integrate traditional project management principles with modern digital solutions. The findings offer actionable insights for industry practitioners seeking to improve schedule performance in EPC projects in Malaysia. By understanding how engineering management practices influence project outcomes, the study supports the adoption of more efficient, risk-aware and technology-enabled strategies to improve EPC project delivery, strengthen industry competitiveness and promote sustainable project execution in the Malaysian oil and gas sector.

EPC contracts encompass the contractor's responsibility for the detailed implementation and integration of all activities from detailed design through procurement, construction and mechanical completion. At this point, facilities are fully equipped and ready for operation. For large-scale, complex industrial projects such as chemical processing facilities, mining developments, power plants, and oil and gas installations, the EPC framework provides a comprehensive structure for managing technical, commercial and execution risks.

Under this arrangement, the EPC contractor holds primary responsibility and liability for engineering and construction. At the same time, the owner is accountable for defining the basic project requirements, design standards and performance criteria before tendering, and for supervising the contractor's work to maintain a system of checks and balances. Project inputs developed during the FEED stages, such as specifications, design basis, conceptual solutions, cost estimates and performance guidelines, serve as the baseline for assessing the adequacy and acceptability of the EPC contractor's deliverables.

Management of EPC projects under lump-sum contracts, particularly in petrochemical and similar facilities, therefore necessitates careful planning to avoid delays and cost overruns while preserving contractor profitability (Sangroungrai *et al.*, 2018). Within this approach, engineering outputs drive the progress of the entire project by enabling procurement decisions, construction work fronts and commissioning activities. Multidisciplinary engineering teams, typically including process, mechanical, electrical, civil/structural and instrumentation disciplines, operate under a coordinated engineering management structure that must align technical solutions with project objectives, contractual requirements and safety. Engineering is inherently iterative, moving from planning and design through construction to operation, and thus requires highly skilled professionals who remain accountable for their outputs and responsive to issues arising in subsequent phases.

Effective engineering management integrates these technical disciplines with clients, procurement and construction teams, and key suppliers, promoting collaboration from bidding through execution. Empirical evidence shows that project outcomes are significantly influenced by the quality of engineering management and the strength of stakeholder collaboration.

When engineering is mismanaged, the consequences include poor-quality designs, extensive rework and deterioration of overall project performance.

At the same time, construction industry pressures for lower costs and shorter delivery times continue to expose tensions between engineering and construction. Engineering teams tend to prefer design maturity before release, while construction teams seek earlier work fronts. To reconcile these competing priorities, Mejlænder-Larsen (2018) advocates parallelism between engineering and construction, supported by digital solutions such as project execution models (PEM) and BIM, to better integrate schedules and compress overall execution timelines. The inherent complexity of EPC contracts, arising from strong interdependence between project phases, reinforces the need for sophisticated and coherent project and engineering management practices (Ghanbari, 2024).

Within this context, the study presented in this chapter focuses on engineering management during the EPC phase and its influence on schedule performance, recognising that engineering decisions and practices have cascading effects on procurement and construction, and therefore on timely project delivery. EPC projects are characterised by technical complexity, multidisciplinary involvement and high interdependence between activities, making structured coordination and collaboration essential. Project management theory (PMT) provides a useful lens for examining these projects, as it conceptualises project delivery as the organised application of knowledge, skills, tools and techniques to meet objectives related to time, cost and quality. Contemporary contributions to PMT emphasise the development of mid-range theories that bridge disciplines and connect conceptual insights with practical project environments.

The PMBOK guide (PMI, 2021) reframes project management through a principles-based approach, stressing creativity, agility and proactivity in tailoring processes to specific project contexts. It identifies core behavioural principles and links them to performance domains, including stakeholders, teams, development approach, lifecycle, planning, project work, delivery, measurement and uncertainty. These domains provide a useful framework for understanding how engineering management practices influence EPC project outcomes, particularly schedule performance.

Digitalisation further enriches this theoretical basis. Tools such as BIM, Primavera and similar scheduling systems, along with emerging technologies like blockchain, enhance the monitoring, coordination and control of time, cost and deliverables. They offer mechanisms to integrate engineering and project management processes and improve visibility across the project lifecycle.

Building on PMT and PMBOK, this study developed a knowledge-management-based model to examine the application of key engineering management principles, specifically stakeholder engagement, resource management, schedule management, integration management, quality management and risk management and their contribution to EPC schedule performance in a Malaysian context. Digitalisation is considered an additional factor that supports integration and schedule management during the engineering stage, potentially amplifying its positive effects on schedule outcomes.

Stakeholder engagement is a central element of engineering management in EPC projects because multiple parties, including clients, contractors, subcontractors, suppliers, regulators, financiers and affected communities, each have distinct interests and influences on project outcomes. Stakeholder engagement involves identifying these parties, understanding their expectations and managing their influence to maintain alignment with project objectives. In complex EPC environments, ineffective engagement often leads to miscommunication, delays, cost escalation and disputes, especially when stakeholder interests diverge or conflict.

During the engineering stage, stakeholder decisions on specifications, performance standards and design solutions significantly affect procurement and construction. Early stakeholder involvement has been shown to reduce the likelihood of costly changes later in the project by incorporating expectations during the planning and design stages (Ezeh *et al.*, 2024; Shaukat *et al.*, 2022). Effective communication also promotes trust and transparency, which are crucial in large projects.

Structured stakeholder engagement frameworks, such as those proposed in the PMBOK, guide stakeholder identification, influence analysis, the development of engagement strategies and continuous monitoring (Faraji *et al.*, 2022). Tools such as stakeholder mapping and power–interest grids help prioritise engagement with key decision-makers (Zhu *et al.*, 2024). Proactive conflict management and clear mechanisms for dispute resolution further help prevent stakeholder disagreements from escalating and disrupting the schedule. Systematic contract review and clear documentation of obligations, performance criteria and contingencies also support compliance and accountability.

In this study, stakeholder engagement was conceptualised as a critical dimension of engineering management that affects schedule adherence in EPC projects. Resource management in EPC projects refers to the planning, allocation and control of workforce, equipment, materials and financial resources to ensure project objectives are met efficiently and without unnecessary waste. Successful resource management supports timely completion, cost control and quality compliance. It begins at the planning stage and continues across engineering, procurement and construction (Ahuja *et al.*, 1994). Engineering requires high-calibre professionals from diverse disciplines, procurement coordinates the availability of materials and equipment, and construction transforms designs into physical assets (Benton and McHenry, 2010).

Inadequate allocation of skilled engineering staff can lead to design errors, late deliverables and delays in procurement and construction schedules. Poor procurement planning may cause material shortages and site disruptions. To mitigate such risks, accurate forecasting of resource requirements based on project scope and schedule is necessary. Flexibility in resource allocation allows project teams to respond to unforeseen delays and constraints (Faria *et al.*, 2020), while focusing on critical path activities helps protect key milestones. Lean practices further support waste reduction and value maximisation, and digital tools such as enterprise resource planning (ERP) systems enable real-time resource tracking and coordination.

Schedule management is fundamental to engineering management because delays translate directly into financial losses and potential contractual penalties. Empirical studies suggest that changes or inefficiencies in the engineering schedule can have disproportionately large effects on overall project duration when compared to other phases (Nazimko and Zakharova, 2023). To address this, structured design schedule management processes have been proposed, including clear milestone definition, progress tracking, productivity monitoring, interface management, vendor document control and work front management.

Human resource capability, robust communication platforms and regular performance assessment further contribute to timely project delivery (Ade and Santoso, 2023). Design changes, often arising from insufficient strategic planning, design errors or onsite modifications, are frequent sources of delay in EPC projects (Habibi *et al.*, 2019). Early involvement of EPC engineering teams during FEED, including participation in design reviews, safety studies and value engineering workshops, can reduce rework and compress overall schedules.

Tools such as the detail engineering completion rating index system (DECRIS) help monitor engineering progress and support the optimisation of construction schedules. Integrating traditional project management techniques such as the program evaluation and review technique (PERT) and the critical path method (CPM) with simulation software has also been shown to reduce project duration and increase completion certainty. Timely decision-making and effective communication between project participants remain essential to improving schedule performance (Habibi *et al.*, 2019).

Integration management in EPC projects focuses on coordinating multidisciplinary engineering teams and aligning engineering outputs with procurement and construction activities. Engineering disciplines such as process, mechanical, electrical, civil/structural and instrumentation each contribute unique expertise, and their outputs must be coherent and consistent to avoid discrepancies during construction, late procurement or misaligned interfaces (Habibi *et al.*, 2019). Effective integration strengthens decision-making, streamlines processes and aligns team efforts with shared project goals.

Key elements of integration include clear communication channels that ensure relevant information is shared accurately and promptly; centralised project or engineering management roles that coordinate priorities, resolve conflicts and maintain accountability; and collaboration technologies and shared platforms that support real-time updates, clash detection and version control. Integration is particularly crucial when EPC phases overlap in parallel workflows; sufficient design maturity must be achieved before procurement and construction proceed at scale. Knowledge sharing across disciplines also helps anticipate issues and identify practical, cost-effective design solutions.

Given the multidimensional nature of EPC projects, quality assurance (QA) and quality control (QC) are crucial for ensuring that outcomes meet client expectations, regulatory requirements and industry standards. QA refers to the systematic processes used to maintain quality throughout the project, while QC comprises specific inspections, tests and verifications to ensure compliance with defined criteria (Qamar *et al.*, 2024). Robust QA/QC reduces failures, rework and delays by ensuring that engineering designs, procured materials and construction activities conform to specified requirements.

Engineering quality is especially important because design flaws propagate downstream and are more costly to rectify later. Applying established quality frameworks, such as ISO 9001 (ISO, 2015) and structured quality management systems, enables systematic planning, documentation and control of quality processes. Comprehensive quality management plans (QMPs) clarify procedures, roles, responsibilities and inspection protocols (PMI, 2021). Integration of QA/QC across engineering, procurement and construction (e.g. joint reviews of material specifications before purchase) helps prevent non-conformances during installation. Transparent communication between stakeholders further supports compliance (Balzarova and Castka, 2012).

Risk management is another central engineering management principle and involves the systematic identification, assessment and mitigation of uncertainties that may affect project success. Potential risks are first identified through the review of project documentation, consultations with stakeholders and analysis of historical data from similar projects. Common categories include commercial, technical, safety, environmental and schedule-related risks. In EPC projects, examples include procurement delays, design errors, equipment failures, labour shortages and unforeseen site conditions (Kabirifar and Mojtahedi, 2019).

Risk assessment evaluates the likelihood and consequences of each risk, often using tools such as risk assessment matrices to prioritise high-impact threats. High-priority risks that might severely affect the schedule or cost are addressed first. Mitigation strategies generally follow four broad approaches: avoidance, reduction, transfer and acceptance. Avoidance eliminates a risk, for example, by changing suppliers. Reduction lowers the probability or impact, such as by strengthening quality control. Transfer shifts risk to another party through insurance or subcontracting (Banks, 2004), while acceptance acknowledges the risk and prepares contingency plans.

Risk management is an iterative process: new risks emerge, existing risks evolve. The effectiveness of mitigation measures must be reassessed through regular reviews, reports and audits. In EPC environments, proactive, continuous risk management enhances resilience and increases the likelihood of on-time project completion. Technological advancement has significantly transformed engineering management by enabling more integrated, data-driven and collaborative approaches. Digital tools such as simulation software, BIM and enterprise information platforms support improved planning, coordination and control across project phases. BIM, in particular, facilitates real-time information sharing, multidisciplinary coordination and lifecycle management from design through operation and decommissioning. It enhances stakeholder collaboration, supports clash detection and promotes a holistic understanding of project constraints.

Enterprise-level platforms support the integrated management of design, manufacturing and assembly, particularly for prefabricated or modular projects, by centralising data and aligning workflows. Machine-learning-based tools, such as cost estimation platforms, improve forecasting accuracy and support decision-making throughout the EPC lifecycle (Choi *et al.*, 2021). Customised ICT platforms and relational databases have also been shown to improve information flow, standardisation and process control in large-scale EPC.

Prior empirical research demonstrates that engineering management practices significantly influence EPC project performance, particularly in relation to schedule and cost outcomes (Ciric Lalic *et al.*, 2022; Shahzad *et al.*, 2025). Studies have identified crucial areas such as planning and scheduling, resource mobilisation, procurement, quality management, monitoring and control, commissioning and stakeholder coordination as central to successful delivery (Bhatti and Nazir, 2024; Shen *et al.*, 2025). Several investigations highlight the disproportionate influence of the engineering and design phase on project schedules and costs (Metwally *et al.*, 2025; Prasad, 2025). Kabirifar and Mojtahedi (2019) emphasised the importance of detailed design and planning.

Effective communication, timely decision-making and Agile project management have been found essential for responding to client-driven changes and minimising schedule and cost impacts. Habibi *et al.* (2019) demonstrated that project performance is significantly affected by design management efficiency and designer competency, and that strong contract management and collaboration with stakeholders enhance design quality and reduce delays. Application of PMBOK-based project management principles in engineering consulting environments has been associated with substantial reductions in project delivery time and risk exposure (Hinostroza *et al.*, 2018; Yilmaz *et al.*, 2025).

The literature also underlines the interdependence between engineering output and procurement performance (Kermani *et al.*, 2024; Nyamah *et al.*, 2023). Late or incomplete engineering information can lead to unsuitable equipment purchases, cost escalation, reduced plant performance and schedule delays (Javed *et al.*, 2022; Moretto *et al.*, 2022). Studies recommend early release of controlled engineering data, pre-approved vendor documentation and structured weightage systems to accelerate procurement without compromising quality (Granheimer *et al.*, 2024; Gupta *et al.*, 2025). Incorporating sustainability criteria in procurement has been shown to support cost reduction, risk mitigation and reputational benefits (Soares *et al.*, 2024).

In the construction phase, efficient site management, contingency planning, resource mobilisation and adherence to safety standards are crucial to avoid delays and disruptions. QC practices play a vital role in ensuring conformance to design and standards.

Guided by project management theory and the principles set out in the PMBOK guide (PMI, 2021), this study conceptualises engineering management as a set of interrelated principles that influence the performance of EPC project schedules (Maslennikov *et al.*, 2020). Six core principles (stakeholder engagement, resource management, schedule management, integration management, quality management and risk management) are treated as independent variables. The EPC schedule performance of EPC projects constitutes the dependent variable. In addition, the study recognises digitalisation (technology adoption) as a moderating factor that can enhance the effectiveness of scheduling and integration management during the engineering stage. Digital tools and platforms are expected to strengthen coordination, improve information flow and facilitate real-time control, thereby amplifying the positive impact of sound engineering management practices on schedule outcomes.

The conceptual framework, presented in Figure 9.1, therefore posits that better implementation of engineering management principles will be associated with improved EPC schedule performance, and that the adoption of digital technologies can further optimise schedule delivery. This framework provides the basis for developing the study's research questions and hypotheses and for empirically examining engineering management practices within a Malaysian EPC organisation.

Drawing on project management theory, PMBOK principles and prior empirical evidence, this study proposed a set of hypotheses linking engineering management practices to EPC project schedule performance, with technological adoption considered as a moderator.

The study hypothesised that stakeholder engagement during engineering has a significant impact on EPC schedule performance, as effective communication, clear decision-making and alignment of expectations reduce delays and rework. Efficient engineering resource management is expected to enhance schedule outcomes by ensuring that the right skills, materials and financial resources are available when needed. Engineering schedule management is

Figure 9.1 Conceptual framework

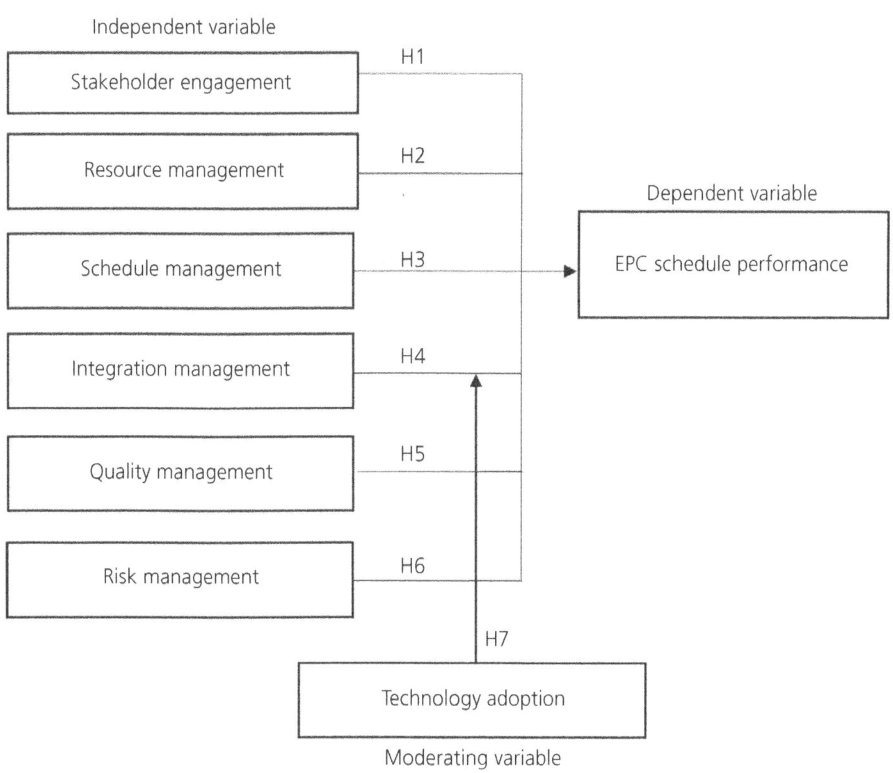

expected to strongly influence schedule delivery, given the cascading effects of engineering delays on procurement and construction.

The study further hypothesised that effective integration management among multidisciplinary engineering teams improves EPC schedule performance by reducing inconsistencies, improving coordination and supporting parallel workflows. Quality management during the engineering stage is expected to have a positive relationship with schedule outcomes by preventing non-conformances and rework. Robust risk management was hypothesised to improve schedule performance by identifying, prioritising and mitigating threats before they materialise. Finally, digitalisation is expected to strengthen the effect of integration and schedule management on EPC schedule performance by enhancing visibility, collaboration and control. Through testing these hypotheses in a Malaysian EPC context, the study deepens understanding of how engineering management practices and technological adoption jointly contribute to schedule optimisation. The findings support practitioners and decision-makers in designing more effective management strategies to improve EPC project delivery.

9.2. Methodology

9.2.1 Research design

This study adopts a quantitative, cross-sectional survey to examine the relationship between engineering management practices and schedule performance in Malaysian EPC projects. Quantitative research proceeds through clear stages of variable identification, data collection and hypothesis testing and allows statistical generalisation of findings based on responses from relevant industry practitioners. A structured questionnaire ensures standardised data collection, strengthening objectivity and reproducibility. Through hypothesis testing, the study evaluates the relationships between engineering management practices (independent variables), schedule performance (dependent variable) and technological integration (moderating variable). The cross-sectional survey provides a practical snapshot of current practices and experiences without requiring long-term observation, making this design appropriate given time and access constraints and suitable for identifying trends and correlations aligned with the research questions.

9.2.2 Population and sampling procedures

This study uses probability-based random sampling, which ensures that every individual in the population has an equal chance of selection, thereby reducing bias and enabling reliable statistical inference. Random sampling is appropriate for this research because it strengthens objectivity in assessing the relationship between engineering management practices and schedule performance in EPC projects. The target population comprises professionals involved in Malaysian EPC projects, including engineers, technical specialists, planners and project managers. Due to the absence of publicly available data on the exact number of EPC professionals in Malaysia and given the study's limited access and timeframe, a single experienced Malaysian EPC contractor, referred to as Company A, was selected. Company A, with approximately 200 staff involved directly or indirectly in EPC activities, serves as the population base for this research.

The sample size was calculated using Yamane's formula for a population of 200 individuals, producing a required minimum of 133 respondents at a 95% confidence level and a 5% margin of error (Sathyanarayana *et al.*, 2024). This aligns closely with Krejcie and Morgan's sample size table, which also recommends at least 132 participants for a population of 200 (Rahman, 2023). Therefore, the study randomly sampled at least 133 professionals from Company A with experience in EPC project execution. This sample size ensures adequate representation of the population and supports the reliability and generalisability of the study's findings.

9.2.3 Data collection method

Data for this study were collected through an online survey distributed to Company A employees involved in EPC projects. Online surveys developed in Google Forms were selected for their speed, convenience, cost-effectiveness and accessibility. The questionnaire, adapted from validated studies and tailored to the Malaysian EPC context, comprised nine sections with 45 items covering demographic details, six engineering management practice variables, a moderating variable (technology adoption) and the dependent variable (schedule performance), all measured using a five-point Likert scale. With permission from Company A's human resources manager, the survey link was shared with targeted participants through WhatsApp, accompanied by an introduction outlining the study's purpose, assurances of confidentiality and voluntary participation. The online format supported anonymity, efficient data management and seamless export into the statistical package for the social sciences (SPSS). The survey remained open for four weeks, from 1 to 30 April 2025, ensuring adequate response time for participants.

9.2.4 Operationalisation and measurement

Defining research variables is essential for transforming abstract concepts into measurable elements. In this study, three types of variables were used: six independent variables representing engineering management practices, one dependent variable reflecting EPC schedule performance, and a moderating variable capturing digitalisation in schedule and integration management. All constructs are measured using a standardised five-point Likert scale. The independent variables include stakeholder engagement, resource management, schedule management, integration management, quality management and risk management, each assessed using specific indicators that reflect practical engineering activities such as communication, workforce planning, schedule integration, coordination across disciplines, adherence to quality standards and risk mitigation procedures. The dependent variable, schedule performance, evaluates timeliness of engineering deliverables, rework frequency, milestone achievement and adherence to deadlines. Digitalisation functions as the moderating variable, assessing the extent to which tools such as BIM, 3D modelling and real-time collaboration platforms strengthen the relationship between engineering practices – particularly integration management – and EPC schedule outcomes.

Reliability and validity are critical to ensuring the credibility of the study's measurements. Reliability was assessed using Cronbach's alpha (α) to confirm internal consistency; a value of 0.70 or higher indicates acceptable reliability. Content and construct validity were established by adapting previously validated items, aligning questionnaire content with theoretical frameworks and obtaining expert reviews from industry professionals and academic supervisors. Statistical validity was further assessed through an exploratory factor analysis (EFA), using Kaiser-Meyer-Olkin (KMO) and Bartlett's test to assess data suitability and strong factor loadings (> 0.5). By combining expert judgement with statistical verification, the study ensured its measures were both robust and appropriate for analysing the influence of engineering management practices on EPC schedule performance.

9.2.5 Data analysis techniques

This study follows a structured data analysis procedure that includes data cleaning, coding and the application of appropriate statistical tests using software tools, ultimately producing findings aligned with the research aims. Descriptive analysis was first employed to summarise respondent characteristics and overall data behaviour using measures such as mean, frequency, percentage and standard deviation, supported by visual representation where relevant. Correlation analysis was then used to determine the strength and direction of relationships between the independent variables and the dependent variable, using Pearson's correlation coefficient, which is well-suited to assessing linear associations.

To further examine the influence of engineering management practices on schedule performance, multiple regression analysis was conducted to evaluate the combined and individual contributions of the independent variables, while accounting for potential overlaps. In addition, moderation analysis was performed to investigate whether digitalisation strengthens the relationship between integration management and project schedule performance. This was carried out using hierarchical regression in SPSS, involving mean-centring of variables, creating interaction terms and assessing changes in R^2 and beta (β) significance to determine the presence of a moderating effect. Hypothesis testing was subsequently conducted to statistically evaluate the seven hypotheses formulated in the study, using correlation, multiple regression and moderation results to determine whether each hypothesis is supported or rejected at the chosen significance level. All analyses were conducted using SPSS, which offers robust capabilities for data entry, statistical computation, visualisation and interpretation, making it well-suited for the descriptive and inferential analyses required in this research.

9.3. Results and analysis

9.3.1 Response rate

This section presents the demographic characteristics of the respondents to contextualise the study's findings, as perspectives on engineering management may differ according to an individual's role, experience, industry exposure and project environment. A total of 140 responses were received through the Google Form survey distributed within Company A and, after data cleaning, 139 valid responses were retained for analysis using SPSS.

9.3.2 Demographic profile

The respondents demonstrated wide industry exposure: although most were involved in oil and gas projects (79.3%), many had also worked in petrochemical, power and energy, and construction or infrastructure sectors. This diversity reflects the multi-industry nature of EPC work and strengthens the study's credibility by incorporating perspectives from professionals accustomed to varied project requirements and stakeholder landscapes. Respondents also represented a broad range of job roles, with 75.8% on the engineering team, including engineering managers, lead engineers, discipline engineers and drafters, whose insights are particularly valuable given their direct involvement in detailed design activities. The remaining participants, including project managers, construction managers, planners, procurement staff and safety or quality personnel, provided an external perspective on engineering performance, adding balance and objectivity to the dataset.

The respondents were generally experienced, with 43.9% having more than 10 years of EPC exposure and 31.4% having 5–10 years, indicating that most participants had mature professional judgement and strong familiarity with the realities of EPC projects. Their involvement across different project scales further enhanced data robustness: more than half had worked on medium- to high-value projects between US$10 million and US$500 million, and 17.1% had experience with mega projects exceeding US$500 million. Additionally, 61% had participated in projects of multiple scales, suggesting an ability to compare engineering management practices across varying levels of complexity and risk. Geographic exposure also contributed to the dataset's richness, as 68.3% had worked on Peninsular Malaysia projects, 11.5% on projects in Sabah or Sarawak, and 20.1% on projects in both regions. This distribution indicates that respondents were familiar with differing logistical, regulatory and stakeholder environments across Malaysia. Collectively, the demographic insights reveal a diverse and experienced respondent group, strengthening the relevance and reliability of the findings in representing engineering management practices and schedule performance within Malaysian EPC project settings. The geographic distribution of respondents' project experience is shown in Figure 9.2.

9.3.3 Reliability analysis

This section presents the reliability and validity assessment to ensure the precision and consistency of the survey instruments before proceeding to hypothesis testing. Reliability was evaluated using Cronbach's alpha, with a threshold of 0.70 adopted as the minimum acceptable level. The reliability results for all constructs are summarised in Table 9.1. All constructs recorded strong internal consistency, with alpha values ranging from 0.780 to 0.908 for stakeholder engagement, resource management, schedule management, integration management, quality management, risk management, technology adoption and schedule performance. These results confirm that the items within each construct are highly reliable and closely aligned in measuring their intended concepts.

Figure 9.2 Geographic location of EPC projects executed by the participants

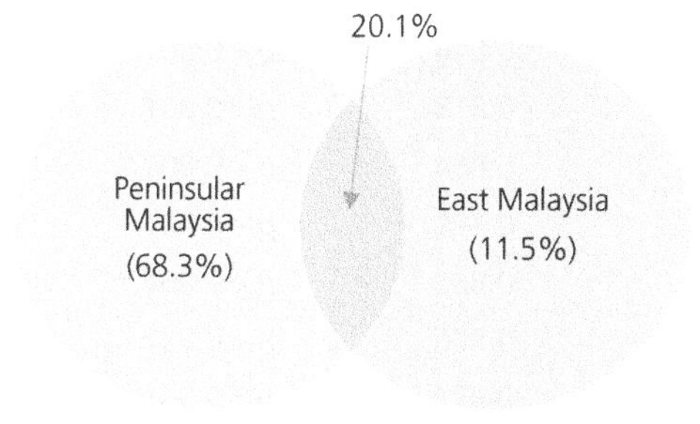

Table 9.1 Summary of reliability test

Construct	Cronbach's alpha, α	Result
Stakeholder engagement	0.780	Acceptable
Resource management	0.876	Acceptable
Schedule management	0.860	Acceptable
Integration management	0.908	Acceptable
Quality management	0.856	Acceptable
Risk management	0.895	Acceptable
Technology adoption	0.860	Acceptable
Schedule performance	0.834	Acceptable

Each construct was further examined for validity using exploratory factor analysis (EFA) with principal component analysis in SPSS to confirm that the measurement items aligned with a single underlying factor as theoretically expected. The validity results, including Kaiser-Meyer-Olkin (KMO) values and Bartlett's test outcomes, are presented in Table 9.2. The KMO values for all constructs exceeded 0.60, and Bartlett's test of sphericity was significant at $p < 0.001$, indicating adequate sampling and suitability for factor analysis. For each construct, a single factor with an eigenvalue greater than 1 was extracted, accounting for more than 50% of the total variance and confirming that items loaded cleanly onto a single dimension without cross-loadings. These results demonstrate strong construct validity and confirm that the instruments appropriately measure the intended concepts, enabling the study to proceed confidently to subsequent hypothesis testing. Together, these findings establish a sound foundation for conducting subsequent correlation, regression and moderation analyses.

Table 9.2 Summary of validity test results using EFA

Construct	KMO	Bartlett's test	Variance explained
Stakeholder engagement	0.730	$p < 0.001$	53.90%
Resource management	0.844	$p < 0.001$	67.15%
Schedule management	0.813	$p < 0.001$	64.26%
Integration management	0.864	$p < 0.001$	73.38%
Quality management	0.791	$p < 0.001$	64.11%
Risk management	0.826	$p < 0.001$	71.02%
Technology adoption	0.776	$p < 0.001$	64.24%
Schedule performance	0.824	$p < 0.001$	61.93%

9.3.4 Descriptive analysis

Descriptive analysis was conducted to summarise respondents' perceptions of each construct using five-point Likert scale data. The descriptive statistics for all research variables are shown in Table 9.3. The results show that all independent variables (stakeholder engagement, resource management, schedule management, integration management, quality management and risk management) received positive mean scores above 4.0, indicating general agreement that these engineering management practices are effectively implemented in EPC projects. However, the wide range between minimum and maximum scores across constructs suggests that respondents' views vary depending on their project experience, role, industry exposure and the differing complexities of EPC environments. Technology adoption also received a high mean score (4.16), reflecting favourable perceptions of the support that digital tools provide for engineering activities.

In contrast, schedule performance recorded a much lower mean of 3.15, indicating that despite effective engineering management, achieving consistent schedule adherence remains challenging in EPC projects. This variability is further reflected in the high standard deviation (SD) for schedule performance (0.90), suggesting diverse perceptions influenced by project complexity and differing stakeholder viewpoints. Lower variation in stakeholder engagement responses (SD = 0.64) indicates greater consistency in how respondents experience engagement practices. Overall, the descriptive results indicate that strong engineering management does not automatically translate into strong schedule performance, reinforcing the need for further inferential analysis, correlation, regression and moderation to examine relationships between variables and the potential role of digitalisation in enhancing EPC schedule outcomes.

9.3.5 Correlation analysis

Pearson's correlation analysis was conducted to examine the relationship between the six engineering management practices and EPC schedule performance. The correlation results

Table 9.3 Descriptive analysis for research variables (n = 139)

Variable	Min.	Max.	Mean	SD
Stakeholder engagement (SE)	1.40	5.00	4.1842	0.6402
Resource management (RE)	1.60	5.00	4.0101	0.7422
Schedule management (SM)	1.00	5.00	3.9799	0.7743
Integration management (IM)	1.00	5.00	4.0129	0.8116
Quality management (QM)	1.00	5.00	4.0906	0.7383
Risk management (RM)	2.00	5.00	3.9957	0.7733
Technology adoption (TA)	1.20	5.00	4.1626	0.7486
Schedule performance (SP)	1.00	5.00	3.1482	0.8994

Table 9.4 Pearson's correlation test between engineering management practices and schedule performance of EPC projects ($n = 139$)

Variable	Pearson's correlation with SP	*Sig. (2-tailed)*	Interpretation of hypothesis
Stakeholder engagement (SE)	0.345	$p < 0.001$	Significant moderate correlation, H1 supported
Resource management (RE)	0.577	$p < 0.001$	Significantly strong correlation, H2 supported
Schedule management (SM)	0.613	$p < 0.001$	Significantly strong correlation, H3 supported
Integration management (IM)	0.558	$p < 0.001$	Significantly strong correlation, H4 supported
Quality management (QM)	0.579	$p < 0.001$	Significantly strong correlation, H5 supported
Risk management (RM)	0.602	$p < 0.001$	Significantly strong correlation, H6 supported

are presented in Table 9.4. All variables showed statistically significant positive correlations ($p < 0.001$), supporting hypotheses H1–H6. Stakeholder engagement showed a moderate positive relationship with schedule performance ($r = 0.345$), indicating that effective communication, clear expectations and timely stakeholder involvement help reduce delays in complex EPC environments. Resource management showed a strong correlation ($r = 0.577$), highlighting the importance of allocating skilled workers, using appropriate tools, balancing workloads and adaptive resource planning to prevent bottlenecks and maintain schedule adherence. Engineering schedule management exhibited the strongest relationship with schedule performance ($r = 0.613$), reinforcing the critical role of baseline scheduling, milestone tracking and continuous progress monitoring, in line with PMBOK principles. Integration management also demonstrated a strong correlation ($r = 0.558$), reflecting the importance of coordinated multidisciplinary workflows, harmonised deliverables and timely conflict resolution. Quality management showed a similarly strong correlation ($r = 0.579$), underscoring how robust QA procedures, design verification and effective non-conformity management help minimise rework and delays. Risk management displayed a strong positive correlation ($r = 0.602$), confirming that early risk identification, thorough assessment, contingency planning and integrated mitigation are essential for preventing schedule disruptions in high-risk EPC settings. Overall, the results provide substantial empirical evidence that all six engineering management practices significantly influence EPC schedule outcomes, forming a solid foundation for subsequent multiple regression analysis to assess their combined effects. The linear relationships between the independent variables and schedule performance are visually illustrated in Figure 9.3.

Figure 9.3 Scatter plots illustrating the correlation between engineering management practices and schedule performance of EPC projects

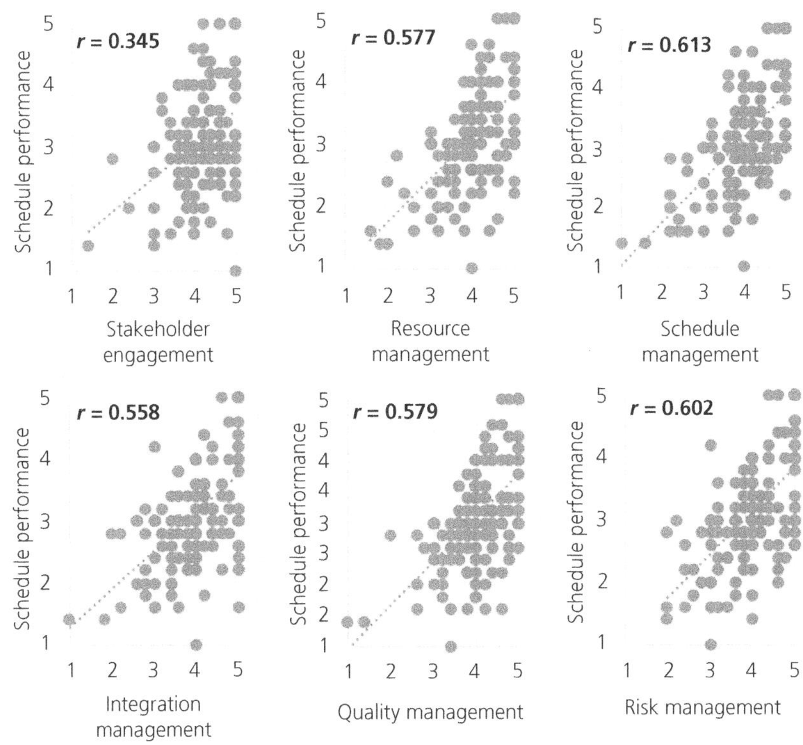

9.3.6 Multiple regression analysis

The overall significance of the regression model was evaluated using analysis of variance (ANOVA), which compares the fitted model against a null model with no predictors. A statistically significant F-value ($p < 0.05$) indicates that the model fits better than would be expected by chance. As shown in Table 9.5, the ANOVA results indicate that the model is statistically significant, demonstrating that the engineering management practices meaningfully contribute to predicting schedule performance in EPC projects.

Coefficient analysis was used to determine the individual predictive contribution of each engineering management practice to schedule performance. As summarised in Table 9.6, only schedule management and risk management showed significant effects, with positive standardised coefficients ($\beta = 0.272$, $p = 0.043$ and $\beta = 0.288$, $p = 0.005$, respectively), thereby supporting hypotheses H3 and H6. The remaining variables, namely stakeholder engagement, resource management, integration management and quality management, did not demonstrate statistically significant predictive power within the multivariate model, despite their

Table 9.5 ANOVA

Model	Sum of squares	*df*	Mean square	*f*	Sig.
Regression	50.067	6	8.344	17.892	< 0.001
Residual	61.561	132	0.466	—	—
Total	111.627	138	—	—	—

Note: dependent variable: SP_avg; predictors: (constant), RM_avg, SE_avg, IM_avg, RE_avg, QM_avg, SM_avg.

Table 9.6 Coefficients and interpretation of hypotheses

Variable	Standardised coefficient, β	*t*	*Sig. (p-value)*	Interpretation of hypothesis
Stakeholder engagement (H1)	–0.116	–1.296	0.197	Not significant, H1 unsupported
Resource management (H2)	0.107	0.839	0.403	Not significant, H2 unsupported
Schedule management (H3)	0.272	2.047	0.043	Significant, H3 supported
Integration management (H4)	0.038	0.298	0.766	Not significant, H4 unsupported
Quality management (H5)	0.121	0.947	0.346	Not significant, H5 unsupported
Risk management (H6)	0.288	2.885	0.005	Significant, H6 supported

earlier positive correlations with schedule performance. This indicates that when all variables operate simultaneously, the influence of schedule and risk management is dominant, overshadowing the unique contribution of the other practices.

Based on the analysis, the change in R^2 for Model 2, reflecting the effect of the interaction term between integration management (IM) and technology adoption (TA), is 0.036 ($p = 0.006$; < 0.05), indicating that including this interaction significantly improves the model. The hierarchical regression model summary is shown in Table 9.7. This finding confirms that technology adoption moderates the relationship between integration management and schedule performance. The coefficient estimates for this moderation analysis are provided in Table 9.8.

Table 9.8 shows a positive, significant interaction term (IM × TA), indicating that technology adoption further strengthens the influence of integration management on the timely delivery

Table 9.7 Model summary of moderation analysis using hierarchical multiple regression

Model	R^2 change	*F* change	*df*1	*df*2	Sig. *F* change
1	0.332	33.866	2	136	< 0.001
2	0.036	7.648	1	135	0.006

Note: Model 1 predictors: (constant), TA_centred, IM_centred; Model 2 predictors: (constant), TA_centred, IM_centred, IM × TA.

Table 9.8 Coefficients of moderation analysis

Predictor	*B*	*β*	*t*	*Sig. (p-value)*	Interpretation of hypothesis
Integration management (IM_centered)	0.537	0.484	5.873	< 0.001	IM has a significant positive impact
Technology adoption (TA_centered)	0.291	0.243	2.821	0.006	TA has a significant positive impact
Interaction term (IM x TA)	0.205	0.207	2.766	0.006	The moderation effect is statistically significant, H7 supported

of EPC projects. This outcome supports hypothesis H7, which posits that digitalisation initiatives such as BIM modelling, integrated communication platforms and cloud-based information sharing enhance the effectiveness of integration management within multidisciplinary engineering environments, thereby improving EPC schedule performance.

9.3.7 Hypothesis testing

A summary of hypothesis testing results is provided in Table 9.9. The correlation analysis confirmed that all six engineering management practices significantly influence EPC schedule performance; however, the multiple regression results revealed that only schedule management and risk management have statistically significant predictive effects, supporting hypotheses H3 and H6. This aligns with project management literature, which consistently identifies schedule planning, sequencing, monitoring and proactive risk mitigation as central to controlling delays in complex EPC environments. Their strong predictive power suggests that these functions act as primary control mechanisms that reduce schedule slippage despite contractual rigidity, multidisciplinary involvement and inherent project uncertainties.

Conversely, stakeholder engagement, resource management, integration management and quality management did not demonstrate significant unique contributions in the multivariate regression model, despite their positive correlations with schedule performance. This indicates that their effects may be overshadowed by the dominant influence of schedule and

Table 9.9 Summary of hypothesis testing

Hypothesis	Inferential analysis		Moderation analysis
	Pearson correlation test	Multiple regression test	Hierarchical regression test
H1: stakeholder engagement during engineering execution has a significant impact on the EPC schedule performance	Significant moderate correlation, H1 supported	Not significant, H1 unsupported	–
H2: efficacy in engineering resource management enhances EPC schedule performance	Significantly strong correlation, H2 supported	Not significant, H2 unsupported	–
H3: the scheduled delivery of the EPC project is highly affected by the efficiency of engineering schedule management	Significantly strong correlation, H3 supported	Significant, H3 supported	–
H4: EPC project schedule performance is positively impacted by effective integration management	Significantly strong correlation, H4 supported	Not significant, H4 unsupported	–
H5: there is a positive relationship between quality management in the engineering stage and the outcomes of the EPC project schedule	Significantly strong correlation, H5 supported	Not significant, H5 unsupported	–
H6: EPC schedule performance is enhanced by effective risk management	Significantly strong correlation, H6 supported	Significant, H6 supported	–
H7: digitalisation enhances the effect of integration management in optimising EPC schedule performance	–	–	The moderation effect is statistically significant, H7 supported

risk management or may operate indirectly. For example, integration management, although strongly correlated with schedule performance, lost significance in the regression model, suggesting its influence may be enhanced or mediated through other factors. This was confirmed in the moderation analysis, which showed that technology adoption significantly strengthened the relationship between integration management and schedule performance, supporting hypothesis H7. Digital tools such as cloud collaboration platforms, BIM and 3D modelling enhance coordination, information flow and real-time problem-solving in multidisciplinary settings, reinforcing the relevance of project management theory in addressing project complexity.

9.4. Practical insights

The findings of this study provide several practical insights for EPC practitioners, project managers and organisational decision-makers involved in complex engineering-driven projects. Most notably, the results show that while multiple engineering management practices contribute positively to schedule performance, schedule and risk management are the most critical determinants of timely EPC project delivery. This highlights the need for EPC organisations to prioritise these two functions as core control mechanisms rather than treating them as supporting activities.

From a practical perspective, EPC contractors should strengthen engineering schedule governance by establishing realistic design milestones, enforcing disciplined progress monitoring and ensuring the timely release of engineering deliverables to support procurement and construction activities. Integrating engineering schedules with procurement and construction timelines is essential to prevent downstream bottlenecks. Dedicated schedule controllers, frequent look-ahead planning and structured design-freeze strategies can significantly reduce schedule slippage originating in the engineering phase.

Risk management also requires heightened managerial attention. The strong predictive impact of risk management underscores the importance of early risk identification during engineering, particularly risks related to design maturity, interface coordination, vendor documentation and scope ambiguity. EPC organisations should institutionalise formal risk registers, conduct regular risk reviews and assign clear ownership for mitigation actions. Proactive contingency planning and early warning systems can help prevent minor engineering risks from escalating into major schedule disruptions.

Although stakeholder engagement, resource management, integration management and quality management did not show significant direct effects in the multivariate regression, their strong correlations with schedule performance indicate that these practices remain essential enablers. Practically, this suggests that their influence is often indirect or conditional. For instance, effective integration management becomes significantly more impactful when supported by technology adoption, as evidenced by the moderation analysis. This finding reinforces the strategic value of digitalisation in EPC projects.

Digital tools such as BIM, cloud-based collaboration platforms, integrated document management systems and real-time communication tools should therefore be leveraged to enhance interdisciplinary coordination, improve information transparency and accelerate decision-making. By digitalising integration processes, EPC organisations can reduce misalignment between engineering disciplines, minimise rework and strengthen schedule adherence, particularly in parallel execution environments.

Overall, the practical implications of this study suggest that EPC organisations should adopt a focused, technology-enabled engineering management strategy, prioritising schedule and risk control while leveraging digitalisation to enhance integration effectiveness. Such an approach can significantly improve schedule performance, reduce execution inefficiencies and enhance the overall competitiveness of EPC projects in the Malaysian context.

9.5. Conclusion and recommendations

This section highlights the key findings of the study, then discusses their theoretical and practical implications. The study's limitations are then examined, providing the basis for recommendations for future research and industry applications. Overall, this section provides a comprehensive synthesis of the analytical outcomes in relation to the research objectives and hypotheses. It also offers insights that deepen understanding of project management practices, particularly during the engineering phase of EPC projects, yielding benefits for both academic and industrial contexts.

9.5.1 Theoretical contributions

The findings of this study provide meaningful theoretical and practical insights into engineering management practices that drive high schedule performance in EPC projects within the Malaysian context. By empirically validating schedule performance factors, particularly schedule- and risk-based control mechanisms, the study reinforces the relevance of PMT. The results highlight that effective planning, monitoring and proactive risk mitigation, as outlined in PMBOK and similar standards, are essential for achieving optimal schedule outcomes in EPC environments.

The study also extends PMT by demonstrating that technology adoption moderates the impact of integration management on schedule performance. Confirming this interaction suggests that digital tools enhance collaboration, increase project visibility and improve timely decision-making, thereby promoting better schedule compliance. This underscores the evolving role of digitalisation in shaping contemporary project management frameworks.

9.5.2 Limitations

Although this study provides valuable insights into the influence of engineering management practices on EPC schedule performance, several limitations must be acknowledged. First, the sample was drawn from a single EPC organisation, yielding 139 valid responses. While statistically adequate, this limits the generalisability of the findings to the broader Malaysian EPC context due to the absence of national data on the actual population of EPC professionals and constraints related to time and access.

The demographic scope was also simplified, with project locations grouped into only two regions: Peninsular Malaysia and East Malaysia. Although this distinction captures major regulatory and logistical differences, other context-specific factors such as local practices, site conditions and geographical constraints may also influence schedule outcomes. Additionally, the diversity of project sectors and scales within the dataset introduces varying complexities that were not statistically examined, as demographic information was analysed descriptively. This leaves potential uncertainties unaddressed.

Moreover, the study explored the moderating role of technology adoption only in the relationship between integration management and schedule performance. Given that digital technologies increasingly affect multiple management domains such as scheduling, risk management and stakeholder coordination, examining their moderating influence across other engineering management practices would provide a more holistic understanding of technology's impact.

9.5.3 Recommendations and future research

Several recommendations are proposed to advance understanding of how engineering management practices influence EPC schedule performance. First, future studies should broaden the research population beyond a single EPC organisation. Although national statistics on EPC professionals are limited, identifying and engaging key EPC players across Malaysia would enable a larger, more representative sample, strengthening the reliability and generalisability of the findings.

Second, to address uncertainties arising from varied demographic profiles, future research should statistically examine how factors such as project type, scale and location influence engineering management practices and schedule outcomes. This would provide deeper insight into schedule behaviour across different project contexts.

Third, the moderation effect of technology adoption should be expanded beyond integration management. Investigating its influence on other engineering management functions, such as scheduling, risk management and stakeholder engagement, would offer a more comprehensive understanding of the role of digitalisation in enhancing project performance. Additionally, incorporating objective project performance metrics (e.g. progress trends, schedule variance, recorded delays) into future studies would strengthen the empirical evidence base, enabling more accurate and realistic assessments of schedule performance. Together, these recommendations can support more comprehensive research that contributes meaningfully to project management theory and practice, particularly by improving the effectiveness of engineering management in successful EPC schedule delivery.

REFERENCES

Ade AF and Santoso S (2023) Delay factor analysis and dynamic system modeling on the EPC project of liquid petroleum gas tank in Kupang. *Journal Research of Social Science, Economics, and Management (JRSSEM)* **2(6)**: 1187–1202.

Ahmed YA, Shehzad HMF, Khurshid MM *et al.* (2024) Examining the effect of interoperability factors on building information modelling (BIM) adoption in Malaysia. *Construction Innovation* **24(2)**: 606–642.

Ahuja HN, Dozzi SP and Abourizk SM (1994) *Project Management: Techniques in Planning and Controlling Construction Projects*. Wiley, Hoboken, NJ, USA.

Balzarova MA and Castka P (2012) Stakeholders' influence and contribution to social standards development: the case of ISO 26000. *Journal of Business Ethics* **111**: 265–279.

Banks E (2004) *Alternative Risk Transfer: Integrated Risk Management Through Insurance, Reinsurance, and the Capital Markets*. Wiley, Hoboken, NJ, USA.

Benton WC and McHenry LF (2010) *Construction Purchasing and Supply Chain Management*. McGraw-Hill, New York, NY, USA.

Bhatti MAA and Nazir MU (2024) The impact of project process management on sustainable project success in the construction sector: the moderating role of risk management practices. *Bulletin of Business and Economics* **13(2)**: 1065–1072.

Choi SW, Lee EB and Kim JH (2021) The engineering machine-learning automation platform (EMAP): a big-data-driven AI tool for sustainable management solutions. *Sustainability* **13(18)**: 10384.

Ciric Lalic D, Lalic B, Delić M, Gračanin D and Stefanovic D (2022) How project management approach impacts project success? From traditional to agile. *International Journal of Managing Projects in Business* **15(3)**: 494–521.

Ezeh MO, Ogbu AD, Ikevuje AH and George EPE (2024) Stakeholder engagement and influence: strategies for successful energy projects. *International Journal of Management & Entrepreneurship Research* **6(7)**: 2375–2395.

Faraji A, Rashidi M, Perera S and Samali B (2022) Applicability-compatibility analysis of PMBOK seventh edition from the perspective of the construction industry distinctive peculiarities. *Buildings* **12(2)**: 210.

Faria J, Araújo M, Demeulemeester E and Tereso A (2020) Project management under uncertainty: using flexible resource management. *European Journal of Industrial Engineering* **14(5)**: 599–631.

Ghanbari L (2024) *Key Successful Elements of EPC Projects*. Vaasan University of Applied Sciences, Vaasa, Finland.

Granheimer K, Karrbom Gustavsson T and Eriksson PE (2024) Public procurement of engineering services: the influence of task characteristics on organisational control. *Engineering, Construction and Architectural Management* **31(13)**: 208–222.

Gupta H, Mishra MVM, Grover N and Chourasia R (2025) Integrating project management with supply chain and big data engineering using AI methodologies for enhanced sustainability. In *AI-Enabled Sustainable Innovations in Education and Business* (Azar AS, Gupta H and Alhamaty F (eds)). IGI Global, New York, NY, USA, pp. 319–352.

Habibi M, Kermanshachi S and Rouhanizadeh B (2019) Identifying and measuring EPC key performance indicators. *Infrastructures* **4(10)**: 1–19.

Hinostroza M, Chavez P, Nuñez V and Raymundo C (2018) Application of PMBOK to improve the deadline of projects in SMEs engineering consultancies. In *Brazilian Technology Symposium*. Springer, Switzerland, pp. 487–494.

Huang Y, Bai F, Shang M and Ahmad M (2023) On the fast track: the benefits of ESG performance on commercial credit financing. *Environmental Science and Pollution Research* **30(35)**: 83961–83974.

ISO (International Organization for Standardization) (2015) ISO 9001: *Quality Management Systems – Requirements*. ISO, Geneva, Switzerland.

Javed S, Hussain MI, Al Aamri AM and Akhtar J (2022) Investigation on factors causing construction delay and their effects on the development of Oman's construction industry. *EUREKA: Physics and Engineering* **3**: 33–44.

Kabirifar K and Mojtahedi M (2019) Impact of engineering, procurement and construction (EPC) phases on project performance. *Buildings* **9(1)**: 1–15.

Kermani MAMA, Maghsoudi M and Darzi E (2024) Enhancing procurement performance in project-oriented organizations: a process analysis approach. *IEEE Access* **12**: 134340–134354.

Maslennikov VV, Popova EV and Lyandau YV (2020) Project management based on PMBOK 7.0. In *International Scientific and Practical Conference*. Springer, Cham, Switzerland, pp. 283–289.

Mejlænder-Larsen Ø (2018) Improving collaboration between engineering and construction in detail engineering using a project execution model and BIM. *Journal of Information Technology in Construction* **23**: 324–339.

Metwally YB, Elsaid AM, Elhamaida SM and Abd Elsamad MAE (2025) Project navigating: a guide to successful construction project management. *Advanced Sciences and Technology Journal* **2(1)**: 1–27.

Moretto A, Patrucco AS, Walker H and Ronchi S (2022) Procurement organisation in project-based setting: a multiple case study of engineer-to-order companies. *Production Planning & Control* **33(9–10)**: 847–862.

Nazimko VV and Zakharova LM (2023) Project schedule expediting under structural and parametric uncertainty. *Engineering Management Journal* **35(1)**: 29–49.

Nyamah EY, Feng Y, Yeboah Nyamah E, Opoku RK and Ewusi M (2023) Procurement process risk and performance: empirical evidence from manufacturing firms. *Benchmarking: An International Journal* **30(1)**: 75–101.

Peansupap V and Ly R (2015) Evaluating the impact level of design errors in structural and other building components in building construction projects in Cambodia. *Procedia Engineering* **123**: 370–378.

PMI (Project Management Institute) (2021) *A Guide to the Project Management Body of Knowledge (PMBOK Guide)*, 7th edn. Project Management Institute, Newtown Square, PA, USA.

Prasad A (2025) Effective strategies in monitoring, control, commissioning, and handover. In: *Handbook of Construction Project Management*. Springer, Singapore, pp. 749–785.

Qamar SZ, Al-Hinai N and Márquez FPG (2024) *Quality Control and Quality Assurance: Techniques and Applications*. Books on Demand, Norderstedt, Germany.

Rahman MM (2023) Sample size determination for survey research and non-probability sampling techniques: a review and set of recommendations. *Journal of Entrepreneurship, Business and Economics* **11(1)**: 42–62.

Sangroungrai S, Sukchareonpong P and Witchakul S (2018) Engineering, procurement and construction (EPC) project management for reducing cost and time: a case study of petrochemical plant. In *2018 5th International Conference on Business and Industrial Research (ICBIR)*. IEEE, Piscataway, NJ, USA, pp. 402–406.

Sanni-Anibire MO, Mohamad Zin R and Olatunji SO (2022) Causes of delay in the global construction industry: a meta-analytical review. *International Journal of Construction Management* **22(8)**: 1395–1407.

Sathyanarayana S, Mohanasundaram T, Pushpa BV and Harsha H (2024) Selecting the right sample size: methods and considerations for social science researchers. *International Journal of Business and Management Invention* **13(7)**: 152–167.

Shahzad A, Terziovski M and Sandhu MA (2025) Benchmarking an integrative project management model to enhance the effectiveness of EPC projects through the adoption of digital technology. *Benchmarking: An International Journal*. Advance online publication, pp. 1–27.

Shaukat MB, Latif KF, Sajjad A and Eweje G (2022) Revisiting the relationship between sustainable project management and project success: the moderating role of stakeholder engagement and team building. *Sustainable Development* **30(1)**: 58–75.

Shen ZW, Wei X, Siraj A *et al.* (2025) Relationship between quality governance model and performance in EPC projects. *International Journal of Productivity and Performance Management* **74(6)**: 2183–2210.

Soares I, Fernandes G and Santos JM (2024) Sustainability in project management practices. *Sustainability* **16(10)**: 4275.

Wan S, Liu Y, Ding G, Runeson G and Er M (2023) Risk allocation for energy performance contracts from the perspective of incomplete contracts: a study of commercial buildings in China. *International Journal of Climate Change Strategies and Management* **15(4)**: 457–478.

Yilmaz S, Kumar D, Hada S *et al.* (2025) A PMBOK-based construction cost management framework for BIM integration in construction projects. *International Journal of Construction Management* **25(8)**: 861–875.

Zhu Q, Xi J, Hu X *et al.* (2024) Stakeholder mapping and analysis of off-site construction projects. *Buildings* **14(9)**: 2865.

Advancing Project Management: Frameworks, Digital Transformation, Sustainability and Governance

Mohammed Reyasudin Basir Khan
ISBN 978-1-83742-799-4
https://doi.org/10.1108/978-1-83742-798-720261013

Chapter 10
Sustainable project management: balancing triple constraints with environmental goals

Thong Jin Cheng and Mohammed Reyasudin Basir Khan

Tun Razak Graduate School, Universiti Tun Abdul Razak (UNIRAZAK), Kuala Lumpur, Malaysia

10.1. Introduction

Traditional project management has long been shaped by the triple constraints model, which prioritises balancing scope, time and cost to ensure efficient delivery. While effective for guiding short-term project outcomes, this model often overlooks environmental and social considerations that have become increasingly important amid escalating climate concerns, regulatory pressures and evolving stakeholder expectations. The idea of sustainability, rooted in the 1987 Brundtland Report, stresses meeting present needs without compromising the ability of future generations to meet theirs (Reiser, 2023). This principle has expanded sustainability into a multidimensional concept that integrates economic, environmental and social objectives into organisational strategies and project practices.

The global shift toward sustainability, driven by initiatives such as the United Nations Sustainable Development Goals and the Paris Agreement, emphasises responsible production, climate action and the transition to clean energy. Despite this momentum, organisations continue to struggle to embed sustainability within project management due to a lack of structured methodologies that reconcile long-term sustainability goals with traditional performance metrics. Sustainable project management requires balancing short-term constraints with long-term value creation through resource optimisation, stakeholder engagement and continuous organisational learning. Although industries such as construction and energy have adopted practices such as leadership in energy and environmental design (LEED) certification and lifecycle assessments to reduce environmental impact, the practical integration of sustainability remains inconsistent.

Studies highlight persistent challenges, including tensions between immediate project deliverables and broader sustainability outcomes, difficulties measuring sustainability performance, conflicting stakeholder interests and resistance to change among organisations accustomed to traditional success metrics (Ali and Haapasalo, 2023; Santos and de Carvalho, 2022).

These issues underscore the need for a more comprehensive framework that bridges sustainability principles and project management practices. The conceptual framework proposed in the study presented in this chapter builds on the triple bottom line (TBL) and project sustainability management models, incorporating elements such as corporate governance, resource management, stakeholder participation and organisational learning to embed sustainability in project workflows without compromising performance.

Although sustainability is increasingly recognised as essential for long-term organisational success, many companies continue to rely on the traditional triple constraints model as the primary indicator of project performance. While this model ensures efficiency in cost, schedule and scope, frameworks such as TBL offer a more comprehensive perspective that extends evaluation to environmental and social dimensions. Yet, most organisations struggle to operationalise these concepts because they lack structured, practical methodologies that link sustainability objectives to conventional project indicators. Research shows that only a minority of firms have successfully integrated sustainability into project processes, largely because sustainability efforts often require higher upfront investments, may extend delivery timelines and are sometimes met with resistance from stakeholders who perceive them as complex or disruptive (Marcelino-Sádaba *et al.*, 2014; Turner *et al.*, 2010).

Moreover, limited empirical evidence exists on the roles of mediating and moderating factors, such as stakeholder engagement and organisational adaptability, in supporting sustainable project management. Without a deeper understanding of how these factors influence performance outcomes, organisations struggle to incorporate sustainability without undermining project efficiency. To address this gap, the study presented in this chapter proposes an integrated framework that aligns sustainability practices with the triple constraints model and examines how corporate policies, resource allocation, stakeholder engagement and organisational learning interact to influence project success.

This study examines how sustainability practices can be effectively integrated into project management frameworks while maintaining performance on cost, time and scope. It explores the influence of corporate policies and resource management on project outcomes, the mediating role of stakeholder engagement, the moderating impact of organisational learning and the combined effects of these factors on project success. By investigating these relationships, the study identifies challenges and proposes strategies for balancing sustainability goals with traditional performance expectations.

The primary objective was to develop and validate a conceptual framework that embeds sustainability principles into project practices without compromising performance metrics. This involves assessing how governance structures and policy enforcement influence sustainable project implementation, examining the effectiveness of resource optimisation strategies, evaluating stakeholder communication and collaboration as mediating mechanisms, investigating the contribution of organisational learning to adaptive sustainability practices, and determining how sustainability initiatives affect adherence to cost, time and scope requirements.

The study contributes theoretically and practically to sustainable project management. Theoretically, it advances existing models by integrating TBL and key sustainability theories, such as stakeholder theory and sustainable development theory, into project management discourse. This creates a more holistic understanding of project success that goes beyond cost efficiency and embraces social responsibility and environmental stewardship. In practice, the study provides project managers with a structured, actionable framework to implement sustainability practices while maintaining performance standards. It emphasises resource efficiency, transparent stakeholder collaboration and robust compliance with environmental regulations, thereby enhancing organisational reputation and long-term value creation. The findings also offer policy insights for organisations and regulators seeking to develop governance mechanisms and performance measurement tools that ensure accountability and promote sustainable project outcomes. By highlighting the role of organisational learning in fostering adaptability and innovation, the study underscores the importance of developing sustainability as a core component of project management practice.

Project management has long been guided by the triple constraints theory, which emphasises scope, time and cost to achieve project efficiency. Although this approach supports timely and cost-effective project delivery, its focus on cost, time and scope creates a structural tension with long-term sustainability objectives, marginalising environmental and social considerations in project decision-making (Al-Emran, 2023). This limitation has driven the evolution of sustainability-oriented theories to incorporate broader ecological and societal concerns into project decision-making.

Sustainable development theory, rooted in the Brundtland Report, promotes balancing economic, environmental and social needs to safeguard future generations (Sakalasooriya, 2021). This principle aligns with contemporary project management trends that advocate embedding sustainability into planning and execution. Similarly, the TBL framework extends performance evaluation across economic, environmental and social dimensions, emphasising organisational accountability and stakeholder value creation (Khalifeh *et al.*, 2020).

The project sustainability management (PSM) framework operationalises TBL by integrating sustainability into project lifecycles through governance, resource optimisation and stakeholder engagement (Khalifeh *et al.*, 2020). Approaches such as lifecycle assessments (LCA) provide systematic methods for evaluating environmental impacts from initiation to closure (Rumayor *et al.*, 2024). Stakeholder theory further reinforces the need for transparent communication and collaborative processes to enhance acceptance and align project outcomes with community and environmental priorities (Mahajan *et al.*, 2023). Complementing these is contingency theory, which emphasises adaptability in dynamic conditions, such as regulatory changes and evolving stakeholder expectations, common challenges in the energy and construction industries (Sabini *et al.*, 2019).

Integrating these theories supports the development of a comprehensive model that balances traditional project constraints with sustainability goals. These theories collectively underpin a framework that examines how corporate policies, resource management and stakeholder engagement influence project performance, with organisational learning as a moderating factor. Corporate policies ensure governance and compliance; resource management promotes efficiency and cost reduction; stakeholder engagement improves acceptance; and organisational learning supports continuous adaptation of sustainability practices.

Empirical studies on sustainable project management (SPM) demonstrate mixed findings regarding cost, timeline and performance implications. While sustainability integration often increases initial costs due to green technologies and compliance requirements (Srivastava *et al.*, 2026), long-term benefits, such as improved resource efficiency and operational savings, help offset upfront investments. Conversely, many organisations still perceive adopting sustainability as burdensome due to additional costs, extended timelines and limited expertise (Escoto *et al.*, 2022). These contrasting findings reflect the trade-off between short-term constraints and long-term sustainability value.

Tools such as the sustainability balanced scorecard (SBSC) and LCAs have become essential for evaluating multidimensional sustainability indicators. Research identified 82 indicators across economic, social and environmental dimensions to structure performance assessments in construction projects (Khalifeh *et al.*, 2020). Despite these advancements, a systematic review of 770 publications found fragmentation and a lack of standardisation in sustainable project management practices, calling for unified governance frameworks and consistent evaluation mechanisms (Stanitsas *et al.*, 2021). Top management plays a crucial role in embedding sustainability into project governance. Leadership teams with sustainability expertise enhance organisational capacities and improve TBL performance. Projects, particularly in energy and infrastructure, achieve better outcomes when sustainability officers are part of decision-making structures. Stakeholder engagement also emerges as a strong mediator in sustainability adoption. Effective communication, collaborative decision-making and transparent processes increase accountability, improve policy alignment and enhance project success (Khalifeh *et al.*, 2020). However, gaps remain in understanding scalability and long-term effects across industries (Sabini *et al.*, 2019).

The conceptual framework, as shown in Figure 10.1, integrates sustainability principles into project management processes to assess their impact on project success, measured by

Figure 10.1 Conceptual framework

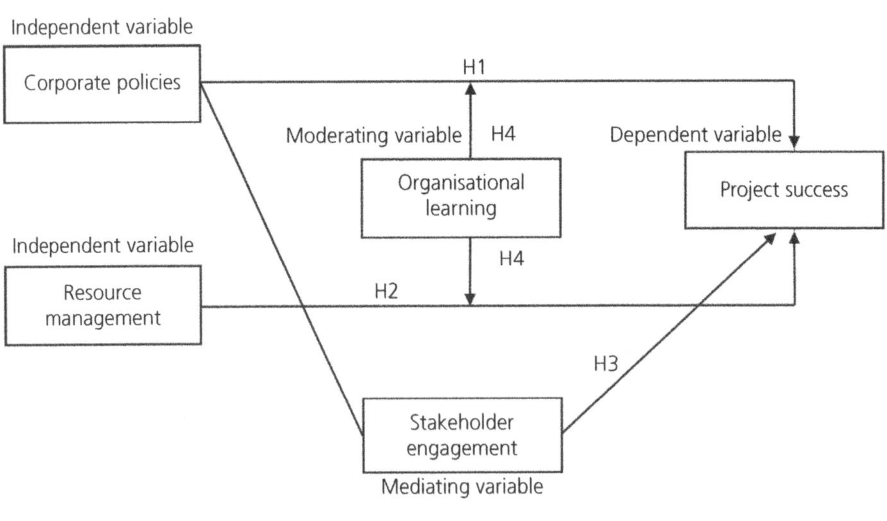

Table 10.1 Distribution of questionnaire sections

Section	Purpose
Demographic information	Captures data on participant's role, industry and years of experience to contextualise findings
Sustainability practices	Evaluates the adoption of sustainability strategies, including corporate policies and resource management
Performance indicators	Measures project success based on adherence to cost, time and scope
Stakeholder engagement and organisational learning	Assesses mediating and moderating influences related to collaboration and adaptability

scope, time and cost. Grounded in stakeholder theory and sustainable development theory, the framework incorporates empirical insights by examining relationships between independent variables (corporate policies and resource management), the mediating variable (stakeholder engagement), the moderating variable (organisational learning) and the dependent variable (project success). Table 10.1 summarises the variables and their operational meanings. Stakeholder engagement is positioned as a mediator, enhancing communication, resolving conflicts and aligning interests toward sustainability objectives. Organisational learning moderates the relationships by enabling continuous improvement, knowledge sharing and adaptability to environmental or regulatory changes.

Drawing from the conceptual framework and theoretical foundations, four hypotheses were proposed to test the relationships between sustainability practices and project performance. These hypotheses were informed by TBL, stakeholder theory and contingency theory, and evaluate how governance, resource optimisation, collaboration and learning influence project outcomes.

10.2. Methodology

10.2.1 Research design

This study employs a quantitative research design to test the proposed hypotheses and examine the relationships between sustainability practices and project success. Quantitative methods were selected for their ability to generate measurable data and enable statistical validation, supporting the generalisability of findings across different project environments. Guided by a deductive approach, the study draws on established theories, including triple constraints theory, stakeholder theory and sustainable development theory, and validates them through empirical analysis. A survey strategy was adopted because it is efficient at collecting data from a large sample within a defined timeframe. The structured questionnaire, developed from the literature, measures key variables including corporate policies, resource management, stakeholder engagement, organisational learning and project success indicators (cost, time and scope). This approach ensures alignment with the conceptual framework. It enables a rigorous assessment of how sustainability practices influence traditional project management outcomes, contributing to a practical model for sustainable project execution.

10.2.2 Data collection method

Primary data were collected through a structured questionnaire distributed electronically through Google Forms. The instrument consists of closed-ended questions to facilitate quantitative analysis and ensure systematic measurement of the study variables. A pilot test involving ten participants was conducted to assess clarity, reliability and validity. Feedback from this phase helped refine ambiguous items before distributing the final questionnaire to the full sample of 200 respondents.

10.2.3 Operationalisation and measurement

To ensure consistency and reliability in data collection, all variables were measured on a five-point Likert scale from 1 (strongly disagree) to 5 (strongly agree). This standardised measurement approach enabled accurate assessment of respondents' perceptions and practices, supporting rigorous statistical analysis.

10.2.4 Framework

The first independent variable, corporate policies, measures the presence and effectiveness of governance mechanisms that support sustainability. This includes assessing whether organisations have formal sustainability policies and whether leadership prioritises sustainability in project planning, reflecting how well sustainability principles are embedded into decision-making structures. The second independent variable, resource management, evaluates strategies related to efficient resource allocation and waste minimisation. This involves examining the extent to which energy-efficient technologies, recycling initiatives and resource-optimisation practices are incorporated into project activities to reduce environmental impact.

The mediating variable, stakeholder engagement, captures the level of collaboration and communication used to align stakeholder interests with project objectives. It assesses whether stakeholders are actively consulted on sustainability goals and whether their feedback is incorporated into project planning, demonstrating the importance of participatory processes in achieving sustainable outcomes. The moderating variable, organisational learning, examines the organisation's ability to adapt and apply lessons learned to enhance sustainability practices. This includes evaluating whether insights from previous projects are used to improve new initiatives and whether innovation is encouraged to strengthen sustainable project management.

Finally, the dependent variable, project success, is measured by cost, time and scope performance, while accounting for sustainability considerations. This involves assessing whether sustainability practices are compatible with budget constraints and whether project deliverables are achieved without compromising environmental or social goals.

10.3. Results and analysis

10.3.1 Demographic profile

This section presents the demographic profile of the 200 respondents who participated in the study. Understanding the characteristics of the sample is essential for contextualising the findings and ensuring that the insights gathered accurately reflect the perspectives of professionals within Malaysia's engineering and project management sectors.

As shown in Table 10.2, the sample comprises 118 male respondents (59%) and 82 female respondents (41%), reflecting the typical gender distribution in technical and engineering sectors. Most participants are mid-career professionals: 38% are aged 31–40, 28% are aged 41–50, 22% are aged 20–30 and 12% are over 50. In terms of education, nearly half of the respondents hold a Bachelor's degree (49.5%), followed by those with a Master's degree (41%) and a smaller group with a diploma (9.5%). This indicates a well-educated sample suitable for providing informed insights on sustainable project management.

Regarding professional roles, 40% of respondents are project managers, 21% are engineers and 9% are sustainability officers; the remaining 30% hold related roles, such as consultants and administrative staff. This diversity provides a comprehensive perspective across project functions. Regarding industry experience, most have more than 6 years in the field. Specifically, 35.5% have 6–10 years of experience, 29% have 11–15 years and 24.5% have more than 15 years, while only 11% have less than 5 years. This indicates that most respondents bring substantial practical experience, enhancing the reliability of insights on sustainability integration in project lifecycles.

Table 10.2 Respondents' demographic profile

Demographic variable	Category	Frequency, *n*	Percentage: %
Gender	Male	118	59.0
	Female	82	41.0
Age group	20–30 years	44	22.0
	31–40 years	76	38.0
	41–50 years	56	28.0
	Over 50 years	24	12.0
Educational qualification	Diploma	19	9.5
	Bachelor's degree	99	49.5
	Master's degree	82	41.0
Current position	Project manager	80	40.0
	Engineer	42	21.0
	Sustainability officer	18	9.0
	Other (e.g. consultant, administrator)	60	30.0
Years of experience	Less than 5 years	22	11.0
	6–10 years	71	35.5
	11–15 years	58	29.0
	More than 15 years	49	24.5

Table 10.3 Cronbach's alpha across variables

Construct	Number of items	Cronbach's alpha, α	Interpretation
Corporate policies	5	0.864	Very good reliability
Resource management	4	0.821	Good reliability
Stakeholder engagement	6	0.882	Very good reliability
Organisational learning	5	0.851	Very good reliability
Project success	7	0.896	Excellent reliability

10.3.2 Reliability analysis

In general, a Cronbach's alpha (α) above 0.70 indicates acceptable reliability, although some authors recommend values between 0.90 and 0.95. As shown in Table 10.3, all constructs in this study demonstrate high reliability, exceeding the 0.70 threshold. Project success recorded the highest alpha (0.896), reflecting excellent internal consistency. Stakeholder engagement also showed strong reliability (0.882), confirming the coherence of items related to stakeholder involvement and communication. Corporate policies (0.864) and organisational learning (0.851) both achieved very good reliability, supporting their roles as aligned governance and learning mechanisms within organisations. Resource management produced an alpha of 0.821, indicating good reliability and satisfactory consistency among its items. Overall, these results confirm the robustness of the measurement instrument. With reliability established, the next step was to test the hypotheses using inferential statistical techniques, as discussed in the following sections.

10.3.3 Normality analysis

The normality test was conducted to assess whether the data for each variable approximates a normal distribution, an essential assumption for parametric analyses such as correlation, regression and structural equation modelling (SEM). Normality was evaluated using skewness and kurtosis, with acceptable values falling within ±2.0 (Orcan, 2020). As shown in Table 10.4, the skewness values for corporate policies (–0.192), resource management

Table 10.4 Normality analysis

Variable	Skewness	Kurtosis
Corporate policies	–0.192	–0.861
Resource management	–0.618	0.519
Stakeholder engagement	–0.825	0.699
Organisational learning	–1.306	1.495
Project success	–0.845	0.753

(–0.618), stakeholder engagement (–0.825) and project success (–0.845) fall well within ± 1, indicating mild skewness and reasonably normal distributions. Organisational learning shows a slightly higher skewness of –1.306, which remains within acceptable limits and does not indicate a severe departure from normality. Kurtosis values for all variables range from –0.861 to 1.495, remaining below the ±2 threshold. This confirms that none of the distributions is overly peaked or flat. Although organisational learning has the highest kurtosis (1.495), it remains within acceptable bounds and reflects only a mildly leptokurtic distribution. Overall, the skewness and kurtosis results, supported by the Q-Q plots illustrated in Figures 10.2 to 10.5, indicate that all variables (corporate policies, resource management, stakeholder engagement, organisational learning and project success) are approximately normally distributed. This confirms that parametric statistical procedures are appropriate for subsequent analyses.

Figure 10.2 Q-Q plot for corporate policies

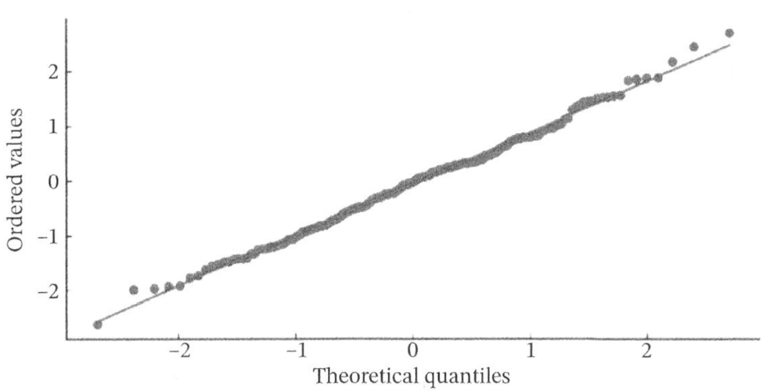

Figure 10.3 Q-Q plot for resource management

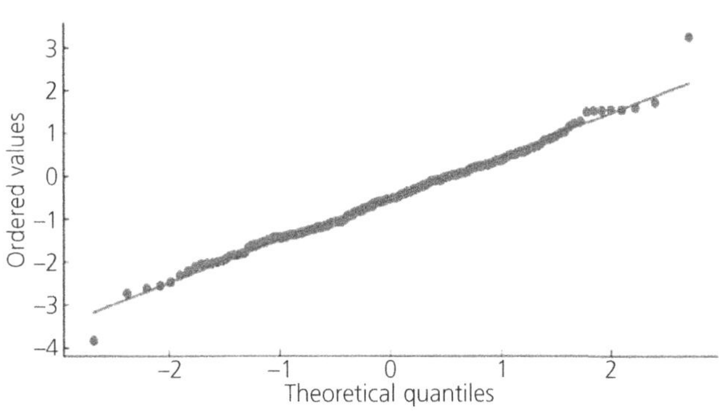

Figure 10.4 Q-Q plot for stakeholder engagement

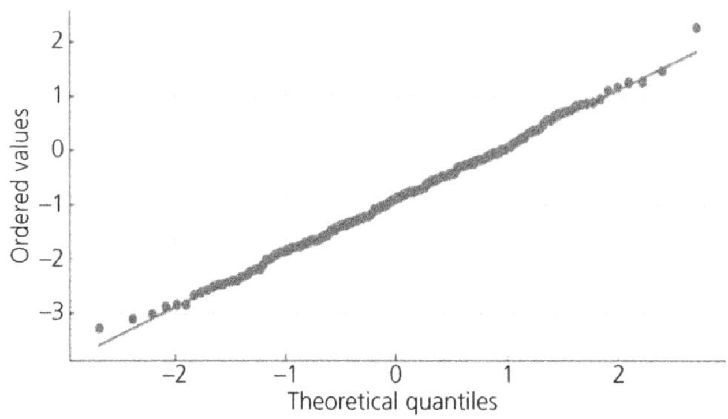

Figure 10.5 Q-Q plot for organisational learning

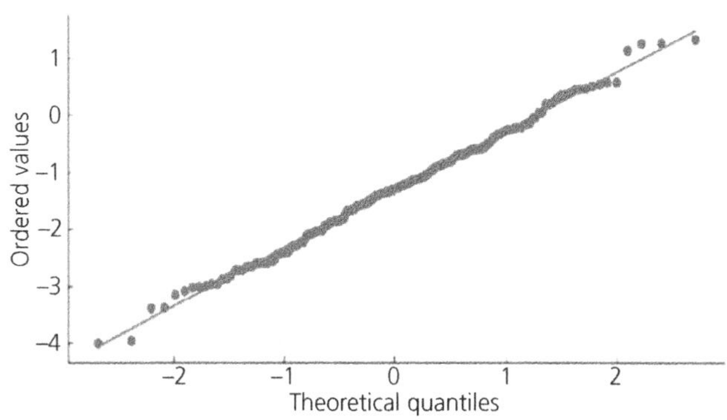

10.3.4 Descriptive analysis

The mean (m) and standard deviation (SD) values for each construct are summarised in Table 10.5. The analysis shows that project success achieved the highest mean score ($m = 4.05$), indicating that respondents generally perceive their projects as well-managed and effective at balancing cost, time and scope with environmental goals. Stakeholder engagement ($m = 3.92$) and organisational learning ($m = 3.88$) also performed strongly, suggesting that collaborative decision-making, transparent communication and continuous improvement practices are well embedded within organisations.

Corporate policies scored slightly lower ($m = 3.85$), implying that although sustainability governance structures are in place, further enhancement may be needed in areas such as policy enforcement, strategic alignment and regulatory compliance. Resource management

Table 10.5 Descriptive statistics of key constructs ($n = 200$)

Construct	Number of items	Mean, *m*	SD
Corporate policies	5	3.85	0.65
Resource management	4	3.78	0.70
Stakeholder engagement	6	3.92	0.60
Organisational learning	5	3.88	0.68
Project success	7	4.05	0.55

Figure 10.6 Likert distribution across resource management variable

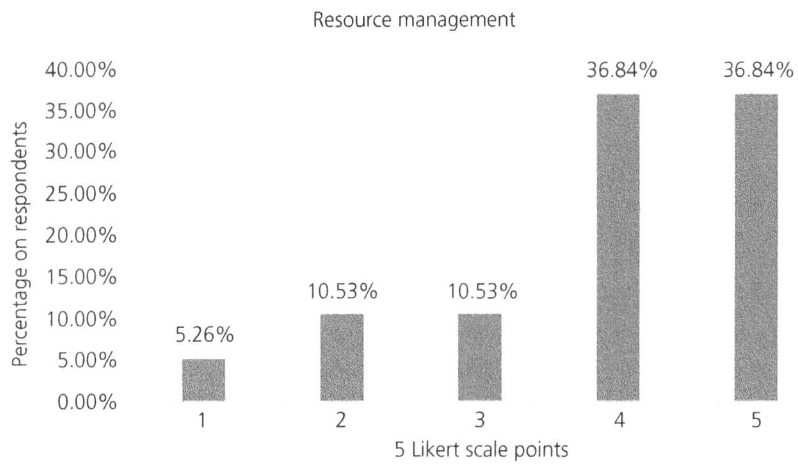

recorded the lowest mean ($m = 3.78$), highlighting challenges in sustainable resource utilisation, potentially due to financial constraints, slow adoption of green technologies or inefficiencies in supply chain practices. The detailed distribution of responses across the Likert scale for resource management is illustrated in Figure 10.6.

10.3.5 Pearson correlation analysis

Pearson correlation analysis was used to assess the strength and direction of relationships between the independent variables (corporate policies, resource management), the mediating variable (stakeholder engagement), the moderating variable (organisational learning) and the dependent variable (project success). As illustrated in Table 10.6, all variables show statistically significant positive correlations with project success, indicating that sustainable project management practices are closely associated with improved performance outcomes.

Corporate policies ($r = 0.651$) demonstrate a moderate-to-strong relationship with project success, suggesting that sustainability-oriented governance supports effective project delivery.

Table 10.6 Summary of Pearson correlation coefficient results

Variable	1	2	3	4	5
Corporate policies	1				
Resource management	0.633 ($p < 0.01$)	1			
Stakeholder engagement	0.602 ($p < 0.01$)	0.645 ($p < 0.01$)	1		
Organisational learning	0.573 ($p < 0.01$)	0.611 ($p < 0.01$)	0.658 ($p < 0.01$)	1	
Project success	0.651 ($p < 0.01$)	0.705 ($p < 0.01$)	0.684 ($p < 0.01$)	0.722 ($p < 0.01$)	1

Resource management shows the strongest correlation ($r = 0.705$), highlighting the critical role of efficient resource use in achieving successful project results. Stakeholder engagement ($r = 0.684$) also exhibits a strong positive relationship, reinforcing the value of active involvement and communication throughout the project lifecycle. Organisational learning ($r = 0.722$) further contributes positively, indicating that continuous improvement and knowledge sharing enhance project sustainability performance.

10.3.6 Hypothesis testing

A summary of hypothesis testing results is shown in Table 10.7. The summary shows a moderate positive correlation ($r = 0.651$) between corporate policies and project success, indicating that well-defined sustainability policies, such as environmental compliance frameworks and governance structures, positively influence project outcomes. This finding aligns with those of Alsudiri *et al.* (2013), who emphasised that strong policy alignment is essential for the effective integration of sustainability in project environments. Resource management shows a stronger relationship with project success ($r = 0.705$), suggesting that the efficient use of energy, materials and time significantly improves project delivery. This supports Stanitsas *et al.* (2021), who found that environmentally responsible resource allocation enhances

Table 10.7 Summary of hypothesis testing

Hypothesis	Correlation coefficient, r	Result
H1: corporate policies positively affect project success	0.651	Supported
H2: resource management positively affects project success	0.705	Supported
H3: stakeholder engagement mediates the relationship between corporate policies and project success	0.684	Supported
H4: organisational learning moderates the relationship between sustainability practices and project success	0.722	Supported

project efficiency and reduces operational costs. Stakeholder engagement, with a correlation of $r = 0.684$, further underscores its mediating role, as effective communication and inclusive decision-making help ensure that sustainability goals align with project execution. This is consistent with Khalifeh *et al.* (2020), who noted that stakeholder involvement reduces resistance, manages expectations and strengthens project accountability. Finally, organisational learning shows a strong correlation with project success ($r = 0.722$), supporting its moderating role; organisations that actively implement lessons learned and adapt based on past experiences tend to achieve higher rates of sustainability-aligned project success. These results reinforce the findings of Ahsan (2025), which emphasised that learning-oriented cultures enhance innovation, adaptability and overall project performance.

10.4. Practical insights

The findings of this study provide several important practical insights for project managers, organisational leaders and policy makers seeking to integrate sustainability into project management without compromising traditional performance objectives related to cost, time and scope. The results clearly indicate that corporate policies and resource management are foundational drivers of project success within sustainable project environments, highlighting the need for structured governance and disciplined operational practices.

From a managerial perspective, organisations should embed sustainability explicitly within corporate project governance frameworks, ensuring that sustainability objectives are not treated as optional or secondary considerations. Clear sustainability policies, leadership commitment and compliance mechanisms help align project decisions with environmental and social goals while maintaining accountability for cost and schedule performance. Project managers should be supported by top management through clear mandates, performance indicators and reporting structures that integrate sustainability into routine project reviews.

Efficient resource management emerged as the strongest contributor to project success, underscoring the importance of optimising energy use, materials and human resources. In practice, this suggests that organisations should invest in sustainable resource planning practices, such as waste minimisation, energy-efficient technologies and lifecycle thinking. While such initiatives may involve higher upfront costs, the findings suggest that improved efficiency, reduced rework and long-term operational savings can offset these investments and enhance overall project outcomes.

The mediating role of stakeholder engagement provides a critical insight for practice. Sustainability initiatives are more likely to succeed when stakeholders, such as clients, contractors, regulators and communities, are actively involved in decision-making. Transparent communication, early consultation and collaborative problem-solving help manage conflicting expectations and reduce resistance to sustainability measures that may otherwise be perceived as disruptive to cost or schedule targets. Project managers should therefore adopt structured stakeholder engagement strategies throughout the project lifecycle rather than limiting engagement to compliance-driven interactions.

The moderating effect of organisational learning further underscores the importance of building adaptive, knowledge-driven project organisations. Organisations that systematically capture lessons learned, encourage innovation and apply insights from previous projects are better positioned to integrate sustainability without undermining performance. Practically, this implies the need for continuous training, knowledge-sharing platforms and post-project reviews that explicitly address sustainability outcomes alongside traditional performance metrics.

Overall, the practical insights from this study suggest that sustainable project management is achievable when organisations adopt a balanced, integrated approach that strengthens governance through corporate policies, optimises resources, engages stakeholders effectively and fosters organisational learning. By doing so, project-based organisations can move beyond the traditional triple constraints mindset and deliver projects that achieve both operational excellence and long-term environmental and social value.

10.5. Conclusion and recommendations

This study examined the key determinants of project success within sustainable project management, focusing on the Malaysian engineering industry. The research explored how corporate policies and resource management influence project success, both directly and through the mediating role of stakeholder engagement and the moderating effect of organisational learning. Using a quantitative approach, structured questionnaires were distributed to 200 engineering professionals across various project-based roles. The descriptive analysis showed that most respondents were experienced practitioners from diverse organisational backgrounds. Reliability testing indicated strong internal consistency across all constructs, confirming the survey instrument's robustness. Descriptive results also revealed that while sustainability practices are moderately implemented across organisations, greater emphasis is placed on corporate policies and resource management compared to stakeholder engagement.

Correlation results demonstrated significant positive relationships between all variables, with corporate policies and resource management showing strong direct links to project success. Stakeholder engagement was positively associated with both independent variables and project outcomes, supporting its mediating function. Organisational learning further strengthened these relationships, confirming its moderating role. Overall, the findings provide empirical support for the integrated sustainable project management model. They highlight the need for structured sustainability strategies within project governance and emphasise the crucial roles of stakeholder engagement and organisational learning in enhancing project performance. These insights align with existing scholarship while offering valuable guidance for practitioners and future research on sustainability-driven project management.

10.5.1 Theoretical contributions

The findings of this study offer important theoretical and practical contributions to sustainable project management. Theoretically, the research advances an integrated framework that links traditional project constraints of cost, time and scope with TBL sustainability principles. By incorporating stakeholder engagement as a mediator and organisational learning as a moderator, the study extends existing models. It provides a more holistic understanding of the factors that drive project success in sustainability-focused environments. The results reinforce the idea that sustainability has become a strategic priority in project management rather than a peripheral concern. Strong relationships between corporate policies, resource management and project success highlight the essential role of governance structures and operational efficiency in implementing sustainable projects. The mediating influence of stakeholder engagement further underscores the need for inclusive and transparent communication. At the same time, the moderating effect of organisational learning shows that continuous improvement and knowledge sharing amplify these positive outcomes.

10.5.2 Limitations

Although this study offers valuable insights, several limitations should be acknowledged. First, the research was confined to the Malaysian engineering sector, which may limit the generalisability of the findings to other industries or regions. Sectors such as healthcare, manufacturing and IT may have different sustainability priorities, so the results should be applied cautiously outside the study's context. Second, the use of a cross-sectional design restricts the ability to draw causal conclusions. While strong associations were identified between the variables, the study cannot establish causality, and longitudinal research would provide a clearer understanding of how these relationships develop over time.

Third, reliance on self-reported questionnaires introduces potential biases, including social desirability bias and varying interpretations of survey items. Although validated scales and pilot testing were employed to enhance accuracy, self-report limitations remain. Lastly, the study examined four key constructs – corporate policies, resource management, stakeholder engagement and organisational learning – but did not account for other potentially influential factors, such as technological innovation, regulatory pressures, leadership commitment or organisational culture. Future research incorporating these elements would provide a more comprehensive understanding of sustainability in project management.

10.5.3 Recommendations and future research

Building on this study's findings and limitations, several recommendations are proposed for future research. First, the scope should be expanded beyond Malaysia's engineering sector to include construction, manufacturing, IT and public infrastructure. Comparative studies across sectors and countries would improve generalisability and reveal how cultural, economic or regulatory differences shape sustainable project management practices. Second, future research should incorporate longitudinal designs to track changes over time. This would enable stronger causal inferences and provide deeper insight into how corporate policies, stakeholder engagement and organisational learning influence long-term project success.

Third, additional mediating and moderating variables such as leadership style, organisational culture and regulatory pressures should be examined to build more comprehensive models of sustainability integration. Finally, adopting mixed-methods approaches would enrich understanding by complementing quantitative findings with qualitative insights from interviews or case studies. Such triangulation would help capture contextual factors and enhance the practical relevance of future research.

REFERENCES

Ahsan MJ (2025) Cultivating a culture of learning: the role of leadership in fostering lifelong development. *The Learning Organization* **32(2)**: 282–306.

Al-Emran M (2023) Beyond technology acceptance: development and evaluation of technology-environmental, economic, and social sustainability theory. *Technology in Society* **75**: 102383.

Ali F and Haapasalo H (2023) Development levels of stakeholder relationships in collaborative projects: challenges and preconditions. *International Journal of Managing Projects in Business* **16(8)**: 58–76.

Alsudiri T, Al-Karaghouli W and Eldabi T (2013) Alignment of large project management process to business strategy: a review and conceptual framework. *Journal of Enterprise Information Management* **26(5)**: 596–615.

Escoto X, Gebrehewot D and Morris KC (2022) Refocusing the barriers to sustainability for small and medium-sized manufacturers. *Journal of Cleaner Production* **338**: 130589.

Khalifeh A, Farrell P and Al-Edenat M (2020) The impact of project sustainability management (PSM) on project success: a systematic literature review. *Journal of Management Development* **39(4)**: 453–474.

Mahajan R, Lim WM, Sareen M, Kumar S and Panwar R (2023) Stakeholder theory. *Journal of Business Research* **166**: 114104.

Marcelino-Sádaba S, Pérez-Ezcurdia A, Lazcano AME and Villanueva P (2014) Project risk management methodology for small firms. *International Journal of Project Management* **32(2)**: 327–340.

Orcan F (2020) Parametric or non-parametric: skewness to test normality for mean comparison. *International Journal of Assessment Tools in Education* **7(2)**: 255–265.

Reiser D (2023) Brundtland report. In *Encyclopedia of Sustainable Management.* Springer, Cham, Switzerland, pp. 378–381.

Rumayor M, Margallo M, Pinedo J and Albo J (2024) Learning by doing using the life cycle assessment tool: LCA projects in collaboration with industries. *Education for Chemical Engineers* **48**: 44–52.

Sabini L, Muzio D and Alderman N (2019) 25 years of 'sustainable projects': what we know and what the literature says. *International Journal of Project Management* **37(6)**: 820–838.

Sakalasooriya N (2021) Conceptual analysis of sustainability and sustainable development. *Open Journal of Social Sciences* **9(3)**: 396–414.

Santos PDO and de Carvalho MM (2022) Exploring the challenges and benefits for scaling agile project management to large projects: a review. *Requirements Engineering* **27(1)**: 117–134.

Srivastava A, Krishan A and Srivastava D (2026) Environmental regulations and compliance. In *Advances in Green Engineering for Sustainable Industrial Developments* (Gupta A, Tripathy DB and Ghosal A (eds.)). Elsevier, Amsterdam, the Netherlands, pp. 41–60.

Stanitsas M, Kirytopoulos K and Leopoulos V (2021) Integrating sustainability indicators into project management: the case of construction industry. *Journal of Cleaner Production* **279**: 123774.

Turner R, Ledwith A and Kelly J (2010) Project management in small to medium-sized enterprises: matching processes to the nature of the firm. *International Journal of Project Management* **28(8)**: 744–755.

Advancing Project Management: Frameworks, Digital Transformation, Sustainability and Governance

Mohammed Reyasudin Basir Khan
ISBN 978-1-83742-799-4
https://doi.org/10.1108/978-1-83742-798-720261014

Chapter 11
Project cost performance: economic, organisational and project-level influences

Zaharullail Abdul Rashid[1], Mohammed Reyasudin Basir Khan[1], Chong Peng Lean[2] and Ahmad Anwar Zainuddin[3]

[1] *Tun Razak Graduate School, Universiti Tun Abdul Razak (UNIRAZAK), Kuala Lumpur, Malaysia*
[2] *MILA University, Nilai, Negeri Sembilan, Malaysia*
[3] *Kulliyyah of Information and Communication Technology, IIUM, Gombak, Malaysia*

11.1. Introduction

Effective project cost performance remains one of the most enduring challenges in project management practice and research. Across both developed and developing economies, projects frequently exceed approved budgets, leading to financial losses, delivery delays, contractual disputes and diminished stakeholder confidence. Empirical evidence indicates that cost overruns are not isolated incidents but systemic phenomena embedded in project environments and governance structures (Flyvbjerg *et al.*, 2002). In Malaysia, cost overruns have been particularly prominent in construction, infrastructure and public-sector development projects, where escalating costs strain public finances and reduce private investment returns.

The Malaysian project environment is characterised by economic volatility, regulatory complexity and multi-stakeholder arrangements. Projects operate in fluctuating macroeconomic conditions, including changes in inflation, interest rates and exchange rates, which directly influence the prices of materials, labour and financing. At the same time, projects are subject to institutional and regulatory requirements related to approvals, compliance and procurement, which may introduce delays and additional costs. These external conditions interact with organisational capabilities and project-specific characteristics, creating complex cost dynamics that cannot be adequately explained through technical estimation accuracy alone.

Although cost estimation and control techniques are widely used in Malaysian projects, cost overruns continue to occur frequently (Hendrickson and Au, 2024). Traditional approaches to cost control often emphasise detailed planning, monitoring and reporting mechanisms.

While these practices are necessary, they are insufficient when broader contextual factors are not adequately accounted for. Studies have shown that even projects with detailed cost plans may experience significant overruns due to unforeseen economic changes, weak organisational coordination or project complexity (Shehu *et al.*, 2014a). This suggests that project cost performance is a multidimensional outcome shaped by interacting influences at multiple levels.

Existing research on project cost performance has contributed valuable insights into specific cost drivers. Prior studies have examined issues such as contractor cash flow problems, design changes, material price escalation and procurement strategies. While these studies enhance understanding of mechanisms, they often adopt narrow analytical lenses. Many focus on single categories of factors, such as technical or managerial variables, without sufficiently accounting for the broader environment in which projects are embedded. As a result, the literature remains fragmented and findings are often difficult to generalise across different project contexts.

In the Malaysian context, this fragmentation is particularly problematic. The country's development trajectory has involved large-scale infrastructure investments, public–private partnerships and rapid urbanisation, all of which amplify cost risks. Projects frequently involve multiple public agencies, private contractors, consultants and financiers, each with distinct objectives and incentives. The interaction of these actors within a regulated and economically dynamic environment creates conditions in which factors beyond project-level decisions influence cost performance. Understanding these interactions requires an integrative analytical approach.

This chapter addresses this need by examining project cost performance through a multilevel perspective that integrates macroenvironmental factors, organisational context and project-specific characteristics. Rather than treating cost overruns as isolated technical failures, the chapter conceptualises cost performance as an outcome shaped by economic conditions, institutional arrangements, organisational processes and project execution practices. By adopting this perspective, the chapter seeks to provide a more comprehensive explanation of cost behaviour in Malaysian projects.

The focus on Malaysia provides valuable context for examining cost performance in a developing economy with a mixed public–private project landscape. Malaysia's reliance on imported construction materials, exposure to global financial markets and structured regulatory environment make it particularly sensitive to macroeconomic and institutional influences. At the same time, variations in organisational maturity, technological adoption and project management practices across firms create heterogeneity in cost outcomes. Examining these factors together allows for deeper insight into why some projects perform better than others under similar external conditions.

This chapter contributes to the project management literature in several ways. First, it integrates multiple categories of determinants within a single analytical framework, addressing limitations of prior fragmented studies. Second, it provides empirical evidence from Malaysian projects, enriching the understanding of cost performance in a context that has received comparatively less attention in international literature. Third, by examining both significant and weak relationships between determinants, the chapter offers a nuanced understanding of which factors matter most for cost performance and under what conditions.

From a practical perspective, the findings of this chapter have implications for project managers, organisational leaders and policy makers. Recognising that cost performance is influenced by external economic conditions underscores the importance of proactive monitoring and adaptive strategies. Understanding the role of organisational coordination and decision-making highlights areas where managerial interventions can improve cost outcomes. Identifying project-specific characteristics associated with cost risk supports better planning and execution decisions.

Overall, this chapter advances understanding of project cost performance by moving beyond narrow technical explanations and emphasising the importance of multilevel determinants. The subsequent sections build on this foundation by reviewing relevant literature, outlining theoretical perspectives, developing hypotheses and presenting empirical findings that collectively explain cost performance in Malaysian projects.

11.1.1 Literature review

11.1.1.1 Project cost performance and cost overruns

Project cost performance is widely regarded as a core dimension of project success and a primary indicator of managerial effectiveness. It reflects the extent to which a project is completed within its approved budget while achieving its intended scope and quality objectives (Kerzner, 2017). Poor cost performance often manifests as cost overruns, budget deviations or inefficient use of financial resources, all of which can undermine stakeholder confidence and organisational sustainability.

Empirical studies consistently demonstrate that cost overruns are pervasive across industries and geographic contexts. Flyvbjerg *et al.* (2002) provide seminal evidence showing that large public projects frequently exceed their initial budgets, suggesting that cost overruns are systematic rather than exceptional. Their findings challenge the assumption that improved forecasting accuracy alone can resolve cost performance problems, instead pointing to deeper behavioural, organisational and institutional causes.

In Malaysia, cost overruns have been repeatedly reported in construction and infrastructure projects, particularly in public-sector projects. Chan (2001) identifies time–cost trade-offs and institutional constraints as major contributors to cost escalation in Malaysian public projects. Similarly, Shehu *et al.* (2014a) highlight contractor cash flow issues, delayed payments and weak financial management as significant causes of cost overruns. These studies underscore that a combination of economic conditions, organisational practices and project execution challenges shapes cost performance.

Despite these insights, much of the existing literature remains fragmented. Many studies focus on isolated cost drivers, such as design changes or procurement strategies, without sufficiently examining how multiple factors interact to influence cost outcomes. This limitation has prompted calls for more integrative approaches that recognise the multidimensional nature of project cost performance.

11.1.1.2 Macroenvironmental factors affecting project cost performance

Macroenvironmental factors form the external context within which projects are planned and executed. These factors are largely beyond project teams' direct control but significantly influence cost outcomes. In the Malaysian context, key macroenvironmental factors include inflation, exchange rate fluctuations, interest rates, economic growth and regulatory requirements.

INFLATION

Inflation is widely recognised as a major driver of project cost escalation. Rising inflation increases the prices of construction materials, labour wages, transportation and equipment, directly affecting project budgets. Chan (2001) emphasises that inflationary pressures can distort initial cost estimates and reduce the reliability of long-term budgeting, particularly for projects with extended durations. In periods of high inflation, even well-managed projects may experience cost overruns if escalation provisions are inadequate.

Empirical findings from Malaysian projects suggest that inflation is among the most consistently cited external cost drivers (Omran *et al.*, 2023; Shehu *et al.*, 2014b). Contractors and clients alike face challenges in managing price volatility, especially when contracts lack appropriate adjustment mechanisms. The strong association between inflation and project cost performance reported in prior studies supports the hypothesis that inflation significantly affects cost outcomes (Musarat *et al.*, 2021; Unegbu *et al.*, 2022).

EXCHANGE RATE FLUCTUATIONS

Exchange rate volatility is particularly relevant in Malaysia due to the country's reliance on imported construction materials, machinery and specialised equipment. Depreciation of the local currency increases the cost of imports, placing upward pressure on project budgets. Exchange rate fluctuations can also disrupt procurement schedules and contractual arrangements, especially in projects involving foreign suppliers or contractors.

While exchange rate risks are often acknowledged in international projects, their impact on domestic projects in developing economies has received less attention. The Malaysian experience suggests that even domestically focused projects are exposed to currency risks through global supply chains. This exposure underscores the importance of accounting for exchange rate volatility as a determinant of project cost performance.

INTEREST RATES

Interest rates influence project cost performance by affecting financing costs. Higher interest rates increase borrowing costs for both public- and private-sector projects, placing additional strain on project budgets. Projects that rely heavily on debt financing are particularly sensitive to interest rate movements, as increased financing costs can reduce financial flexibility and increase overall expenditure.

Studies examining project financing in Malaysia indicate that interest rate fluctuations can significantly alter project viability, particularly for long-term infrastructure projects. Rising interest rates may lead to cost-cutting measures, scope adjustments or delays, all of which can affect cost performance. These findings support the inclusion of interest rates as a key macroenvironmental determinant.

ECONOMIC GROWTH

Economic growth influences project cost performance in complex ways. On one hand, strong economic growth stimulates investment and increases demand for construction and infrastructure development. On the other hand, heightened demand can lead to resource shortages, increased competition for skilled labour and rising material prices, all of which contribute to cost escalation.

In Malaysia, periods of rapid economic growth have often coincided with increased construction activity and cost pressures. While growth creates development opportunities, it also intensifies competition and reduces cost predictability. The literature therefore suggests a positive association between economic growth and project costs, particularly in capacity-constrained environments (Galli, 2018).

REGULATORY REQUIREMENTS

Regulatory frameworks and institutional processes play a critical role in shaping project cost performance. Approval procedures, compliance requirements, procurement regulations and contractual standards influence project timelines and administrative costs. Chan (2001) highlights that regulatory delays and institutional inefficiencies can significantly increase project costs in the Malaysian public sector.

Alashwal and Chew (2017) further note that regulatory complexity can hinder the adoption of innovative management practices and digital tools, indirectly affecting cost performance. In highly regulated environments, project teams may face limited flexibility in responding to cost pressures, reinforcing the importance of institutional context in cost performance analysis.

11.1.1.3 Organisational context and cost performance

Organisational context refers to the internal capabilities, processes and cultures that shape how projects are managed. Unlike macroenvironmental factors, organisational factors are partially controllable and represent key levers for improving cost performance. Prior studies identify several organisational attributes that influence cost outcomes, including cost management expertise, communication and coordination, risk management culture, technology adoption and decision-making processes (Ameer *et al.*, 2022; Fernandes *et al.*, 2015).

COST MANAGEMENT EXPERTISE

Cost management expertise encompasses the skills, experience and professional competence of personnel responsible for budgeting, monitoring and controlling project costs. Conventional wisdom suggests that higher expertise should lead to better cost performance. However, empirical evidence is mixed. While some studies find positive associations between expertise and cost control, others report weak or inconsistent effects (Ahmadi Eftekhari *et al.*, 2022; Kaufmann and Kock, 2022).

This inconsistency may reflect the fact that expertise alone is insufficient without supportive organisational systems and authority structures. Organisational policies, limited data access or fragmented decision-making processes may constrain skilled professionals. The mixed findings in the literature justify further empirical examination of this factor.

COMMUNICATION AND COORDINATION

Effective communication and coordination between project stakeholders are widely regarded as critical for managing cost risks. Poor communication can lead to misunderstandings, rework, duplication of effort and delayed decision-making, all of which contribute to cost overruns. Fobiri *et al.* (2022) emphasise that stakeholder engagement and information sharing are essential for aligning expectations and resolving conflicts.

In Malaysian projects, communication challenges are often exacerbated by complex stakeholder networks involving public agencies, private contractors, consultants and financiers. Effective coordination mechanisms are therefore essential for maintaining cost control across organisational boundaries.

RISK MANAGEMENT CULTURE

Risk management culture reflects the extent to which organisations proactively identify, assess and respond to uncertainties. Projects with strong risk management practices are better equipped to anticipate cost-related risks and implement mitigation measures. However, empirical findings suggest that risk management practices are unevenly implemented across organisations (Bhatti and Nazir, 2024; Crispim *et al.*, 2019).

In some Malaysian organisations, risk management may be treated as a formal requirement rather than an integrated managerial practice. This may explain why the relationship between risk management culture and cost performance is sometimes weaker than expected.

TECHNOLOGY ADOPTION

The adoption of digital tools and technologies has been promoted to improve cost accuracy, transparency of information and coordination. Alashwal and Chew (2017) examine the use of simulation and digital tools in Malaysian construction projects and find mixed evidence regarding their impact on cost performance. These findings suggest that technology adoption alone does not guarantee improved outcomes and that organisational readiness and integration are critical.

DECISION-MAKING PROCESSES

Efficient decision-making processes enable timely responses to cost deviations and emerging risks. Organisations with clear authority structures and streamlined approval mechanisms are better positioned to implement corrective actions. Conversely, bureaucratic delays and fragmented governance can exacerbate cost overruns. Prior studies highlight decision-making efficiency as an important but often overlooked determinant of cost performance (Secundo *et al.*, 2022; Tsiga and Emes, 2022).

11.1.1.4 Project-specific factors influencing cost performance

Project-specific factors represent the immediate operational conditions under which projects are executed. These factors directly influence resource utilisation, uncertainty and cost behaviour. Key project-specific determinants include complexity, scope clarity, planning and scheduling effectiveness, material quality and resource allocation efficiency.

PROJECT COMPLEXITY

Project complexity refers to the degree of technical difficulty, interdependence and uncertainty associated with a project. Complex projects require greater coordination and are more susceptible to unforeseen events. Empirical evidence consistently shows that higher complexity is associated with increased cost risk and poorer cost performance (Bakhshi *et al.*, 2025; Kaufmann and Kock, 2022).

SCOPE CLARITY AND STABILITY

A clear and stable project scope is essential for effective cost control. Poorly defined scopes or frequent scope changes disrupt planning and lead to cost escalation. In Malaysian projects, scope changes are often driven by evolving client requirements or regulatory changes, underscoring the importance of scope management.

PLANNING AND SCHEDULING EFFECTIVENESS

Effective planning and scheduling support cost control by enabling accurate forecasting, efficient activity sequencing and proactive monitoring. Kerzner (2017) emphasises that planning quality is a foundational element of project success. Weak planning increases uncertainty and limits the ability to manage cost deviations.

MATERIAL QUALITY AND RESOURCE ALLOCATION

Material quality affects both initial costs and long-term value. While higher-quality materials may increase upfront expenditure, they can reduce maintenance and rework costs. Efficient allocation of labour, equipment and materials further supports productivity and budget adherence.

11.1.1.5 Summary of literature and research gaps

The literature demonstrates that project cost performance is influenced by a wide range of factors operating at macroenvironmental, organisational and project-specific levels. However, existing studies often examine these factors in isolation, limiting their explanatory power. There is a clear need for integrative analyses that capture the combined effects of multiple determinants within a coherent framework, particularly in the Malaysian context.

11.1.2 Theoretical foundation

Understanding project cost performance requires a robust theoretical grounding that accounts for the complex interactions between economic conditions, organisational behaviour and project execution practices. No single theory is sufficient to explain the persistent occurrence of cost overruns across diverse project environments. Consequently, this section draws on multiple complementary theoretical perspectives to explain how and why project cost performance is shaped by macroenvironmental, organisational and project-specific determinants in the Malaysian context.

Agency theory provides a foundational lens for examining cost performance in projects characterised by multiple stakeholders with differing objectives. Jensen and Meckling (1976) argue that agency problems arise when principals and agents have misaligned incentives and unequal access to information. In project settings, clients, contractors, consultants and subcontractors often pursue competing goals related to cost, quality and time. These misalignments can result in opportunistic behaviour, cost padding, inadequate transparency and weak cost control. In Malaysia, where many projects involve public sector clients and complex contractual arrangements, agency-related issues are particularly salient and may contribute to cost inefficiencies.

Resource dependence theory further explains how external constraints influence organisational behaviour and project outcomes. Organisations depend on critical external resources, such as materials, labour and financing, and this dependence shapes strategic and operational decisions (Öztürk and Bağış, 2025). In the Malaysian project environment, dependence on imported materials, specialised equipment and external financing exposes projects to inflationary pressures, exchange rate volatility and interest rate fluctuations. These dependencies limit managerial control over costs and reinforce the importance of macroenvironmental conditions in shaping cost performance.

Institutional theory provides insight into how regulatory frameworks, professional norms and cultural expectations influence project practices. Organisations often conform to institutional pressures to gain legitimacy, even when such conformity may reduce efficiency (Qiu and Chen, 2023). In Malaysia, regulatory requirements related to procurement, approvals and compliance exert strong influences on project processes. Institutional constraints may delay decision-making, restrict flexibility and increase administrative costs, thereby affecting project cost performance. Institutional theory thus explains why projects operating under similar technical conditions may exhibit different cost outcomes depending on regulatory and normative environments.

Contingency theory complements these perspectives by emphasising that there is no universally optimal approach to managing projects or controlling costs. Organisational effectiveness depends on the alignment between management practices and contextual conditions (Gallagher and Coleman Gallagher, 2012). Applied to project cost performance, this suggests that cost control strategies must be tailored to specific macroenvironmental conditions, organisational capabilities and project characteristics. The Malaysian project environment, characterised by diversity in project size, sector and governance arrangements, underscores the relevance of a contingency-based approach.

Systems thinking offers a holistic perspective, viewing project cost performance as an emergent outcome of interactions between multiple subsystems. Complex problems cannot be understood by examining individual components in isolation (Naveed and Khan, 2022). Cost overruns often result from feedback loops involving economic changes, organisational responses and project execution decisions. Systems thinking, therefore, supports an integrative analytical framework that captures interdependencies across levels.

Organisational learning theory further enriches the analysis by highlighting the role of experience, knowledge acquisition and continuous improvement. Senge (1990) argues that organisations that systematically learn from past experiences are better equipped to adapt to changing environments. In project-based organisations, the ability to capture lessons learned from previous projects and apply them to future endeavours can significantly influence cost performance. Weak learning mechanisms may lead to repeated cost overruns despite prior experience.

Collectively, these theoretical perspectives justify a multilevel approach to examining project cost performance. Macroenvironmental factors shape the external constraints and opportunities faced by projects, organisational context determines managerial responses to these conditions, and project-specific characteristics influence the execution of cost control practices. The integration of these theories provides a comprehensive foundation for the hypotheses developed in this chapter.

11.1.3 Hypothesis development

Drawing on the literature and theoretical foundations discussed above, this section develops a set of hypotheses linking macroenvironmental factors, organisational context and project-specific characteristics to project cost performance. These hypotheses operationalise the conceptual framework and provide a basis for empirical examination.

11.1.3.1 Macroenvironmental factors

As explained earlier, macroeconomic and institutional conditions influence project cost performance by shaping input prices, financing costs and administrative processes. Inflation increases the cost of materials, labour and services, thereby exerting upward pressure on project budgets. Exchange rate volatility affects projects that rely on imported inputs, while interest rate changes affect financing costs. Economic growth, although generally positive, may intensify competition for resources and drive cost escalation. Regulatory requirements and policy frameworks further affect timelines and administrative costs.

Based on these considerations, the following hypotheses were proposed.

- H1a: exchange rate fluctuations significantly influence project cost performance.
- H1b: inflation rates significantly increase construction material prices and overall project costs.
- H1c: government regulations and policy requirements significantly affect project timelines and costs.
- H1d: economic growth is positively associated with increased project costs.
- H1e: interest rate changes significantly affect project financing costs and budget performance.

11.1.3.2 Organisational context

Organisational attributes shape how projects respond to external pressures and manage internal processes. Cost management expertise reflects personnel's ability to plan and control costs, while communication and coordination influence information flow and stakeholder alignment. Risk management culture affects the anticipation and mitigation of uncertainties, technology adoption supports data accuracy and integration, and decision-making processes determine the speed and effectiveness of corrective actions.

Theoretical perspectives suggest that stronger organisational capabilities should enhance cost performance, although empirical findings indicate that some factors may exert weaker or conditional effects (Rungi, 2014; Yazici, 2020). Accordingly, the following hypotheses were proposed.

- H2a: higher levels of cost management expertise are associated with better project cost performance.
- H2b: a strong organisational risk management culture improves control over unforeseen project costs.
- H2c: effective communication and coordination between project stakeholders positively influence cost control.

- H2d: adoption of advanced technologies and digital tools positively influences project cost performance.
- H2e: clear and efficient decision-making processes are positively associated with budget adherence.

11.1.3.3 Project-specific factors

Project-level characteristics directly influence uncertainty, resource utilisation and execution complexity. Complex projects require greater coordination and are more susceptible to unforeseen events, increasing cost risk. Clear and stable project scope supports effective planning and cost control, while strong planning and scheduling practices enable proactive monitoring. Material quality influences both initial costs and long-term value, and efficient resource allocation supports productivity and budget adherence.

Based on these considerations, the following hypotheses were proposed.

- H3a: higher project complexity is associated with increased project costs.
- H3b: clear and stable project scope positively influences budget adherence.
- H3c: effective project planning and scheduling significantly improve project cost performance.
- H3d: the quality of materials used influences project costs and long-term value.
- H3e: efficient allocation of resources positively influences project cost performance.

The conceptual framework integrating macroenvironmental, organisational and project-specific determinants of project cost performance is illustrated in Figure 11.1.

11.2. Methodology

11.2.1 Research design

This study adopted a quantitative, cross-sectional research design to examine the relationships between the proposed determinants and project cost performance. Quantitative methods are appropriate for testing hypotheses and identifying statistically significant

Figure 11.1 Conceptual framework

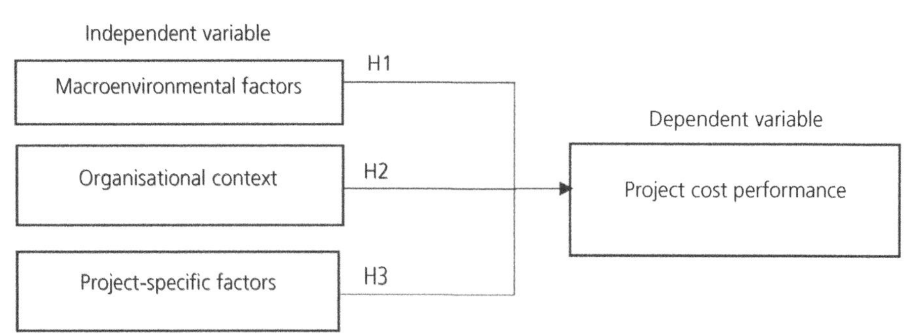

relationships between variables across a broad sample of practitioners. The cross-sectional approach provides a snapshot of perceptions and experiences regarding cost performance in Malaysian projects.

11.2.2 Population and sample

The target population comprises professionals involved in project management activities in Malaysia, particularly within the construction and related sectors. This includes project managers, quantity surveyors, contractors, consultants and other stakeholders responsible for planning, executing and controlling project costs. Given the diversity of project types and organisational contexts, a stratified sampling approach was adopted to enhance representativeness.

A total of 85 valid responses were obtained. While the sample size is modest, it is consistent with prior exploratory studies in project management research and sufficient for correlation and variance-based analyses. Respondents represented a range of organisational types, project roles and experience levels, providing a heterogeneous sample suitable for examining determinants of cost performance.

11.2.3 Data collection

Data were collected using an electronic, structured questionnaire. The questionnaire consisted of multiple sections designed to capture respondents' perceptions of macroenvironmental conditions, organisational practices, project-specific characteristics and project cost performance. Likert-scale items were used to assess the extent of agreement with each construct's statements. Demographic information was also collected to contextualise the responses.

Using an electronic survey facilitated access to geographically dispersed respondents and ensured consistent data collection. Participation was voluntary and respondents were assured of confidentiality to encourage honest and accurate responses.

11.2.4 Measurement and reliability

Measurement items were adapted from established project management literature and tailored to the Malaysian context. Constructs related to macroenvironmental factors included items on inflation, exchange rates, interest rates, economic growth and regulatory requirements. Organisational context constructs captured cost management expertise, communication and coordination, risk management culture, technology adoption and decision-making processes. Project-specific constructs measured complexity, scope clarity, planning and scheduling effectiveness, material quality, and resource allocation efficiency.

Reliability was assessed using Cronbach's alpha to measure internal consistency. All major constructs achieved acceptable reliability levels, with alpha values exceeding commonly accepted thresholds. This indicates that the measurement items consistently captured their intended constructs and were suitable for subsequent analysis.

11.2.5 Data analysis techniques

Data analysis was conducted using the statistical package for the social sciences (SPSS). Descriptive statistics were used to summarise respondent characteristics and variable distributions. Pearson correlation analysis was employed to examine relationships between independent variables and project cost performance. Analysis of variance (ANOVA) was used to assess the explanatory power of different groups of determinants and to examine whether the integrated framework explained greater variance in cost performance than partial models.

11.3. Results and analysis

This section presents and interprets the study's empirical findings based on quantitative data collected from 85 project professionals in Malaysia. The results are structured according to the conceptual framework, which conceptualises project cost performance as influenced by macroenvironmental factors, organisational context and project-specific characteristics. Statistical evidence is drawn from descriptive statistics, reliability analysis, Pearson correlation and analysis of variance, with explicit reference to the relevant tables and figures.

11.3.1 Respondent profile

The demographic profile of respondents is summarised in Table 11.1 and illustrated in Figure 11.2. The sample is predominantly male, reflecting the gender composition of the Malaysian construction and project management sectors. Respondents represent a broad range of professional roles, including project managers, quantity surveyors, contractors, consultants and senior management personnel. This diversity ensures that the findings reflect multiple perspectives on project cost performance rather than the views of a single stakeholder group.

Most respondents report moderate to extensive professional experience, indicating substantial exposure to project cost planning, monitoring and control activities. Organisational representation spans small, medium and large firms, as well as public- and private-sector organisations. This variation enhances the relevance of the findings across different project environments and strengthens the applicability of the conceptual framework.

Table 11.1 Demographic

Item	Mean	Median	SD	Skewness	Kurtosis	*n* (valid)
Gender	1.21	1.0	0.411	1.436	0.064	85
Geographic location	1.29	1.0	0.458	0.92	−1.182	85
Age	3.66	4.0	1.129	−0.508	−0.6	85
Education	3.31	3.0	1.012	−0.44	0.481	85
Position	3.13	3.0	1.926	0.14	−1.572	85
Organisation type	2.16	2.0	0.986	0.346	−0.938	85
Organisation size	2.38	3.0	0.845	−0.814	−1.104	85
Years of experience	3.46	4.0	1.532	−0.388	−1.4	85
Industry sector	2.68	1.0	2.161	0.893	−0.672	85

Figure 11.2 Industry sector

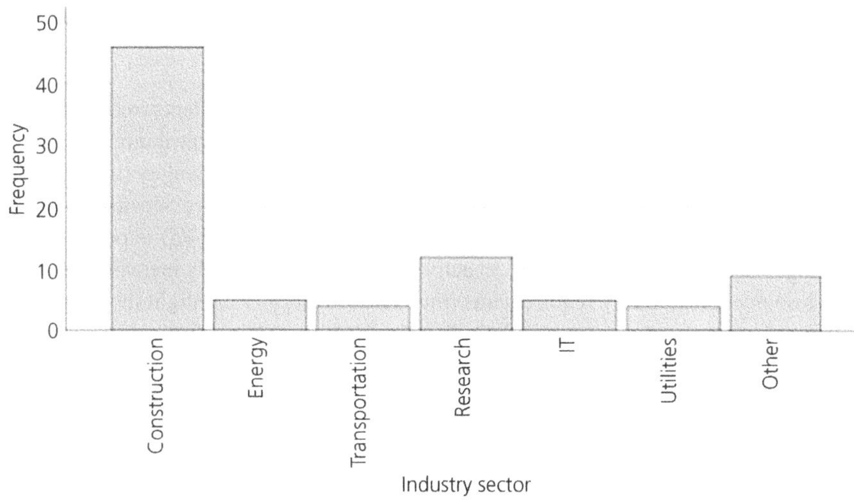

11.3.2 Reliability of measurement scales

Reliability analysis was conducted to assess the internal consistency of the questionnaire's constructs. As reported in Table 11.2, the Cronbach's alpha (α) values for macroenvironmental factors ($\alpha = 0.824$), organisational context ($\alpha = 0.783$) and project-specific factors ($\alpha = 0.735$) exceed the commonly accepted threshold of 0.70. These results indicate satisfactory internal consistency and confirm that the items within each construct reliably measure their intended dimensions.

The strong reliability of the macroenvironmental construct suggests that respondents consistently perceive external economic and regulatory conditions as coherent influences on project cost performance. Similarly, acceptable reliability values for organisational and project-specific factors indicate that internal management practices and project characteristics are perceived as distinct yet internally consistent dimensions within the conceptual framework.

Table 11.2 Summary of reliability statistics

Item	Cronbach's alpha, α
Macroenvironmental factors	0.824
Organisational context	0.783
Project-specific factors	0.735

11.3.3 Descriptive analysis of macroenvironmental factors

Descriptive statistics for macroenvironmental factors are presented in Table 11.3, with visualisations in Figures 11.3 and 11.4. Overall, respondents strongly agree that macroenvironmental conditions substantially influence project cost performance.

Inflation records the highest mean score (4.27), indicating that respondents overwhelmingly perceive rising prices for materials, labour and services as a primary contributor to cost escalation. Exchange rate fluctuations (mean = 4.20) and interest rate changes (mean = 4.16) also receive high mean scores, indicating that Malaysian projects are sensitive to global financial conditions and financing costs. Economic growth (mean = 3.95) is perceived as a double-edged factor that stimulates project activity while simultaneously increasing competition for resources. Regulatory requirements (mean = 4.06) further highlight the institutional dimension of cost performance, particularly for projects subject to public-sector oversight.

The negatively skewed distributions across these variables indicate strong clustering of responses at the upper end of the scale, suggesting broad consensus among respondents on the importance of macroenvironmental influences. These findings provide initial descriptive support for hypotheses H1a–H1e and reinforce the macroenvironmental component of the conceptual framework.

11.3.4 Analysis of variance and framework validation

Analysis of variance (ANOVA) was conducted to assess the explanatory power of the integrated conceptual framework. As reported in Table 11.4, the full model incorporating macroenvironmental, organisational and project-specific variables is statistically significant ($F = 2.958$, $p < 0.001$). The sequential inclusion of variable groups demonstrates that

Table 11.3 Statistics: macroenvironmental

Item	Mean	Median	SD	Skewness	Kurtosis	*n* (valid)
Fluctuations in exchange rates significantly affect our project costs	4.2	4.0	0.687	−1.184	4.399	85
Inflation rates directly affect the prices of construction materials in our projects	4.27	4.0	0.762	−1.33	3.313	85
Government regulations and policies significantly affect project timelines and costs	4.06	4.0	0.807	−1.083	2.1	85
Economic growth leads to increased investment in construction projects, which in turn affects project costs	3.95	4.0	0.785	−0.671	4.462	85
Interest rate changes affect the cost of financing projects, thereby influencing the overall budget	4.16	4.0	0.614	−0.423	1.028	85

Figure 11.3 Fluctuation in exchange rates

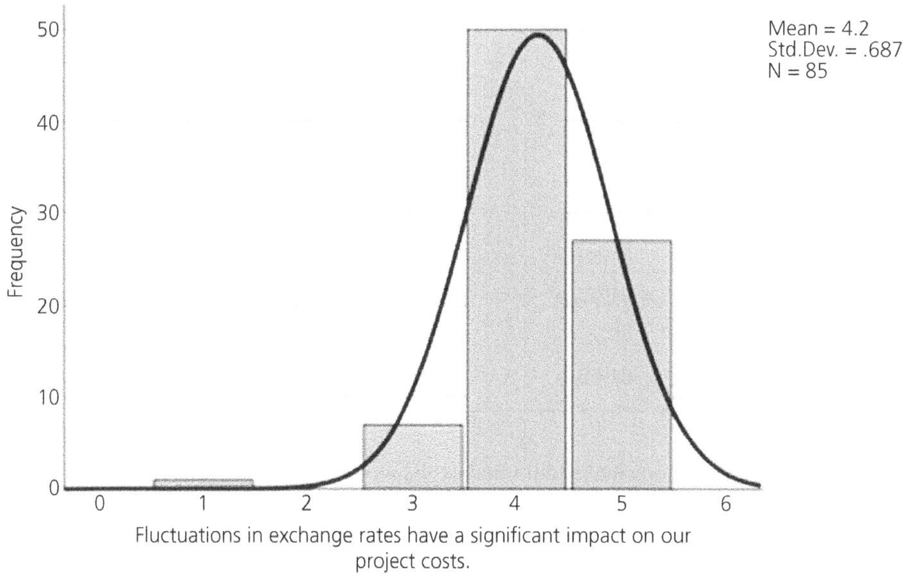

Figure 11.4 Government regulations and policies

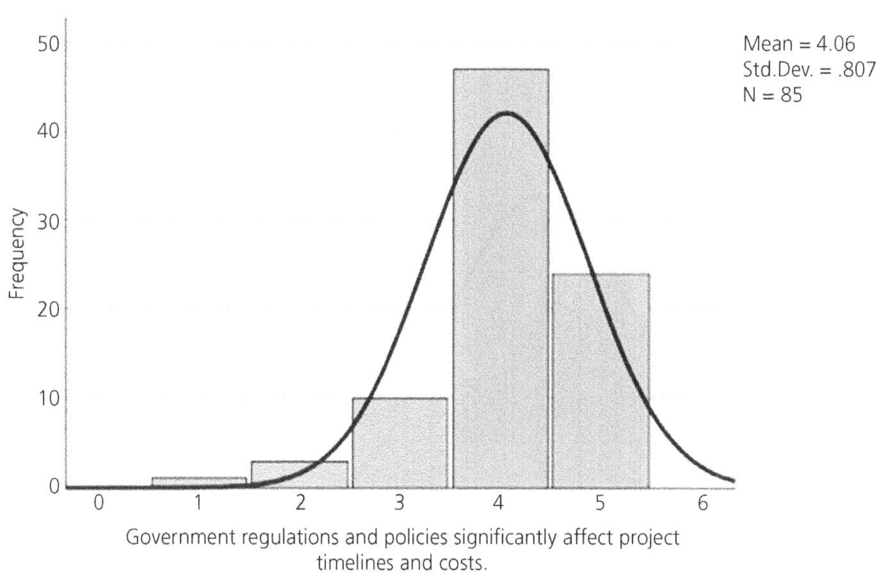

Table 11.4 ANOVA

Model	Source	Sum of squares	*df*	Mean square	*F*	Sig.
1	Regression	18.875	5	3.775	4.747	< 0.001
	Residual	62.819	79	0.795		
	Total	81.694	84			
2	Regression	21.340	10	2.134	2.616	0.009
	Residual	60.355	74	0.816		
	Total	81.694	84			
3	Regression	31.970	15	2.131	2.958	<0.001
	Residual	49.724	69	0.721		
	Total	81.694	84			

project-specific factors contribute additional explanatory power beyond macroenvironmental and organisational variables.

Figures 11.5 and 11.6 illustrate the distribution of the dependent variable and the normal P-P plot of the residuals, which is used to assess model assumptions. These findings confirm that the multilevel conceptual framework offers a more comprehensive explanation of project cost performance than models that focus on isolated determinants.

Figure 11.5 Dependent variable

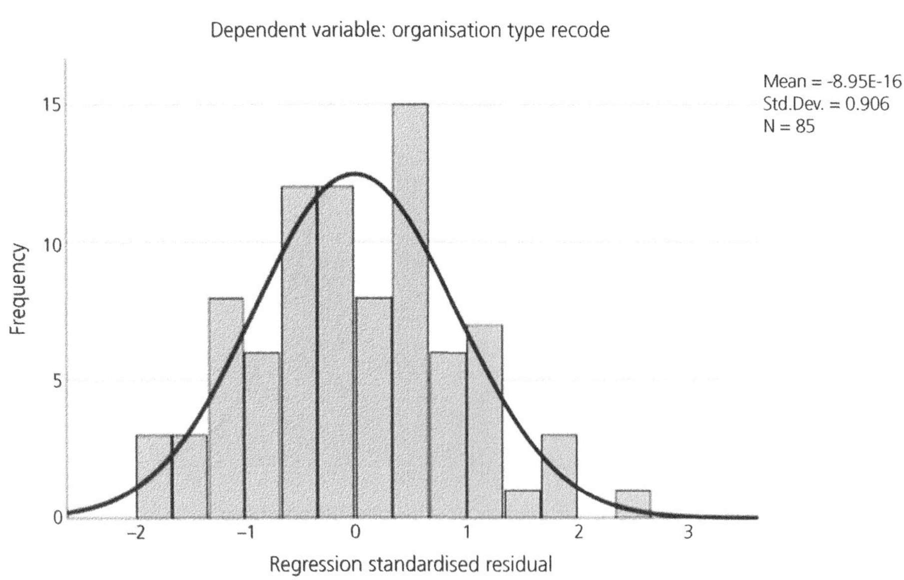

Figure 11.6 Normal P-P plot of regression standardised residual

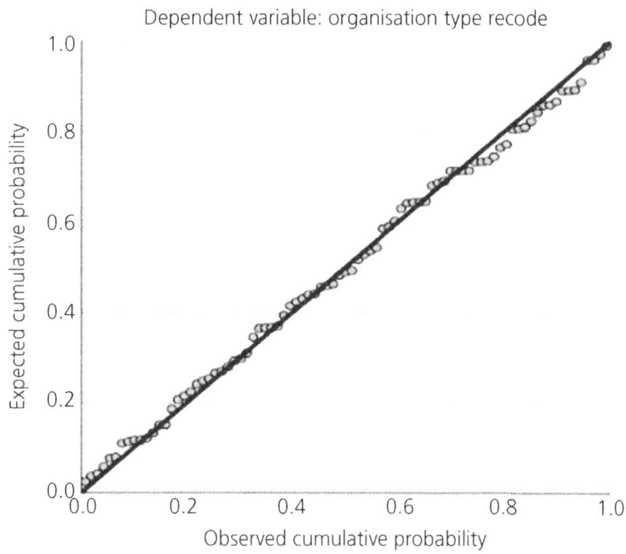

11.3.5 Correlation analysis and hypothesis testing

Pearson correlation analysis was employed to test the hypothesised relationships between the independent variables and project cost performance. The results of hypothesis testing are summarised in Table 11.5 and discussed below in relation to each dimension of the conceptual framework.

Table 11.5 Hypothesis testing (continued on next page)

Hypothesis	Pearson correlation	Sig. (2-tailed)	Result
H1a: exchange rate fluctuations significantly influence project cost performance	0.242	0.026	Supported
H1b: inflation rates significantly increase construction material prices and overall project costs	0.409	<0.001	Supported
H1c: government regulations and policy requirements significantly affect project timelines and costs	0.252	0.020	Supported
H1d: economic growth is positively associated with increased project costs	0.374	<0.001	Supported
H1e: interest rate changes significantly affect project financing costs and budget performance	0.360	<0.001	Supported

Table 11.5 Continued

Hypothesis	Pearson correlation	Sig. (2-tailed)	Result
H2a: higher levels of cost management expertise are associated with better project cost performance	0.186	0.087	Weak positive correlation
H2b: a strong organisational risk management culture improves control over unforeseen project costs	0.171	0.118	Weak positive correlation
H2c: effective communication and coordination between project stakeholders positively influence project cost control	0.244	0.024	Supported
H2d: adoption of advanced technologies and digital tools positively influences project cost performance	0.242	0.026	Supported
H2e: clear and efficient decision-making processes are positively associated with budget adherence	0.230	0.035	Supported
H3a: higher project complexity is associated with increased project costs	0.379	<0.001	Supported
H3b: clear and stable project scope positively influences budget adherence	0.232	0.033	Supported
H3c: effective project planning and scheduling significantly improve project cost performance	0.262	0.016	Supported
H3d: the quality of materials used influences project costs and long-term value	0.340	0.001	Supported
H3e: efficient allocation of resources positively influences project cost performance	0.193	0.077	Weak positive correlation

11.3.5.1 Macroenvironmental factors (H1a–H1e)

All macroenvironmental variables demonstrate statistically significant positive relationships with project cost performance. Inflation exhibits the strongest correlation ($r = 0.409$, $p < 0.001$), providing strong support for H1b and confirming that inflationary pressures significantly undermine cost performance. Economic growth also shows a moderate positive relationship ($r = 0.374$, $p < 0.001$), supporting H1d and indicating that increased demand during growth periods contributes to cost escalation through resource competition.

Interest rate changes are significantly correlated with cost performance ($r = 0.360$, $p < 0.001$), supporting H1e and highlighting the importance of financing conditions. Regulatory requirements ($r = 0.252$, $p = 0.020$) and exchange rate fluctuations ($r = 0.242$, $p = 0.026$) also demonstrate significant relationships, supporting H1c and H1a, respectively.

Collectively, these results confirm that macroenvironmental factors form a dominant layer of influence within the conceptual framework and validate the theoretical emphasis placed on external economic and institutional conditions.

11.3.5.2 Organisational context (H2a–H2e)

The results for organisational factors reveal a more differentiated pattern. Effective communication and coordination are statistically significantly related to project cost performance ($r = 0.244$, $p = 0.024$), providing empirical support for H2c. This finding underscores the importance of information flow and stakeholder alignment in mitigating cost overruns, consistent with agency theory and systems thinking perspectives.

Technology adoption is also significantly associated with cost performance ($r = 0.242$, $p = 0.026$), supporting H2d. This suggests that digital tools improve cost monitoring and coordination, although the moderate strength of the relationship implies that technology's effectiveness depends on complementary organisational processes. Decision-making efficiency shows a weaker but statistically significant relationship ($r = 0.230$, $p = 0.035$), supporting H2e and indicating that timely decisions help maintain budget control.

In contrast, cost management expertise ($r = 0.186$, $p = 0.087$) and risk management culture ($r = 0.171$, $p = 0.118$) do not achieve statistical significance, providing limited support for H2a and H2b. These findings suggest that formal expertise and risk management practices may exert indirect or contingent effects rather than direct influences on cost performance, particularly when constrained by external or institutional factors.

11.3.5.3 Project-specific factors (H3a–H3e)

Project-specific factors demonstrate consistently strong relationships with project cost performance. Project complexity exhibits a moderate positive correlation ($r = 0.379$, $p < 0.001$), supporting H3a and confirming that complex projects are more susceptible to cost escalation. Planning and scheduling effectiveness is significantly associated with cost performance ($r = 0.262$, $p = 0.016$), supporting H3c and reinforcing the role of execution discipline in controlling costs.

Scope clarity ($r = 0.232$, $p = 0.033$) and material quality ($r = 0.340$, $p < 0.001$) also show significant relationships, supporting H3b and H3d, respectively. These results highlight the importance of clearly defined project requirements and quality-oriented decisions in managing cost outcomes. Resource allocation efficiency shows a positive but non-significant relationship ($r = 0.193$, $p = 0.077$), providing limited support for H3e.

11.4. Practical insights

The findings of this chapter reinforce the view that project cost performance is shaped by interacting determinants operating at multiple levels. Macroenvironmental factors emerge as particularly influential, underscoring the vulnerability of Malaysian projects to external economic and institutional conditions. The strong effects of inflation, interest rates and economic growth confirm that cost performance cannot be effectively managed without close attention to the broader economic environment.

The significant impact of regulatory requirements highlights the role of institutional context in shaping cost outcomes. Regulatory delays and compliance costs may limit managerial flexibility and increase administrative burdens, contributing to cost escalation. This finding is consistent with institutional theory, which emphasises the influence of regulatory frameworks on organisational behaviour (Hirsch and Lounsbury, 1997).

Organisational factors exhibit mixed effects. The strong influence of communication and coordination supports agency theory, which emphasises the importance of information sharing and stakeholder alignment (Jensen and Meckling, 1976). Technology adoption also demonstrates a positive relationship with cost performance, although the modest strength of this relationship suggests that technology must be accompanied by appropriate organisational processes to be effective.

The weak effects of cost management expertise and risk management culture warrant careful interpretation. These findings do not imply that expertise and risk management are unimportant; rather, their influence may be indirect or contingent on other factors. For example, skilled personnel may be constrained by organisational structures, regulatory requirements or limited decision-making authority. Similarly, risk management practices may exist formally but lack effective integration into daily project operations.

Project-specific factors play a critical role in shaping cost outcomes. The strong association between project complexity and cost performance supports prior research indicating that complex projects are more vulnerable to uncertainty and cost escalation (Safapour *et al.*, 2018). Effective planning and scheduling are essential for mitigating these risks, consistent with the perspectives of contingency theory and systems thinking.

The findings of this chapter offer several practical implications for project managers and organisational leaders in Malaysia. First, the strong influence of macroenvironmental factors underscores the importance of proactive environmental scanning and adaptive planning. Project teams should regularly monitor economic indicators such as inflation and interest rates and incorporate appropriate contingencies into budgets and schedules.

Second, organisations should prioritise strengthening communication and coordination mechanisms across stakeholder groups. Clear information flows and collaborative decision-making can reduce misunderstandings, minimise rework and support timely cost-control interventions. Investments in digital tools should be accompanied by training and process integration to maximise their effectiveness.

Third, project managers should pay particular attention to project complexity during the planning phase. Complex projects require enhanced coordination, robust planning and realistic budgeting to mitigate cost risks. Ensuring scope clarity and stability can further support cost control by reducing uncertainty and change-related disruptions.

11.5. Conclusion and recommendations

This chapter provides a comprehensive examination of the determinants of project cost performance in Malaysia. By integrating macroenvironmental, organisational and project-specific factors, it offers a holistic explanation of cost behaviour that moves beyond narrow technical perspectives. The findings demonstrate that cost performance is shaped by interacting multilevel determinants and that effective cost control requires alignment across economic conditions, organisational capabilities and project execution practices.

The chapter contributes to the literature by addressing fragmentation in prior research and providing empirical evidence from the Malaysian context. It also offers practical, policy-relevant insights to improve cost performance in projects operating under complex, dynamic conditions.

Despite its contributions, this chapter has limitations. The cross-sectional design captures perceptions at a single point in time and does not account for changes in economic conditions or organisational practices over time. The sample size, while adequate for exploratory analysis, limits the generalisability of the findings. Future research should employ longitudinal designs, larger samples and advanced multivariate techniques to further validate and extend the findings. Comparative studies across countries or sectors could also enhance understanding of contextual influences on project cost performance.

REFERENCES

Ahmadi Eftekhari N, Mani S, Bakhshi J and Mani S (2022) Project manager competencies for dealing with socio-technical complexity: a grounded theory construction. *Systems* **10(5)**: 161.

Alashwal AM and Chew MY (2017) Simulation techniques for cost management and performance in construction projects in Malaysia. *Built Environment Project and Asset Management* **7(5)**: 534–545.

Ameer A, Naz F, Gul Taj B and Ameer I (2022) The impact of manager's personality traits on project success through affective professional commitment: the moderating role of organizational project management maturity system. *Journal of Facilities Management* **20(2)**: 284–305.

Bakhshi J, Golzad H, Martek I, Hosseini MR and Papadonikolaki E (2025) Unveiling the complexity code: navigating BIM-enabled projects with a project management complexity index. *Engineering, Construction and Architectural Management* **32(11)**: 7703–7732.

Bhatti MAA and Nazir MU (2024) The impact of project process management on sustainable project success in the construction sector: the moderating role of risk management practices. *Bulletin of Business and Economics* **13(2)**: 1065–1072.

Chan APC (2001) Time–cost relationship of public sector projects in Malaysia. *International Journal of Project Management* **19(4)**: 223–229.

Crispim J, Silva LH and Rego N (2019) Project risk management practices: the organizational maturity influence. *International Journal of Managing Projects in Business* **12(1)**: 187–210.

Fernandes G, Ward S and Araújo M (2015) Improving and embedding project management practice in organisations – a qualitative study. *International Journal of Project Management* **33(5)**: 1052–1067.

Flyvbjerg B, Holm MS and Buhl S (2002) Underestimating costs in public works projects: error or lie? *Journal of the American Planning Association* **68(3)**: 279–295.

Fobiri G, Musonda I and Muleya F (2022) Reality capture in construction project management: a review of opportunities and challenges. *Buildings* **12(9)**: 1381.
Gallagher KP and Coleman Gallagher V (2012) Organizing for post-implementation ERP: a contingency theory perspective. *Journal of Enterprise Information Management* **25(2)**: 170–185.
Galli BJ (2018) The future of economic decision making in project management. *IEEE Transactions on Engineering Management* **67(2)**: 396–413.
Hendrickson C and Au T (2024) Cost estimation. In *Project Management for Construction (and Deconstruction): Fundamental Concepts for Owners, Engineers, Architects and Builders*. Prentice-Hall, Upper Saddle River, NJ, USA.
Hirsch PM and Lounsbury M (1997) Putting the organization back into organization theory: action, change, and the 'new' institutionalism. *Journal of Management Inquiry* **6(1)**: 79–88.
Jensen MC and Meckling WH (1976) Theory of the firm: managerial behavior, agency costs and ownership structure. *Journal of Financial Economics* **3(4)**: 305–360.
Kaufmann C and Kock A (2022) Does project management matter? The relationship between project management effort, complexity, and profitability. *International Journal of Project Management* **40(6)**: 624–633.
Kerzner H (2017) *Project Management: A Systems Approach to Planning, Scheduling, and Controlling*. Wiley, Hoboken, NJ, USA.
Musarat MA, Alaloul WS and Liew MS (2021) Impact of inflation rate on construction projects budget: a review. *Ain Shams Engineering Journal* **12(1)**: 407–414.
Naveed F and Khan KIA (2022) Investigating the influence of information complexity on construction quality: a systems thinking approach. *Engineering, Construction and Architectural Management* **29(3)**: 1427–1448.
Omran A, Saleh MSH and Gebril AO (2023) Factors causing time and cost overruns of construction projects in Malaysia. *Design, Construction, Maintenance* **3**: 179–186.
Öztürk O and Bağış M (2025) Bridging resource dependence theory and resource-based view: a theoretical synthesis. *Management Decision*, ahead of print.
Qiu Y and Chen H (2023) A systematic review of the knowledge domain of institutional theory in construction project management. *Engineering, Construction and Architectural Management* **30(6)**: 2523–2544.
Rungi M (2014) The impact of capabilities on performance. *Industrial Management & Data Systems* **114(2)**: 241–257.
Safapour E, Kermanshachi S, Habibi M and Shane J (2018) Resource-based exploratory analysis of project complexity impact on phase-based cost performance behavior. In *Construction Research Congress 2018*, pp. 439–448.
Secundo G, Elia G, Margherita A and Leitner KH (2022) Strategic decision making in project management: a knowledge visualization framework. *Management Decision* **60(4)**: 1159–1181.
Senge PM (1990) *The Fifth Discipline: The Art and Practice of the Learning Organization*. Doubleday, New York, NY, USA.
Shehu Z, Endut IR, Akintoye A and Holt GD (2014a) Cost overrun in the Malaysian construction industry projects: a deeper insight. *International Journal of Project Management* **32(8)**: 1471–1480.

Shehu Z, Endut IR and Akintoye A (2014b) Factors contributing to project time and hence cost overrun in the Malaysian construction industry. *Journal of Financial Management of Property and Construction* **19(1)**: 55–75.

Tsiga Z and Emes M (2022) Decision making in engineering projects. *Procedia Computer Science* **196**: 927–937.

Unegbu HCO, Yawas DS and Dan-Asabe B (2022) An investigation of the relationship between project performance measures and project management practices of construction projects for the construction industry in Nigeria. *Journal of King Saud University – Engineering Sciences* **34(4)**: 240–249.

Yazici HJ (2020) An exploratory analysis of the project management and corporate sustainability capabilities for organizational success. *International Journal of Managing Projects in Business* **13(4)**: 793–817.

Part IV

Risk and governance

Mohammed Reyasudin Basir Khan
ISBN 978-1-83742-799-4
https://doi.org/10.1108/978-1-83742-798-720261015

Chapter 12
The role of project management applications in delivering railway infrastructure success

Mohd Suhail Kamari and Mohammed Reyasudin Basir Khan

Tun Razak Graduate School, Universiti Tun Abdul Razak (UNIRAZAK), Kuala Lumpur, Malaysia

12.1. Introduction

In the modern era, technology plays an increasingly critical role across various industries, including banking, medicine, education, manufacturing and construction. Within the construction sector, particularly in large-scale and specialised infrastructure projects such as railway developments in Malaysia, advanced technological tools are essential for enhancing management efficiency, streamlining workflows and improving overall productivity and project performance. These technologies encompass both hardware and software solutions, including smartphones, tablets, laptops, servers, communication systems and project management platforms. In their absence, organisations frequently encounter fragmented communication, poor coordination and project delays.

One of Malaysia's major double-track railway rehabilitation initiatives (hereafter referred to as Project A) clearly illustrates these challenges. Awarded in 2015, Project A involved extensive track rehabilitation works, station upgrades, electrification systems, signalling and communication installations, radio communication infrastructure and enhancements to centralised train control systems. Despite an anticipated 48-month completion timeline, the project experienced persistent delays and multiple extensions of time (EOTs). As of March 2025, cumulative physical progress exceeded 99%, yet completion remained outstanding. The delays, amounting to several years, reflect systemic weaknesses in project deployment, particularly the limited or inconsistent utilisation of project management tools (PMTs) among stakeholders and project teams.

The insufficient adoption of PMTs contributes to ineffective communication, poorly coordinated work programmes, deviation from specifications and concerns regarding material and installation quality. Empirical studies reinforce this observation. Research by Quan *et al.* (2022) in the Malaysian construction sector found that poor communication significantly affects project outcomes, leading to delays, cost escalations, safety incidents and reduced project success. Similarly, studies conducted in East Malaysia show that ineffective communication weakens team cohesion, reduces morale and leads to operational inefficiencies (Ding *et al.*, 2023; Yong and Mustaffa, 2013). Despite the availability of advanced digital communication platforms, organisational hierarchy, unclear objectives and cultural barriers continue to impede effective information exchange. These findings underscore the need for robust communication mechanisms and integrated digital systems to enhance information flow and support informed managerial decision-making.

The research presented in this chapter examines how PMTs, when used systematically, can enhance project deployment across three major railway initiatives, referred to in this study as Project A, Project B and Project C. The challenges observed in Project A have similarly emerged in Project B, which has encountered governance concerns, safety issues and contractual disputes. Project C, a long-distance electrified double-tracking initiative, has also experienced coordination challenges and limited real-time monitoring capabilities. Collectively, these cases reflect broader industry concerns, including fragmented communication, inadequate planning structures and insufficient integration of digital project management solutions.

Although prior research emphasises the importance of communication and planning in construction project success (Akintelu *et al.*, 2023; Gamil and Abd Rahman, 2023), few studies have examined how different categories of PMTs, namely, planning and scheduling tools, communication and collaboration platforms, and resource allocation systems, directly influence project outcomes in Malaysia's railway infrastructure context. This study addresses this gap by focusing on three primary dimensions.

First, planning and scheduling tools such as Primavera and Microsoft Project are essential for establishing realistic timelines, monitoring physical progress and forecasting risks (Bachwani *et al.*, 2023). Their absence or improper use often leads to schedule conflicts, resource misallocation and delayed milestones. Second, communication and collaboration tools, including building information modelling (BIM) platforms, cloud-based dashboards and digital reporting systems, support timely decision-making, real-time updates and effective stakeholder coordination. In railway project environments, delays in decision cycles and information dissemination have repeatedly disrupted progress. Third, resource allocation systems, including enterprise resource planning (ERP) platforms and performance dashboards, enable strategic deployment of materials, equipment and personnel. These systems promote accountability and facilitate multistakeholder coordination, which is critical in complex infrastructure projects.

The main objective of this research was to assess the influence of PMTs on project deployment success within Projects A, B and C. Specifically, the study examines how planning and scheduling systems contribute to timeline accuracy and milestone tracking, how communication tools enhance coordination and decision-making, and how resource allocation platforms improve team efficiency and overall project delivery. In addition to evaluating PMT usage, the research also considers user proficiency, recognising that even advanced systems are ineffective without adequate training and alignment with project objectives.

To guide the investigation, three research questions were formulated.

- To what extent do planning and scheduling tools influence project success?
- What is the role of communication and collaboration tools in improving project outcomes?
- How do stakeholder and resource allocation tools impact project success?

The findings were expected to demonstrate that systematic and comprehensive use of PMT enables more predictable, efficient and timely project execution, ultimately enhancing quality and stakeholder satisfaction.

The significance of this study lies in its potential to strengthen project management practices within Malaysia's railway infrastructure sector and the broader construction industry. Effective adoption of PMTs enables stakeholders to manage planning, execution, monitoring, verification, testing, commissioning and reporting with greater precision. Real-time data access enables faster decision-making, earlier issue detection and timely corrective action. A structured digital approach also enhances team competence, fosters trust among contractors and clients, and reduces disputes and miscommunication. Ultimately, improved coordination and communication contribute to more efficient project delivery, reduced costs and higher-quality outcomes.

This chapter further examines the theoretical and empirical foundations underlying the relationship between PMT and project success. PMTs are closely related to project management information systems (PMIS), which support activities from project initiation to final handover. A project is commonly defined as a temporary endeavour with a defined scope, limited resources and a specific timeline involving multidisciplinary teams. In railway projects, this includes specialised departments such as signalling, track engineering, electrification and project management.

Project management is broadly defined as the integration of knowledge, skills, tools and techniques to plan, direct, coordinate, monitor and control project activities while meeting time, cost, quality and stakeholder requirements. PMIS supports this integration by providing essential information on cost, time and performance parameters, along with their interrelationships. Massimo *et al.* (2025) describe PMIS as a structured system that provides managers with timely, relevant data for effective control. In this study, PMIS is treated as conceptually aligned with PMTs, as both involve systematic management of project information and processes.

Project management information systems are designed to support decision-making across the project lifecycle by collecting, integrating and disseminating project data for planning, execution, monitoring and control (Monteiro *et al.*, 2025; PMI, 2017). Studies show that PMIS can provide managers with essential information on project costs, schedules and performance parameters, thereby enabling more informed decisions and improving overall project control (Raymond and Bergeron, 2014). Empirical research indicates that higher-quality PMIS use is associated with enhanced planning efficiency, improved monitoring practices and timelier decision-making, collectively contributing to better project outcomes (Raymond and Bergeron, 2014).

Figure 12.1 Conceptual framework

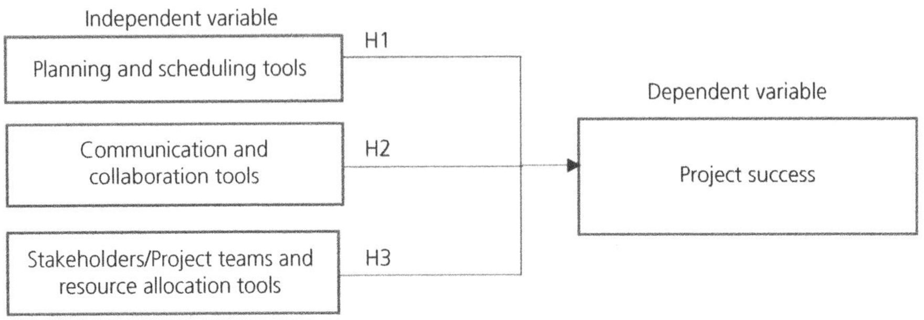

The information systems (IS) success model offers a theoretical foundation for understanding PMIS effectiveness. This model posits that system quality, information quality, user satisfaction and system use influence organisational outcomes and net benefits (DeLone and McLean, 2003). When applied to PMIS, the model suggests that well-designed information systems that deliver high-quality data and satisfy user needs are more likely to yield positive project performance outcomes (DeLone and McLean, 2003). In project contexts, user satisfaction and functional information quality have been shown to correlate with improved planning accuracy and control effectiveness.

Recent systematic studies highlight that PMIS adoption and use vary widely across industries but are increasingly linked to improved coordination, real-time progress visibility and reduced workflow bottlenecks (Onaghinor *et al.*, 2022). These studies support the conceptual linkage between PMIS utilisation and performance outcomes, forming the basis of the framework illustrated in Figure 12.1, which conceptualises PMT usage quality and completeness as primary determinants of perceived usefulness, user satisfaction and overall project success.

In this framework, more comprehensive and effective PMT utilisation enhances planning accuracy, communication flow, coordination and the functions of monitoring and reporting. Questionnaires were designed to measure usage levels, perceived quality and satisfaction, enabling statistical examination of their relationship with project performance.

Based on this framework, three hypotheses were proposed.

- H1: planning and scheduling tools significantly influence project success.
- H2: communication and collaboration tools positively contribute to project success.
- H3: stakeholder and resource allocation tools significantly impact project success.

These hypotheses were tested using quantitative data collected from railway project stakeholders to determine the extent to which PMT influences the success of infrastructure projects.

12.2. Methodology

12.2.1 Research design

The research approach for this study is grounded in established theories and supported by previous studies. This study adopted a quantitative approach to measure the influence of PMT on the success of railway infrastructure projects, using numerical data collected through questionnaires. Quantitative data, expressed as numbers, enables statistical analysis of respondents' perceptions and provides an objective basis for evaluating the relationship between tool usage and project success. This research is descriptive in nature, aiming to present a clear and comprehensive picture of the current issues affecting railway infrastructure projects, particularly delays stemming from communication and management challenges.

In terms of research strategy, a survey was the most suitable method, as it directly addresses the research questions and aligns with prior PMIS studies. The questionnaire was developed based on the conceptual framework to ensure that each item accurately reflects the constructs being measured. Ensuring the questionnaire is both valid and reliable is essential to ensure the data collected accurately reflects the real conditions experienced by stakeholders and project teams.

12.2.2 Population and sampling procedures

Identifying the population is essential for determining an appropriate research strategy. In this study, the population and sample size were clearly defined, as the research focuses on the deployment of railway infrastructure projects among stakeholders and project teams. The national rail operator (owned by the Ministry of Finance and operating under the Ministry of Transport) has more than a century of operational history. The organisation manages railway services across Peninsular Malaysia and Singapore under the Malaysian Railway Act 1991, with an extensive network of approximately 2800 km of track and more than 188 passenger stations.

Malaysia's rail sector has undergone substantial expansion through major double-track and rehabilitation initiatives, including several large-scale corridor upgrades and electrification programmes. National development strategies, such as NIMP 2030, have further accelerated infrastructure growth, increasing demand for skilled project management teams. According to the Department of Statistics Malaysia, the rail industry employed over 14 000 workers in 2022, with continued growth expected as new projects are initiated (Kalhoro *et al.*, 2025).

The national rail operator (NRO) itself employs approximately 6600 staff nationwide, including about 450 personnel in its project management department. These teams are responsible for managing complex, high-value projects, including Project A and Project B, both are critical infrastructure that has been in use for nearly 30 years and is now in need of upgrading. For this study, the population was narrowed to the 450 employees working in the NRO's key project-related departments: signalling and communication, electrification, track engineering and project management. The sample size was calculated based on Morgan's sample size table (Rahman, 2023).

A population of 450 yields a sample size of 207. However, because only 120 of these employees are directly involved in deploying Projects A, B and C, the more accurate reference point is the

120-person active project team. Based on Morgan's table, the corresponding sample size for a population of 120 is 90. Despite this minimum requirement, the study received responses from 114 employees across different departments, roles and technical backgrounds. This exceeds the recommended sample size, thereby strengthening the reliability of the findings.

12.2.3 Data collection method

The data were collected using a set of 26 questionnaire items created through Google Forms and distributed to all selected respondents through email and WhatsApp. The questionnaire consisted of five main sections: demographic information, planning and scheduling tools, communication and collaboration tools, stakeholders/project teams and resource allocation tools, and project success factors. These sections included multiple-choice questions and Likert-scale items designed to capture respondents' perceptions and experiences regarding the use of PMT and project outcomes. Details of the survey are shown in Table 12.1.

12.2.4 Operationalisation and measurement

Cronbach's alpha (α) was used to assess internal consistency across all 26 questionnaire items. Cronbach's alpha ranges from 0 to 1, with higher values indicating greater reliability. This method, widely used in the statistical package for the social sciences (SPSS), is efficient for determining whether the questionnaire items consistently measure their intended constructs.

The study variables were grounded in well-established theories. PMTs are associated with technology adoption and usability models, such as the technology acceptance model (TAM) and the unified theory of acceptance and use of technology (UTAUT), which focus on how accessible, user-friendly and beneficial the tools are for project teams. Project deployment is linked to the DeLone and McLean (2003) information systems success model, which emphasises user engagement, system usage and task performance. Project success aligns with established project management frameworks, such as the project management body of knowledge (PMBOK) and Agile methodologies, and with general IT project success models. It is assessed based on performance, quality and stakeholder satisfaction.

Three independent variables form the core of the research framework. Planning and scheduling tools measure how effectively project activities are structured, monitored and forecasted

Table 12.1 Questionnaire sections

Questionnaire section	No. of questions
Section 1: demographic information	5
Section 2: planning and scheduling tools	5
Section 3: communication and collaboration tools	5
Section 4: stakeholders/project teams and resource allocation tools	5
Section 5: project success	6
Total	26

to prevent delays and support teamwork and productivity. Communication and collaboration tools examine how effectively technology facilitates information sharing, reduces misunderstandings and supports timely decision-making among stakeholders. Stakeholder and resource allocation tools evaluate how well project teams manage resources, track responsibilities and monitor individual and team performance, ensuring informed, timely project decisions.

Project success is the dependent variable and is defined by time, quality, cost and stakeholder satisfaction. In the context of railway projects, success is reflected in the absence of delays, compliance with technical specifications, adherence to installation standards and maintaining costs within approved budgets. Stakeholder satisfaction is also considered a key measure of overall success. The research, therefore, examines how effectively using PMT can prevent delays, ensure high-quality outcomes and support cost-efficient project execution. By providing accurate, real-time data on project status and performance, these tools are expected to enhance satisfaction among NRO stakeholders and project teams and contribute to more successful project delivery.

12.2.5 Data analysis techniques

This study used descriptive analysis to summarise and present the characteristics of the data collected through the questionnaire. Measures such as mean, median, mode, frequency, minimum and maximum are used to describe respondents' perceptions of PMT use. The analysis was conducted using SPSS, which allows efficient data management, statistical testing and visualisation. All 26 questionnaire items were evaluated to ensure validity and reliability. Descriptive analysis provides an overview of the dataset. It produces logical summaries of responses, while SPSS facilitates data cleaning and the calculation of descriptive statistics, correlations, regression analyses, *t*-tests and graphical outputs for deeper interpretation. Inferential analysis is used to infer conclusions about a population from sample data. Regression analysis was used to determine whether PMT significantly influences project success.

12.3. Results and analysis

12.3.1 Response rate

This study involved 114 respondents who completed all 26 questionnaire items, designed to measure the independent variables in relation to the dependent variable.

12.3.2 Demographic profile

The respondents include stakeholders and project team members, including railway project staff involved in fieldwork, project consultants representing both contractors and the client, and personnel from the transport ministry and government agencies. The results show that 66% of respondents have less than 6 years of experience, 8% have between 7 and 9 years, and the remaining 26% have more than 10 years. This distribution may influence the overall analysis of PMT usage and its impact on project success. Figure 12.2 shows the distribution of years of experience in project management.

12.3.3 Descriptive analysis

Descriptive analysis was conducted to identify the overall pattern of the data and to provide meaningful insights for subsequent analysis. To summarise the dataset's characteristics,

Figure 12.2 Years of experience in project management

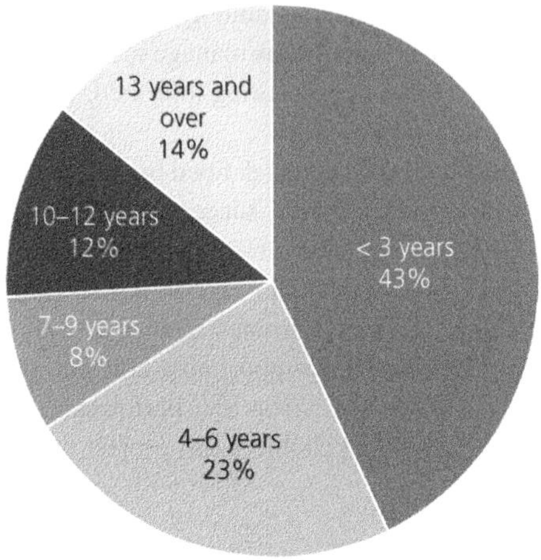

statistical measures such as the mean, median, mode, standard deviation (SD), variance, minimum and maximum values, and frequencies and percentages are presented in Table 12.2.

For the analysis and interpretation of the data, a mean above 3.5 indicates agreement or a positive perception, indicating that respondents believe PMT can enhance productivity,

Table 12.2 Mean, standard deviation, minimum and maximum

Question	Mean	SD	Min	Max
The project was well-planned using scheduling tools	3.67	0.85	1	5
Planning tools helped avoid delays	3.69	0.87	1	5
Planning tools tracked tasks and milestones	3.69	0.85	1	5
Tools improved team productivity	3.76	0.82	1	5
Communication tools enhanced teamwork	3.85	0.79	1	5
Stakeholders/project teams received timely updates	3.44	0.82	1	5
Projects were completed on time (A, B, and C)	2.71	1.02	1	5
Projects met their budget	3.03	0.84	1	5
Deliverables met quality standards	3.12	0.84	1	5
Stakeholders/project teams were satisfied with project outcomes	3.13	0.71	1	5
Projects achieved strategic and original objectives	3.24	0.75	1	5

teamwork and planning. A mean value around 3.0 suggests a moderate or neutral perception, particularly regarding the influence of these tools on project timeliness, quality and budget compliance. Standard deviation values higher than 0.8 indicate greater variability in responses, suggesting differing viewpoints among respondents.

Project success measures, such as budget compliance and stakeholder satisfaction, fall within a neutral-to-slightly-positive range. However, the mean score for timely project completion is below 3.0, indicating disagreement and reflecting the current situation where PMTs are not actively used. Instead, stakeholders and project teams rely on conventional methods such as physical documents, face-to-face meetings, messaging and phone calls, which create communication barriers and hinder efficient project management.

Cronbach's alpha (α) was used to assess the reliability of constructs such as planning and scheduling tools, communication and collaboration tools, and stakeholder/resource allocation tools by measuring the internal consistency of the variables. A value above 0.7 indicates acceptable reliability, values above 0.8 reflect good reliability and values above 0.9 represent excellent internal consistency. Table 12.3 shows the reliability assessment of this study. The lowest Cronbach's alpha recorded is 0.71, demonstrating that all constructs meet the acceptable threshold for internal consistency.

Table 12.3 Cronbach's alpha

Question	Cronbach's alpha, α	Interpretation
The project was well-planned using scheduling tools	0.85	Good internal consistency
Planning tools helped avoid delays	0.87	Good internal consistency
Planning tools tracked tasks and milestones	0.85	Good internal consistency
Tools improved team productivity	0.82	Good internal consistency
Communication tools enhanced teamwork	0.79	Acceptable internal consistency
Stakeholders/project teams received timely updates	0.82	Good internal consistency
Projects were completed on time (Projects A, B and C)	0.88	Good internal consistency
Projects met their budget	0.84	Good internal consistency
Deliverables met quality standards	0.84	Good internal consistency
Stakeholders/project teams were satisfied with project outcomes	0.71	Acceptable internal consistency
Projects achieved strategic and original objectives	0.75	Acceptable internal consistency

12.3.4 Pearson's correlation analysis

Correlation analysis was used to assess the relationship between PMT effectiveness and project success outcomes, based on Likert-scale responses. A correlation coefficient (r) greater than 0.7 indicates a strong positive correlation, while values between 0.5 and 0.7 represent a moderate positive correlation. A positive correlation indicates that greater effectiveness in using PMT is associated with improved project success. Table 12.4 illustrates the correlation analysis of this study.

The results indicate positive relationships between the variables, demonstrating a significant influence on the research objectives. All independent variables (planning and scheduling tools, communication and collaboration tools and stakeholder/resource allocation tools) show meaningful correlations with key components of project success, including timely completion, budget compliance, quality of work and stakeholder or project team satisfaction, particularly with communication effectiveness.

Figure 12.3 presents a Q-Q plot of the distribution of the ordered values relative to the theoretical normal distribution. In a Q-Q plot, if the sample distribution closely follows the straight reference line, the assumption of normality is considered reasonable. In this chart, most observed values lie near the reference line, particularly in the mid-range of the distribution, indicating that the overall distribution does not deviate substantially from normality. Small deviations at the tails, such as the highest values plotted slightly above the line, are common in empirical data and do not necessarily indicate serious non-normality, especially in sample sizes typical of exploratory studies. Although a few points show minor departures from the reference line, the deviations are limited in extent and do not exhibit systematic curvature or clustering that would indicate significant skewness or kurtosis. The Q-Q plot suggests that the variables or residuals are approximately normal for parametric analyses such as Pearson correlation and regression modelling.

Table 12.4 Pearson's correlation

Variable	Correlation coefficient, r	Interpretation
Scheduling tools' effectiveness affects project completion on time	0.65	Moderate positive correlation
Scheduling tools' effectiveness affects project budget compliance	0.58	Moderate positive correlation
Communication tools' effectiveness affects team productivity	0.72	Strong positive correlation
Communication tools' effectiveness affects stakeholder satisfaction	0.68	Moderate positive correlation
Planning tools' effectiveness affects task tracking and milestones	0.75	Strong positive correlation
Planning tools' effectiveness affects no delays	0.63	Moderate positive correlation

Figure 12.3 Q-Q plot of correlation coefficients

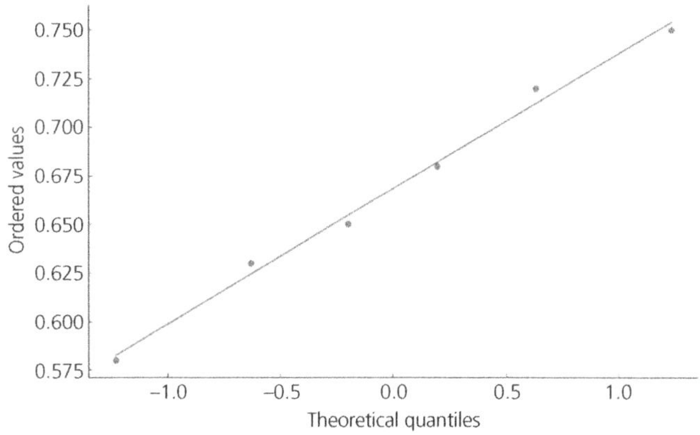

12.3.5 Multiple regression analysis

The multiple regression analysis results shown in Table 12.5 assess the predictive influence of planning and scheduling tools, communication and collaboration tools, and stakeholder and resource allocation tools on project success. This analysis examines the relationship between the dependent variable (DV) and combinations of independent variables (IV) to determine how strongly each predictor contributes to project success. In the multiple regression analysis, the dependent variable is Q26, which measures respondents' overall agreement with the implementation and use of PMT for project success. Items Q21 to Q25 capture broader project success indicators and informed the conceptual framing of the model but were not included as regression variables. The independent variables are Q6 (scheduling), Q7 (delay avoidance), Q9 (productivity), Q11 (communication), Q17 (resource allocation) and Q19 (performance monitoring).

The results show an R^2 value of 0.659, meaning that 65.9% of the variance in respondents' overall agreement with the implementation and use of PMT (DV, Q26) is explained by the combined predictors (independent variables) in the model. This indicates that scheduling, communication and stakeholder/resource allocation tools have a moderate influence on respondents' views of PMT implementation. The remaining 34.1% of the variance is attributed to factors not examined in this study, such as tool features, system functionality or user training. Overall, the findings confirm that PMT use by NRO stakeholders and project teams significantly contributes to project success by reducing delays, improving quality compliance and enhancing stakeholder satisfaction. Nonetheless, some external factors not included in the research may also shape respondents' perspectives.

Table 12.5 Multiple regression analysis

R	R^2	Adjusted R^2	Std. error
0.812	0.659	0.641	0.456

To complement the multiple regression analysis, an analysis of variance (ANOVA) test was conducted to evaluate whether the overall regression model shows a significant relationship among all variables. The results shown in Table 12.6 indicate that the model is highly significant. The ANOVA table shows an F-statistic of 22.653 with degrees of freedom ($df = 6, 107$) and a p-value below 0.001. This means the combination of the six independent variables significantly predicts the dependent variable. In summary, the regression model fits the data well, and the predictors collectively explain a substantial portion of the variation in project success.

Table 12.7 presents the results of the multiple regression analysis examining the influence of project management application variables on project success. A p-value below 0.05 indicates that a variable has a significant effect, whereas a p-value above 0.05 means the variable does not significantly influence the outcome. In the model, Q6, Q7, Q9 and Q11 are significant predictors ($p < 0.05$), whereas Q17 and Q19 are not ($p > 0.05$). Overall, the multiple regression analysis shows that the model is strong, explaining 66% of the variance in PMT usage. The

Table 12.6 ANOVA

Source	Sum of squares	*df*	Mean square	*F*	Sig.
Regression	64.351	6	10.725	22.653	0.000[a]
Residual	33.298	107	0.311		
Total	97.649	113			

[a]$p < 0.01$.

Table 12.7 Coefficients

Variable	Non-standard coefficient, *B*	Std. error	Standard coefficient, β	*t*	Sig. (*p*-value)
(Constant)	0.842	0.320		2.631	0.010
Q6: scheduling	0.201	0.089	0.212	2.258	0.026[a]
Q7: avoid delays	0.185	0.085	0.198	2.176	0.032[a]
Q9: productivity	0.264	0.094	0.243	2.804	0.006[b]
Q11: communication	0.175	0.079	0.194	2.215	0.029[a]
Q17: resource allocation	0.144	0.088	0.143	1.636	0.105
Q19: performance	0.127	0.080	0.130	1.588	0.116

[a]$p < 0.05$; [b]$p < 0.01$.

predictors related to planning, scheduling, timeliness, quality and stakeholder satisfaction have a significant impact on respondents' overall perception of PMT.

12.4. Practical insights

The findings of this study provide several practical insights for project managers, infrastructure stakeholders and public-sector organisations involved in delivering large-scale railway infrastructure. First, the strong positive relationship between planning and scheduling tools and project success highlights the need for systematic adoption of digital scheduling platforms. Project teams should consistently use tools such as integrated schedules, milestone tracking and progress dashboards to improve visibility, anticipate delays and support timely decision-making.

Second, communication and collaboration tools emerged as a critical driver of team productivity and stakeholder satisfaction. The results suggest that fragmented communication practices remain a major contributor to project delays. Railway project organisations should therefore institutionalise the use of shared digital platforms, such as cloud-based reporting systems and real-time dashboards, to ensure consistent information flow, faster approvals and clearer accountability among contractors, consultants and client representatives.

Third, while stakeholder and resource allocation tools showed weaker direct statistical significance, their practical importance remains evident. These tools support transparency in resource deployment, clarify roles and responsibilities, and help manage interdependencies between multiple work packages. Strengthening integration between resource allocation systems and planning tools can improve coordination and reduce inefficiencies.

Finally, the study underscores that the effectiveness of project management applications depends not only on the availability of tools but also on user proficiency and organisational commitment. Continuous training, standardised operating procedures and leadership support are essential to maximise the benefits of digital project management solutions. Embedding these tools within formal project governance frameworks can significantly enhance predictability, quality compliance and overall success in complex railway infrastructure projects.

12.5. Conclusion and recommendations

The study presented in this chapter examined PMTs as essential systems used by NRO stakeholders and project teams for Projects A, B and C. Although these projects have experienced major delays, the aim of the research was to provide valuable insights to help address the challenges project teams face.

Survey data were collected from stakeholders and project personnel through questionnaires covering demographic details such as age, education, years of project management experience, departmental placement and familiarity with PMT. Respondents evaluated the tools' effectiveness in planning, avoiding delays, tracking milestones, boosting productivity and enhancing communication and collaboration. Most participants agreed that the tools support planning, monitoring and productivity, although opinions were mixed on how to improve communication and reduce misunderstandings. The survey also asked respondents to assess

NRO project outcomes on timeliness, budget compliance, quality, stakeholder satisfaction and achievement of strategic goals. Responses varied, reflecting both positive and negative experiences across different projects.

12.5.1 Limitations

The study acknowledges limitations, particularly the exclusion of broader factors such as tool accessibility, system functionality and user training, which could influence the degree of project success. Despite these limitations, the findings show that the six predictor variables explain 65.9% of the variance in respondents' overall agreement with PMT implementation and use ($R^2 = 0.659$, DV: Q26), indicating moderately strong support for the research objectives. The remaining 34.1% of the variance is attributable to factors outside the scope of this model, such as tool features, system functionality and user training. In summary, the survey highlights the perceived benefits of PMT in NRO projects and identifies key areas for improvement to support more effective project delivery.

The PMT referenced in this study may include both free and paid applications currently available in the industry. As such, the questionnaire items were designed around general project management tool functions to establish a baseline understanding of how these tools support and influence overall project deployment across time, cost, quality and stakeholder satisfaction. Global external factors such as market conditions, international conflicts, political influences, regulatory changes, governance issues and project cost fluctuations were excluded from the scope of this research.

12.5.2 Recommendations and future research

Based on the conclusions, the variables used in this study should be further refined to improve accuracy, particularly regarding the accessibility and functionality of PMT. It is recommended that future research build on the current findings to explore additional dimensions of PMTs, including essential system functions; the capacity to store large amounts of technical data, such as drawings and material specifications; integration capabilities with other devices and project management systems; and overall user-friendliness. Enhancing these variables may provide deeper insights, especially in ensuring that new employees or stakeholders can use the tools effectively with minimal training, ultimately supporting more efficient project deployment.

The research focuses on how PMTs influence project success. There are no further details on project management applications, software or programme structures and functionalities. Future research may use this approach to identify the effects of project management applications during ongoing project deployment. This research focused exclusively on NRO projects to narrow the sampling scope and accurately address the research objectives. For future studies, the sample may be broadened to include project deployments across Kuala Lumpur or targeted toward specific construction sectors, such as housing developments or commercial infrastructure projects.

REFERENCES

Akintelu SO, Oyebola AI, Tiamiyu S and Olateju O (2023) The impact of project communication management on successful project delivery in the construction industry: a case study. *International Journal of Development and Sustainability* **12(8)**: 376–386.

Bachwani D, Malek M and Bharadiya R (2023) Project management techniques for planning and scheduling: a general overview. In *Emerging Trends and Innovations in Industries of the Developing World: A Multidisciplinary Approach* (Karras D, Oruganti S and Ray S (eds.)). Taylor and Francis, Abingdon, UK, pp. 186–190.

DeLone WH and McLean ER (2003) The DeLone and McLean model of information systems success: a ten-year update. *Journal of Management Information Systems* **19(4)**: 9–30.

Ding CS, Wong SY, Ting AMY *et al.* (2023) The impacts of project communication skills on the effectiveness of construction project teams in Sarawak, Malaysia. *International Journal of Academic Research in Business and Social Sciences* **13(10)**: 2222–6990.

Gamil Y and Abd Rahman I (2023) Studying the relationship between causes and effects of poor communication in construction projects using PLS-SEM approach. *Journal of Facilities Management* **21(1)**: 102–148.

Kalhoro M, Au Yong HN and SPR CR (2025) Urban mobility and resilient city planning: continuity of rail services and commuters' behavior in Greater Kuala Lumpur. *Continuity & Resilience Review*: 1–22.

Massimo R, Zenezini G, Ottaviani FM and De Marco A (2025) Project management information systems: a systematic review. *Procedia Computer Science* **256**: 1739–1747.

Monteiro A, Varajão J and Santos V (2025) Effects of project management information systems (PMIS) on project manager's performance: the moderating role of PMO. *International Journal of Managing Projects in Business* **18(6–7)**: 777–803.

Onaghinor O, Uzozie OT and Esan OJ (2022) Optimizing project management in multinational supply chains: a framework for data-driven decision-making and performance tracking. *Engineering and Technology Journal* **3(1)**: 907–913.

PMI (Project Management Institute) (2017) *A Guide to the Project Management Body of Knowledge (PMBOK® Guide)*, 6th edn. PMI, Newtown Square, PA, USA.

Quan BLT, Noh NIFM, Lin NJ *et al.* (2022) Effects of poor communication in the construction industry in Klang valley, Malaysia. *Jurnal Kejuruteraan* **34(5)**: 837–841.

Rahman MM (2023) Sample size determination for survey research and non-probability sampling techniques: a review and set of recommendations. *Journal of Entrepreneurship, Business and Economics* **11(1)**: 42–62.

Raymond L and Bergeron F (2014) Impact of project management information systems on project performance. In *Handbook on Project Management and Scheduling* (Schwindt C and Zimmermann J (eds.)), vol. 2. Springer, Cham, Switzerland, pp. 1339–1354.

Yong YC and Mustaffa NE (2013) Critical success factors for Malaysian construction projects: an empirical assessment. *Construction Management and Economics* **31(9)**: 959–978.

Advancing Project Management: Frameworks, Digital Transformation, Sustainability and Governance

Mohammed Reyasudin Basir Khan
ISBN 978-1-83742-799-4
https://doi.org/10.1108/978-1-83742-798-720261016

Chapter 13
Construction challenges and governance realities: a project delivery perspective

Noorazan Morshidi and Mohammed Reyasudin Basir Khan

Tun Razak Graduate School, Universiti Tun Abdul Razak (UNIRAZAK), Kuala Lumpur, Malaysia

13.1. Introduction

The construction sector plays a vital role in Sarawak's economic development, contributing approximately RM20 billion to the state's gross domestic product (GDP) in 2022, or around 10% of the total economy. As a major employer of more than 100 000 workers representing 5% of the state's workforce, the industry is a foundational pillar of Sarawak's socioeconomic progress. Its growth has been fuelled by rising population demands, expanding urbanisation, government-driven infrastructure investments and a robust regional economy. According to projections by the Construction Industry Development Board (CIDB), the sector is expected to contribute RM30 billion to GDP by 2025, underscoring its growing importance (Dehdasht *et al.*, 2022).

Despite this strong growth trajectory, the construction industry in Sarawak continues to face persistent and emerging challenges that hinder project performance and overall industry competitiveness. A major constraint is the shortage of skilled labour, driven by an ageing workforce, limited availability of structured training pathways and the migration of experienced workers to other countries in search of more lucrative employment opportunities. These labour shortages, combined with financial constraints and difficulties securing bank funding, significantly slow project execution. Infrastructure limitations, particularly in remote regions, further exacerbate delays, as transporting materials and equipment becomes more costly, time-consuming and logistically complex.

Other challenges affecting the industry include escalating costs of construction materials and equipment, which strain contractors' financial capacity and erode profit margins. The impacts of climate change, such as increasingly frequent floods, landslides and droughts, pose additional risks to site operations, causing disruptions and increasing the likelihood of project delays. Moreover, technological advancements are reshaping global construction practices, requiring companies in Sarawak to invest in new systems, equipment and workforce training to remain competitive. Together, these challenges create a complex operating environment that threatens project success, profitability and long-term industry sustainability.

Given the multifaceted nature of the issues, this chapter analyses the critical challenges affecting the performance of construction projects in Sarawak and proposes targeted, contextually relevant solutions. According to Morad (2023), construction projects in Sarawak commonly exceed their schedules by an average of 10%, primarily due to labour shortages, financial limitations, underdeveloped infrastructure and site remoteness. Cost overruns of approximately 15% are also widespread and stem from inadequate planning, unforeseen ground conditions and recurrent changes in project scope. Safety remains a pressing concern; construction workers in Sarawak have a 20% higher accident rate than the national average, a trend often attributed to insufficient training, poor safety practices and risky site conditions. Such delays, cost escalations and safety incidents undermine project viability, weaken stakeholder trust and hamper regional economic development.

Considering these issues, the study presented in this chapter provides an in-depth understanding of how project environment, team motivation, project management processes and risk management practices influence project performance across various types of construction projects in Sarawak. The aim of the research was to uncover the root causes of delays, cost overruns and safety risks, and to formulate practical strategies to enhance project planning, execution, financial management, safety compliance and regulatory support. Central to this study is the need for collaborative engagement among project managers, contractors, government agencies and local communities to foster a culture of responsible and efficient project delivery.

The significance of this research lies in its potential to contribute actionable insights that can strengthen construction project outcomes, enhance industry competitiveness and support Sarawak's long-term socioeconomic development. By analysing how environmental factors, team dynamics and risk management approaches interact with project management processes, the study provides a holistic understanding of the challenges confronting the sector. The findings will guide project managers in refining planning and scheduling practices, improving communication and motivation within teams, and adopting more robust risk management frameworks. Contractors can utilise the insights to strengthen cost control mechanisms, improve safety practices and invest in workforce development initiatives that enhance productivity and profitability. For policy makers, the study offers evidence-based recommendations to develop targeted training programmes, streamline regulatory processes, support technological innovation and introduce financial incentives for safer, more sustainable construction practices.

The study focused primarily on construction personnel across Sarawak and used a quantitative survey to examine how project environment, project management processes, team motivation and risk factors influence project performance. This introduction outlines the background, problem statement, objectives, research questions and the study's significance, followed by a comprehensive literature review covering theoretical foundations, empirical studies, the conceptual framework and hypothesis development. The methodology details the research design, sampling strategies, data collection instruments, variable measurement and analytical techniques. The study's findings are then presented and their implications for theory and practice are discussed, proposing recommendations for enhancing project performance in Sarawak's construction industry. This chapter concludes with a summary of key findings, policy and industry recommendations, and suggestions for future research to further advance knowledge in construction project management within developing regional contexts.

Several theoretical frameworks provide the basis for understanding the complexities of managing construction projects. A central reference is the project management triangle, which emphasises the interdependence of time, cost and quality (PMI, 2017). Any adjustment to one dimension inevitably affects the others. For instance, attempts to reduce project costs may require compromises in quality or extended timelines, while accelerating completion often leads to higher costs or increased quality risks. This triadic relationship highlights the trade-offs that project managers must navigate when dealing with resource constraints and performance demands.

The project lifecycle framework, also grounded in the project management body of knowledge (PMBOK: PMI, 2017), further supports understanding of project challenges by viewing projects as progressing through distinct phases: planning, execution and closure. Each phase presents its own set of issues: planning focuses on defining scope, objectives and schedules; execution focuses on coordinating resources and implementing the plan; closure involves finalisation and handover. Problems in the early stages, such as poor scoping or unrealistic scheduling, often manifest as delays, cost overruns or quality issues later in the project. Examining challenges by lifecycle phase helps in identifying where weaknesses in project management practices are most acute.

Risk management frameworks add another crucial dimension. They provide structured processes for identifying, assessing and responding to risks that may affect project outcomes (Shrivastava and Rathod, 2017). Typical steps include risk identification, qualitative and quantitative assessment of likelihood and impact, development of mitigation or contingency strategies, and continuous monitoring throughout the project. Applied to construction projects in Sarawak, such frameworks help explain how unmanaged or poorly managed risks contribute to delays, cost escalations, safety incidents and compromised quality.

Beyond these project management-specific frameworks, broader organisational theories also illuminate the challenges of construction project management (Zarghami, 2024). Systems theory conceptualises a project as a system of interrelated components, including the project team, management processes, environment and risk profile, in which changes in one element can ripple throughout the system. This perspective is particularly useful in Sarawak, where project conditions range widely from remote infrastructure works to relatively straightforward urban or residential developments. It underscores how weaknesses in planning, communication or environmental readiness can compound across multiple phases and actors.

Contingency theory complements this by arguing that there is no single best way to manage projects; instead, effective management must be tailored to contextual factors such as project size, complexity, uncertainty and organisational culture (Patanakul, 2023). In Sarawak, the strategies required to manage a large, complex infrastructure project in a remote area differ substantially from those suitable for smaller, less complex projects in urban centres. Contingency theory, therefore, justifies the need to study project management practices, team dynamics and risk strategies within the specific socioeconomic, regulatory and geographic context of Sarawak.

Together, these theoretical perspectives – that is, the project management triangle, project lifecycle, risk management, systems theory and contingency theory – offer a robust foundation for analysing how environmental conditions, team characteristics, management processes and risk factors interact to shape project outcomes.

Empirical studies on construction in Sarawak and related contexts consistently highlight several recurrent challenges (Esther *et al.*, 2025; Wahab *et al.*, 2025). A prominent issue is the shortage of the workforce. Labour scarcity arises from an ageing pool of experienced workers, limited training opportunities for new entrants and the migration of skilled labour to other regions or countries offering higher wages. This shortage contributes to project delays, higher labour costs, reduced construction quality and increased safety risks, while also placing additional stress on remaining workers and potentially damaging the industry's reputation.

Insufficient financial resources constitute another critical constraint. Government budget constraints limit the scale and pace of public infrastructure investments, while private contractors often face difficulties obtaining bank financing because construction is perceived as high-risk. Uncertainties related to weather, site conditions, material price fluctuations and stakeholder disputes heighten lenders' risk perceptions. As a result, projects may be delayed, scaled down or forced to compromise on quality. Restricted financial capacity also hampers innovation, limits investment in new technologies and training, and undermines the competitiveness of local firms compared to better-financed companies from other regions.

Inadequate infrastructure, particularly in terms of road networks, logistics, and port and airport facilities, further complicates construction activities in Sarawak. Poorly developed or poorly maintained transport links increase the time and cost of moving materials and equipment to and from sites, especially in rural or remote areas. These logistical constraints can cause significant delays, elevate costs, reduce productivity, heighten safety risks and contribute to negative perceptions of the industry among clients and the public.

Regulatory complexity is another commonly cited challenge (Gisong, 2026; Yip *et al.*, 2026). Multiple agencies, overlapping requirements and complex approval procedures make it difficult, especially for small and medium-sized contractors, to understand and comply with relevant regulations fully. This situation leads to delays in obtaining permits and approvals, increased compliance costs and heightened risk of penalties for noncompliance. It may also discourage innovation and deter external firms from participating in the Sarawak construction market, thereby reducing competition. Collectively, these empirical findings point to a challenging project environment characterised by labour, financial, infrastructural and regulatory constraints that directly and indirectly affect project delivery and performance.

Based on theoretical foundations and empirical evidence, this study proposed a conceptual framework linking four key independent variables (project environment, project team, project management process and project risks) to the dependent variable (project performance). Project environment encompasses factors such as infrastructure quality, regulatory stability, availability of skilled labour, economic conditions and political context. Project team refers to the competencies, experience, motivation, communication and leadership capabilities of those responsible for planning and executing the project. Project management process encompasses the effectiveness of planning, execution, monitoring and control. Project risks include both internal and external threats that can affect cost, time, quality and safety outcomes. Figure 13.1 illustrates the conceptual framework linking project management practices, risk factors and governance conditions to construction project performance.

The framework posits that a supportive project environment, characterised by adequate infrastructure, clear and stable regulations and favourable economic conditions, is more likely to facilitate timely, cost-effective and high-quality project delivery. Conversely, environmental constraints are expected to intensify challenges and reduce performance. Similarly, motivated, well-led and well-coordinated project teams are expected to handle complexity more effectively, mitigate problems and improve project outcomes. In contrast, weak teams with poor communication and limited skills are likely to exacerbate delays and cost overruns.

Effective project management processes, including realistic planning, rigorous scheduling, sound procurement strategies and robust monitoring and control, are generally considered positively associated with project performance. Inefficient processes, by contrast, tend to generate misalignment between plans and realities on site, leading to rework, disputes and performance shortfalls. Finally, the systematic identification, assessment and mitigation of project risks are expected to reduce the frequency and severity of adverse events. In contrast, inadequate risk management can lead to serious disruptions, safety incidents and financial losses.

The framework also acknowledges that these four elements are interrelated. Environmental conditions influence team availability and performance, shape the design of management

Figure 13.1 Conceptual framework

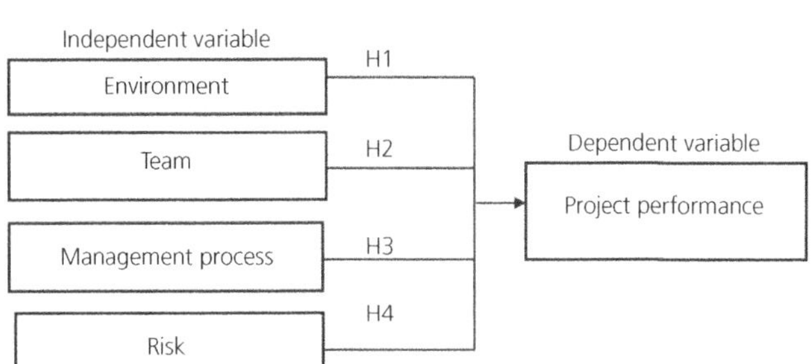

processes and affect project risk profiles. Team capabilities can strengthen or weaken management processes and determine how risks are perceived and addressed. Management processes, in turn, shape how environmental constraints and risks are handled in practice. Understanding these interdependencies is essential for developing comprehensive strategies to improve project performance in Sarawak.

Drawing on the conceptual framework, this study advances four main hypotheses. First, it is proposed that the project environment significantly affects the challenges of managing construction projects in Sarawak and, in turn, overall project performance. Inadequate infrastructure, regulatory uncertainty, economic volatility and political instability are expected to increase delays, raise costs and reduce quality and safety.

Second, the study hypothesises that the project team significantly influences the challenges of project management. The skills, experience, motivation and communication of the project team, especially the project manager and key technical staff, are expected to shape how effectively planning, coordination and problem-solving are carried out and, in turn, affect project outcomes.

Third, it is hypothesised that the quality of the project management process itself significantly affects the challenges encountered. Clear, realistic planning, disciplined execution and systematic monitoring and control are expected to reduce delays, cost overruns and quality and safety issues, whereas weak or ad hoc processes will amplify these problems.

Fourth, the study posits that project risks significantly affect the challenges of managing construction projects. High levels of unmanaged or poorly managed risk, whether internal or external, are expected to lead to schedule slippages, budget overruns and compromised quality and safety.

These hypotheses were tested using a quantitative research design, drawing on survey data from construction project stakeholders in Sarawak. By empirically examining the relationships between project environment, project team, project management processes, project risks and project performance, the study generated evidence-based insights and practical recommendations to help improve the management and outcomes of construction projects in Sarawak.

13.2. Methodology

13.2.1 Research design

The collected data were analysed using appropriate statistical techniques. Descriptive statistics, including frequency, percentage, mean and standard deviation, were employed to summarise respondents' demographic profiles and their perceptions of key variables. Demographic characteristics such as age, job sector, education level, working experience, site location, type of construction work and project cost were presented using frequency distribution and percentage. Meanwhile, the variables related to project environment, project team, project management process, project risks and project performance were analysed using mean score, standard deviation, skewness and kurtosis to assess their distribution and central tendencies.

A quantitative research approach was adopted, using a systematic method to collect and analyse numerical data to test hypotheses and explain relationships between variables (Yin, 2018). A cross-sectional quantitative design was used to investigate the challenges of managing construction projects in Sarawak. Convenience sampling was applied to select professional construction personnel from six construction sites across the state. This sampling method allowed the researcher to access participants, primarily engineers, who were readily available and willing to participate.

Data were collected using closed-ended questionnaires designed to capture respondents' demographic details and measure their perceptions of the project environment, project team, project management processes, project risks and project performance. All items were rated using a five-point Likert scale to capture the respondents' views consistently and objectively (Oppenheim, 2000). Inferential statistical analyses, including correlation and regression, were conducted to examine the relationships between the independent variables (project environment, project team, project management process and project risks) and the dependent variable (project performance). These analytical procedures helped identify significant predictors and determine the extent to which each factor influences project outcomes.

13.2.2 Population and sampling procedures

The target population for this study consists of professional construction personnel working across six major regions in Sarawak: Kuching/Sri Aman, Betong, Sibu/Kapit, Bintulu/Mukah, Miri/Baram and Limbang/Lawas. Convenience sampling was employed due to its practicality, cost-effectiveness and suitability for accessing respondents spread across large geographic areas. As described by Bhardwaj (2019), this approach selects participants based on their accessibility and availability, making it well-suited for efficiently collecting data from personnel at construction sites. While this method may introduce sampling bias and limit generalisability, it remains suitable for exploratory research aimed at identifying challenges in managing construction projects in Sarawak. A total of 100 respondents were selected for this study, based on established sample size guidelines for behavioural and survey research (Lea and Cohen, 2025; Memon *et al.*, 2020).

Participants were approached based on their willingness and availability and briefed on the study's purpose before providing consent. Recruitment was conducted through various channels, including email invitations, social media posts and face-to-face engagement at construction sites. In accordance with Dillman *et al.* (2014), efforts were made to include personnel from different departments, roles and experience levels to ensure a diverse and informative sample within the limits of convenience sampling. Confidentiality and voluntary participation were emphasised throughout the process to encourage honest and reliable responses.

13.2.3 Data collection method

The primary data for this study were collected through a self-administered questionnaire designed to assess the challenges of managing construction projects in Sarawak. The instrument was developed based on the study's objectives and hypotheses and a pilot test with a small group of engineers was conducted to ensure clarity, reliability and overall suitability.

The final questionnaire was distributed electronically through email and social media, accompanied by clear instructions and assurances of confidentiality to encourage participation. Reminders were issued during the data collection period to improve response rates, and all completed responses were stored securely to protect participants' privacy.

Secondary data were obtained from a range of credible sources, including reference books, government economic reports, corporate annual statements, scholarly journal articles, online databases and reputable media publications. Platforms such as Google Scholar, ScienceDirect and ResearchGate were used to access academic literature relevant to construction project management and the Sarawak context. These secondary sources supported the development of the literature review, provided theoretical grounding and supplemented the analysis by offering insights from existing research.

13.2.4 Operationalisation and measurement

This section outlines the operationalisation and measurement of the study variables, which include the independent variables (project environment, project team, project management process, and project risks) and the dependent variable (project performance). Measurement items were adapted from established scales and supplemented with context-specific items relevant to the Sarawak construction industry. All items were incorporated into the questionnaire and measured using a five-point Likert scale ranging from strongly disagree to strongly agree, ensuring standardised data collection and supporting the reliability and validity of the constructs examined.

The independent variables capture key dimensions influencing construction project management. The project environment reflects contextual factors such as infrastructure, political and economic stability, labour availability and regulatory conditions. The project team encompasses the knowledge, skills, communication and collaboration abilities of individuals involved in planning and executing the project. The project management process refers to the effectiveness of planning, execution and control mechanisms, while project risks include internal and external threats that may disrupt project success. The dependent variable, project performance, was assessed through perceptions of cost, schedule, quality and overall project outcomes, serving as a proxy for construction management effectiveness.

Demographic variables in Section 1 of the questionnaire were measured using nominal and ordinal scales to capture respondent characteristics, including age, gender, job sector, years of experience, education level, site location, type of construction work and construction cost. Sections 2 through 6 measured the independent and dependent variables using a five-point Likert scale.

13.2.5 Data analysis techniques

The questionnaire was first reviewed to ensure clarity, accuracy and the absence of errors. All items were designed using simple and comprehensible language. A pilot test was then conducted with a small group of respondents to evaluate the instrument's validity and reliability. Feedback from the pilot test helped refine the questionnaire to ensure it was suitable for full deployment.

Data analysis was conducted using the statistical package for the social sciences (SPSS). Both descriptive and inferential statistical techniques were applied to summarise the data, explore patterns and examine relationships between variables.

Descriptive statistics, including frequency, percentage, mean, median, standard deviation (SD) and range, were used to summarise demographic characteristics and examine the distribution of key variables. These techniques provided an overview of respondent profiles and initial patterns in the data, offering a clear, interpretable summary of the dataset.

Inferential analyses were employed to test hypotheses and assess relationships between variables. Correlation analysis was used to determine the strength and direction of associations, while multiple regression analysis was used to examine the influence of independent variables on project performance. Reliability testing using Cronbach's alpha assessed the internal consistency of the measurement scales and validity checks ensured that the constructs accurately measured their intended concepts. These techniques collectively support robust statistical conclusions and meaningful insights into the challenges of managing construction projects in Sarawak.

13.3. Results and analysis

13.3.1 Response rate

The study targeted engineers working across different regions of Sarawak and data were collected through a Google Form questionnaire. A total of 100 survey links were distributed through social networking platforms such as WhatsApp and Facebook, and respondents were given two weeks to complete the form. At the end of the collection period, 76 completed surveys were received for a 76% response rate.

13.3.2 Demographic profile

The demographic analysis provides an overview of the 76 respondents who participated in the study. The age distribution shows that the majority of respondents (42.1%) are between 41 and 50 years old, indicating that a large portion of the sample comprises mid-career professionals with substantial industry experience. This is followed by those aged 31–40 years (21.1%). Smaller groups include respondents aged 21–30 years and 51–60 years, each representing 14.5% of the sample, while 7.9% are aged 61 and above. This distribution suggests that most respondents have extensive practical experience, thereby strengthening the reliability of the collected data.

In terms of gender, the sample is predominantly male, with 85.5% of respondents identifying as male and 14.5% as female. This reflects the gender imbalance commonly observed in the construction sector. Regarding the job sector, 82.9% of respondents work in the private sector, while 17.1% work in government agencies, indicating greater participation by private-sector construction personnel.

Work experience data show that 39.5% of respondents have more than 21 years of experience in construction project management, indicating that many participants possess deep professional knowledge. Another 19.7% have 0–5 years of experience and the remaining respondents fall within the 6–20-year range. The majority of respondents also hold strong educational qualifications: 50% hold a Bachelor's degree, 27.6% hold a Master's degree and 22.4% hold a Diploma or Doctorate. This reflects a highly educated sample that is likely to understand project management concepts well. Figure 13.2 presents the distribution of respondents' construction site locations.

Most respondents (57.9%) are involved in construction activities located in the Kuching/Sri Aman/Betong region, reflecting the concentration of development in Sarawak's central urban areas. Other respondents are distributed across Miri/Baram, Sibu/Kapit, Limbang/Lawas and Bintulu/Mukah. The types of construction work undertaken by the respondents vary, with building works and road works each accounting for 32.9% of the sample. Mechanical/electrical works and other infrastructure works each represent 11.8%, followed by water supply works and electrical works.

Finally, the analysis of project cost involvement shows that 35.5% of respondents are working on projects valued at over RM100 million, while 23.7% are engaged in projects costing under RM10 million. The remaining respondents are involved in medium-scale projects ranging from RM11 million to RM100 million. This distribution suggests that the sample includes personnel involved in both major and smaller-scale construction projects, providing diverse insights into project management challenges across varying project sizes.

13.3.3 Reliability analysis

Table 13.1 shows that the dependent variable (project performance) demonstrates a high reliability coefficient of 0.838. The independent variables (project environment, project team, project management process and project risks) also exhibit strong reliability, with coefficients

Figure 13.2 Respondents' construction site location

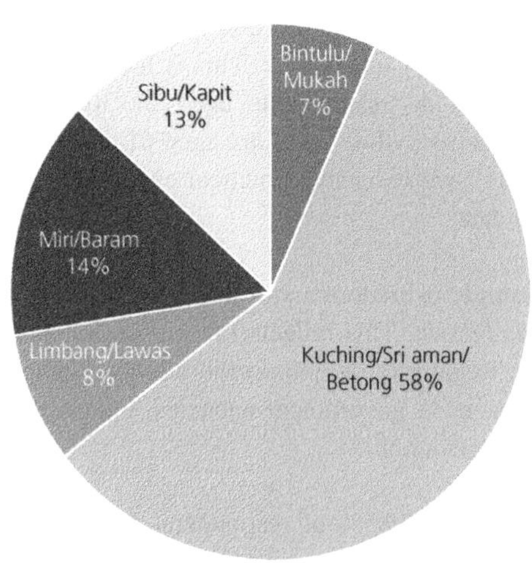

Table 13.1 Reliability analysis

Variable	n	Number of items	Cronbach's alpha, α
Project environment	76	9	0.880
Project team	76	6	0.828
Project management process	76	8	0.804
Project risks	76	8	0.808
Project performance	76	8	0.838

of 0.880, 0.828, 0.804 and 0.808, respectively. As all values exceed the recommended threshold of 0.8, the results indicate excellent internal consistency across all measures. This confirms that the questionnaire items are well-constructed, reliable and suitable for the study.

13.3.4 Descriptive analysis

The descriptive analysis examined the independent variables (project environment, project team, project management process and project risks) and the dependent variable (project performance). Table 13.2 presents the descriptive statistics for the first independent variable, project environment. The highest mean score (4.17) corresponds to the item stating that Sarawak's topography poses challenges for construction projects, followed by the item on political stability with a mean of 3.88. The lowest mean score (2.61) relates to the stability of construction material prices. Overall, mean values range from 2.61 to 4.17, with a grand mean of 3.35 (SD = 0.87), indicating moderate variability in respondents' perceptions of environmental factors affecting construction projects.

Table 13.2 Descriptive statistics of project environment

Descriptive statistics	Mean	SD
The climate and weather in Sarawak are conducive to construction activities	3.39	0.850
The topography and geographic conditions of Sarawak pose challenges for construction projects	4.17	0.681
There is a sufficient supply of construction resources in Sarawak	3.01	0.986
Laws and regulations in Sarawak support construction projects	3.67	0.719
The political environment in Sarawak is stable and conducive to construction projects	3.88	0.816
Corruption is not a major problem in Sarawak's construction industry	2.95	0.978
The economic climate in Sarawak is favourable for construction projects	3.58	0.804
The inflation rate in Sarawak is stable and poses no significant risk to construction projects	2.91	0.996
The fluctuation rate of construction materials in Sarawak is stable and poses no significant risk to construction projects	2.61	0.981

Table 13.3 presents the descriptive statistics for the second independent variable, project team. The highest mean score (4.01) relates to the statement that team members demonstrate strong teamwork and collaboration. Items on team motivation, commitment and conflict resolution share the next-highest mean (3.96). The lowest mean score (3.89) concerns effective communication among team members. Overall, the mean values range from 3.89 to 4.01, with a grand mean of 3.95 (SD = 0.81), suggesting consistently high perceptions of team performance and moderate variability across responses.

Table 13.4 shows the descriptive statistics for the third independent variable, project management process. The highest mean score (4.09) is for a detailed project plan outlining objectives, tasks, timelines and resources, followed by a clear and well-defined project plan with a mean of 3.95. The lowest mean (3.79) concerns the tracking and management of project costs. Overall, mean scores range from 3.79 to 4.09, with a grand mean of 3.90 (SD = 0.76), indicating generally positive assessments of project management practices. The standard deviations, all below 1, indicate moderate variability and suggest minimal differences across items.

Table 13.3 Descriptive statistics of the project team

Descriptive statistics	Mean	SD
The project team members have the necessary skills and experience to manage the project effectively	3.95	0.831
The project team members are motivated and committed to the project	3.96	0.824
The project team members are enthusiastic and passionate about the project	3.91	0.803
The project team members have a strong sense of teamwork and collaboration	4.01	0.792
The project team members communicate effectively with each other	3.89	0.810
The project team members can resolve conflicts effectively	3.96	0.824

Table 13.4 Descriptive statistics of the project management process

Descriptive statistics	Mean	SD
The project has a clear, well-defined plan	3.95	0.630
The project has a detailed project plan that outlines the project's objectives, tasks, timelines and resources	4.09	0.657
The project plan is regularly updated to reflect changes in the project scope or environment	3.89	0.793
The project schedule is realistic and achievable	3.84	0.801
The project activities are being effectively monitored and controlled	3.86	0.844
The project costs are being tracked and managed effectively	3.79	0.869
The project deliverables are being produced to the required quality standards	3.87	0.772
The project team is effectively communicating with stakeholders	3.92	0.707

Table 13.5 presents the descriptive statistics for the fourth independent variable, project risks. The items with the highest mean score, 3.72, were shared between two items: 'The project team has effectively identified all potential risks to the project' and 'The project team is effectively communicating the risks to stakeholders'. The lowest mean score of 3.49 was for the statement 'Stakeholders are aware of the risks and support the project team's efforts to mitigate them'. The mean values range from 3.49 to 3.72, with a grand mean of 3.62 (SD = 0.87), indicating moderate risk management performance. While some aspects are strong, others show potential for improvement. All standard deviations are under 1, suggesting moderate variability within each aspect. Individual strengths and weaknesses may exist within the team.

Table 13.6 presents the descriptive statistics for the dependent variable, project performance. The highest mean score (3.83) corresponds to overall satisfaction with working alongside the project team. In contrast, the next-highest mean (3.78) is shared by items

Table 13.5 Descriptive statistics of project risks

Descriptive statistics	Mean	SD
The project team has effectively identified all potential risks to the project	3.72	0.810
The project team has assessed the likelihood and impact of each risk	3.59	0.912
The project team has prioritised the risks based on their likelihood and impact	3.66	0.825
The project team has developed contingency plans to mitigate the risks	3.58	0.928
The project team is implementing the contingency plans effectively	3.54	0.871
The project team is monitoring risks and adjusting contingency plans as needed	3.63	0.892
The project team is effectively communicating the risks to stakeholders	3.72	0.810
Stakeholders are aware of the risks and support the project team's efforts to mitigate them	3.49	0.931

Table 13.6 Descriptive statistics of project performance

Descriptive statistics	Mean	SD
How satisfied are you with the project's overall performance?	3.57	0.869
How satisfied are you with the project team's communication?	3.70	0.749
How satisfied are you with the project team's responsiveness?	3.76	0.690
How satisfied are you with the overall experience of working with the project team?	3.83	0.681
Has the project met its objectives?	3.75	0.819
Has the project delivered the expected benefits?	3.78	0.858
Has the project been completed on time and within budget?	3.21	1.099
Has the project met the required quality standards?	3.78	0.776

related to expected project benefits and meeting required quality standards. The lowest mean score (3.21) pertains to completing the project on time and within budget. Overall, the mean values range from 3.21 to 3.83, with a grand mean of 3.67 (SD = 0.82), indicating moderate satisfaction with project outcomes. Standard deviations below 1 reflect moderate variability, suggesting some differences in stakeholder perceptions but no extreme divergence.

13.3.5 Correlation analysis

The dependent variable (project performance) and the four independent variables (project environment, project team, project management process and project risks) were examined using Pearson correlation analysis. Table 13.7 presents the correlation results for the study. The findings show that the project environment has a moderate positive relationship with project performance ($r = 0.446$), indicating that a supportive environment contributes to better outcomes. The project team also shows a moderate positive correlation with project performance ($r = 0.502$), suggesting that team dynamics influence performance.

The strongest relationship is between project management process and project performance ($r = 0.642$), indicating a moderately strong positive correlation. Project risks show a moderate positive correlation with project performance ($r = 0.605$), suggesting that managing internal and external risks is an important factor in determining project outcomes.

The Q-Q plot in Figure 13.3 assesses the distribution of Pearson correlation coefficients obtained in this study. The plotted points follow the diagonal reference line reasonably closely, indicating that the correlation coefficients are approximately normally distributed. Minor deviations are observed at both the lower and upper tails. The smallest coefficients fall slightly below the reference line, while the largest coefficients lie somewhat above it. However, these departures are modest and do not indicate substantial skewness or extreme outliers. The overall pattern suggests gradual deviation rather than systematic distortion.

Although normality testing is more commonly applied to raw variable distributions or regression residuals, the Q-Q plot of the correlation coefficients provides additional visual confirmation that no severe distributional abnormalities are present. Therefore, the assumption of approximate normality is considered acceptable, supporting the robustness of subsequent parametric analyses.

13.3.6 Multiple regression analysis

Regression analysis was conducted to examine the causal relationships between the independent variables (project environment, project team, project management process and project risks) and the dependent variable (project performance). The model summary in Table 13.8 shows an R^2 of 0.476, indicating that 47.6% of the variance in project performance is explained by the four predictors, with the remaining 52.4% attributed to external factors not captured in the model.

Table 13.7 Findings of Pearson correlation

		Project performance	Project environment	Project team	Project management process	Project risks
Project performance	Pearson correlation	1	0.446[a]	0.502[a]	0.642[a]	0.605[a]
	Sig. (2-tailed)		0.000	0.000	0.000	0.000
	n	76	76	76	76	76
Project environment	Pearson correlation	0.446[a]	1	0.334[a]	0.439[a]	0.405[a]
	Sig. (2-tailed)	0.000		0.003	0.000	0.000
	n	76	76	76	76	76
Project team	Pearson correlation	0.502[a]	0.334[a]	1	0.718[a]	0.717[a]
	Sig. (2-tailed)	0.000	0.003		0.000	0.000
	n	76	76	76	76	76
Project management process	Pearson correlation	0.642[a]	0.439[a]	0.718[a]	1	0.734[a]
	Sig. (2-tailed)	0.000	0.000	0.000		0.000
	n	76	76	76	76	76
Project risks	Pearson correlation	0.605[a]	0.405[a]	0.717[a]	0.734[a]	1
	Sig. (2-tailed)	0.000	0.000	0.000	0.000	
	n	76	76	76	76	76

[a]Correlation is significant at the 0.01 level (2-tailed).

Figure 13.3 Q-Q plot of Pearson correlation coefficients

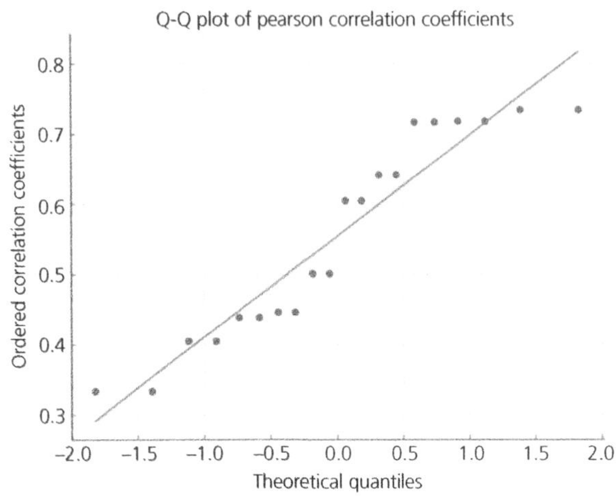

Table 13.8 Model summary

Model summary								
R	R^2	Adjusted R^2	Std. error of the estimate	Change statistics				
				R^2 change	F change	$df1$	$df2$	Sig. F change
0.690	0.476	0.447	0.47153	0.476	16.153	4	71	0.000

Note: predictors: (constant), project environment, project team, project management process, project risks.

Table 13.9 presents the coefficients for each variable. Project management process has the strongest and most significant positive effect on project performance ($\beta = 0.387, p = 0.008$), demonstrating its critical role in improving outcomes. Project risks also show a positive influence ($\beta = 0.271, p = 0.056$), with results approaching significance, suggesting that better risk management may enhance performance. The project environment has a weaker positive impact ($\beta = 0.176, p = 0.073$), indicating a marginal association. Meanwhile, the project team shows no meaningful effect ($\beta = -0.028, p = 0.836$), suggesting that team-related factors did not significantly contribute to performance in this sample.

The analysis of variance (ANOVA) results given in Table 13.10 confirm that the overall regression model is statistically significant ($F = 16.153, p < 0.001$). This indicates that, collectively, the independent variables meaningfully predict project performance, even though their individual contributions vary in strength and significance.

Table 13.9 Coefficients

	Coefficient				
Model	Unstandardised coefficients		Standardised coefficient	*t*	Sig.
	B	Std. error	*β*		
(Constant)	0.592	0.437		1.355	0.180
Project environment	0.227	0.125	0.176	1.821	0.073
Project team	–0.024	0.118	–0.028	–0.208	0.836
Project management process	0.406	0.149	0.387	2.728	0.008
Project risks	0.230	0.118	0.271	1.944	0.056

Note: dependent variable: project performance.

Table 13.10 ANOVA

		ANOVA			
Model	Sum of squares	*df*	Mean square	*F*	Sig.
Regression	14.365	4	3.591	16.153	0.000
Residual	15.786	71	0.222		
Total	30.151	75			

Note: dependent variable: project performance; predictors: (constant), project environment, project team, project management process, project risks.

13.3.7 Hypothesis testing

The analysis indicates partial support for hypothesis 1, as environmental factors showed a weak positive relationship with project performance. However, many aspects received neutral responses; topography and regulatory issues emerged as noteworthy influences, warranting further examination of regulatory and corruption-related challenges. Hypothesis 2 was rejected because team-related factors such as skills, motivation and communication were generally rated positively, suggesting no significant negative impact on performance. However, a deeper investigation into leadership, team dynamics and stakeholder relationships may uncover underlying issues. Hypothesis 3 was strongly supported, with results highlighting that weaknesses in planning clarity, schedule realism and monitoring systems significantly affect project performance, emphasising the need to strengthen project management processes. Hypothesis 4 received moderate support, showing that while basic risk identification and assessment are adequate, shortcomings in risk prioritisation, contingency planning and communication negatively influence performance; thus, improvements in these areas are essential to reduce challenges and enhance project outcomes.

13.4. Practical insights

The findings of this study provide several practical insights for construction practitioners, policy makers and project stakeholders operating in Sarawak and similar regional contexts. First, the strong influence of project management processes on project performance underscores the need to prioritise robust planning, realistic scheduling and systematic monitoring and control. Project managers should place greater emphasis on developing detailed project plans, regularly updating schedules and strengthening cost and progress tracking mechanisms to reduce delays and cost overruns.

Second, the results highlight the importance of proactive risk management. While basic risk identification practices are in place, weaknesses remain in risk prioritisation, contingency planning and ongoing risk communication. Construction organisations should adopt structured risk management frameworks and integrate risk assessment into all stages of the project lifecycle. Regular risk reviews and clearer communication with stakeholders can help minimise disruptions caused by environmental, financial and operational uncertainties.

Third, although team-related factors were generally rated positively and did not show a strong direct statistical effect, their indirect role should not be underestimated. Continuous skills development, leadership training and improved coordination between project team members can enhance the effectiveness of project management processes and risk mitigation efforts.

Finally, environmental challenges such as topographical constraints, regulatory complexity and infrastructure limitations require coordinated responses beyond individual projects. Policy makers and regulatory bodies should streamline approval processes, improve infrastructure support in remote areas and invest in workforce development programmes. Collaborative engagement between government agencies, contractors and local communities is essential to creating an enabling environment for the efficient and sustainable delivery of construction projects in Sarawak.

13.5. Conclusion and recommendations

This study examined the challenges of managing construction projects in Sarawak by exploring how the project environment, project team, project management process and project risks relate to project performance. Using quantitative design and online surveys, the study identified key factors influencing project outcomes in the region. The results show that effective project management processes have the strongest positive impact on project performance, underscoring the importance of robust planning, execution and control. Proactive risk management also plays a significant role in minimising disruptions and improving outcomes. Although the project environment and team factors show some influence, their specific effects on project challenges remain unclear and warrant further investigation. These findings suggest that organisations should prioritise strengthening project management capabilities and integrating risk management into all phases of project delivery.

13.5.1 Limitations

It is important to acknowledge several limitations of this study. Reliance on quantitative data, single-point data collection and potential selection bias highlight the need for future research using mixed-methods and longitudinal approaches. Further studies should explore context-specific dynamics through qualitative methods, case studies and examinations of moderating factors such as project type, size and complexity. Expanding research to other regions may also improve generalisability.

Using an online survey as the sole data collection method restricts participation to individuals with internet access and a willingness to respond, potentially introducing selection bias. Because the survey was conducted at a single point in time, the findings represent only a snapshot of the challenges faced. They may not fully capture how these issues evolve throughout the project lifecycle. In terms of measurement and analysis, some variables, particularly those related to the project environment and team dynamics, may require further refinement to improve interpretation. The study also did not account for other potentially relevant variables or moderating factors, such as project type, size or complexity. These omissions may have limited the depth of the analysis. The generalisability of the findings is also constrained by the sample size and its representativeness, meaning the results may not fully reflect the broader construction sector in Sarawak or apply to other regions. Methodologically, relying solely on quantitative methods limits the richness of insight that qualitative approaches, such as interviews, could provide. Additionally, because the study employed correlational analysis, it cannot establish causal relationships, and future research using longitudinal or experimental designs may be required to validate causal effects.

13.5.2 Recommendations and future research

Future research should broaden the scope of data collection by incorporating qualitative methods such as interviews, focus groups and in-depth case studies to gain deeper insight into the experiences of project managers and team members, as well as the contextual factors shaping project challenges. Longitudinal approaches would also allow researchers to track how challenges evolve and how they relate to project management practices, risk management strategies and environmental conditions. Further studies should explore how project characteristics such as type, size and complexity moderate the relationships between key variables and project challenges, and examine the mechanisms by which factors such as project management processes and team dynamics influence outcomes. Experimental or quasi-experimental designs could help establish causal relationships. Refinement of measurement tools is also essential; future work should enhance the operationalisation of complex variables, such as project environment and team composition, and develop contextually relevant, validated instruments tailored to the Sarawak construction sector. Expanding the sample size and ensuring broader representation across project types and regions would improve generalisability, and comparative studies across different regions or countries could help identify best practices. Additionally, future research could focus on specific categories of challenges, such as delays, cost overruns, safety risks or environmental issues, and develop targeted interventions to address them, with findings shared widely among industry stakeholders to support practical application and knowledge exchange.

REFERENCES

Bhardwaj P (2019) Types of sampling in research. *Journal of the Practice of Cardiovascular Sciences* **5(3)**: 157.

Dehdasht G, Ferwati MS, Abidin NZ and Oyedeji MO (2022) Trends of construction industry in Malaysia and its emerging challenges. *Journal of Financial Management of Property and Construction* **27(2)**: 161–178.

Dillman DA, Smyth JD and Christian LM (2014) *Internet, Phone, Mail, and Mixed-Mode Surveys: The Tailored Design Method*, 4th edn. Wiley, Hoboken, NJ, USA.

Esther WSW, Magdalen P and Stephanie SD (2025) Quantitative study on mitigating foreign worker impacts in Sarawak's construction industry. *Borneo Journal of Social Science & Humanities* **7(2)**: 63–80.

Gisong J (2026) Risk management lessons from the Pan Borneo Highway Project: a comparative case study of public infrastructure delivery in Sabah and Sarawak, Malaysia. *International Journal of Research & Technology* **14(1)**: 33–46.

Lea RB and Cohen BH (2025) *Introductory Statistics for the Behavioral Sciences*. Wiley, Hoboken, NJ, USA.

Memon MA, Ting H, Cheah J-H *et al.* (2020) Sample size for survey research: review and recommendations. *Journal of Applied Structural Equation Modeling* **4(2)**: i–xx.

Morad A (2023) An analysis of global construction projects: causes and implications of cost overruns. *International Journal of Innovative Research in Science Engineering Technology* **8(6)**: 10.

Oppenheim AN (2000) *Questionnaire Design, Interviewing and Attitude Measurement*. Routledge, London, UK.

Patanakul P (2023) Contingency theory and its applications to complex project organizing. In *Research Handbook on Complex Project Organizing* (Winch GM, Brunet M and Cao D (eds.)). Edward Elgar, Cheltenham, UK, pp. 60–69.

PMI (Project Management Institute) (2017) *A Guide to the Project Management Body of Knowledge (PMBOK® Guide)*, 6th edn. PMI, Newtown Square, PA, USA.

Shrivastava SV and Rathod U (2017) A risk management framework for distributed agile projects. *Information and Software Technology* **85**: 1–15.

Wahab A, Bohari SA and Sheng WC (2025) Critical success factors in Malaysian housing projects: examining contractual, process and human management with knowledge sharing moderation. *International Journal of Housing Markets and Analysis* **18(5)**: 1287–1307.

Yin RK (2018) *Case Study Research and Applications: Design and Methods*, 6th edn. Sage, Thousand Oaks, CA, USA.

Yip HL, Wong SY, Ding CS and Low WW (2026) Ethical transgressions in the construction industry in Sarawak, Malaysia: perspectives of quantity surveyors. *Engineering, Construction and Architectural Management* **33(1)**: 435–451.

Zarghami SA (2024) Project schedule contingency planning: building on von Bertalanffy's open systems theory and critical systems practice. *Systems Research and Behavioral Science* **41(2)**: 247–261.

Advancing Project Management: Frameworks, Digital Transformation, Sustainability and Governance

emerald PUBLISHING ice Publishing

Mohammed Reyasudin Basir Khan
ISBN 978-1-83742-799-4
https://doi.org/10.1108/978-1-83742-798-720261017

Chapter 14
Diagnosing construction delays: an empirical study of a major commercial development project

Rozaimoorsyari Abdullah[1], Mohammed Reyasudin Basir Khan[1], Chong Peng Lean[2] and Ahmad Anwar Zainuddin[3]

[1]*Tun Razak Graduate School, Universiti Tun Abdul Razak (UNIRAZAK), Kuala Lumpur, Malaysia*
[2]*MILA University, Nilai, Negeri Sembilan, Malaysia*
[3]*Kulliyyah of Information and Communication Technology, IIUM, Gombak, Malaysia*

14.1. Introduction

Construction project delays are widely recognised as among the most common, costly, complex and risky challenges faced by the construction industry (Sovacool and Ryu, 2025). Delays may stem from numerous factors and inevitably lead to increased costs and extended project duration. The research presented in this chapter focuses on identifying the factors contributing to delays in a high-rise commercial development project in Kuala Lumpur (KL). Contract duration, budgeting and time management often pose significant challenges for construction projects, and failure to meet predetermined targets in these areas typically results in financial losses or project disruptions. Prior studies have emphasised that timely completion is a critical measure of project success, and delays may arise from owners, contractors, consultants or external conditions such as weather, regulations or political instability (Hansen *et al.*, 2022).

Project delays disrupt construction activities, reduce productivity, increase costs and create circumstances for disputes, claims or even abandonment. Because projects rarely complete within the planned timeframe, a clear understanding of the underlying causes of delay is essential to evaluate their significance and reduce their impact. Although delay factors vary across projects, they can be effectively managed through proper planning and coordination between all stakeholders involved in the construction process.

The problem of delay remains one of the most persistent and significant issues in the construction sector (Tariq and Gardezi, 2023). Effective time and cost control are essential, yet many projects still struggle with time management, leading to time extensions and cost overruns. Since the extent of delay varies from project to project, identifying the specific underlying causes is crucial. Previous studies have classified delays into internal and external causes arising from owners, designers, contractors, consultants, government agencies, suppliers or natural conditions (Sanni-Anibire *et al.*, 2022; Welde and Bukkestein, 2022). Several key causes have been frequently highlighted in earlier research, including poor contractor planning, inadequate site management, insufficient contractor experience, financial difficulties, conflicts with subcontractors and shortages of labour, materials and equipment (Bryde and Robinson, 2005; Naji *et al.*, 2022). Because these factors vary across projects, the study presented in this chapter examines the specific conditions that influence project delays, identifies the stakeholders contributing to them and demonstrates how these delays affect project progress. Gaining this understanding is vital in ensuring that projects are completed within the intended timeline and budget.

The main purpose of this study was to identify the factors contributing to construction delays in Malaysia, with a focus on a high-rise commercial development project in KL. The study also compares these findings with previous research to establish how earlier conclusions relate to the current situation. In doing so, the research explored the relationship between delay factors and their impact on the project, in order to determine how these factors contribute to overall project delay and identify the stakeholders who play the most significant roles in affecting project timelines.

To achieve these aims, the study addresses several key questions about the most significant factors influencing project delays and the relationships between financial constraints, contractual changes, site-related matters, communication issues, external conditions and overall project performance. By examining these relationships, the study clarifies which factors most strongly influence project delay.

The significance of this study is twofold. From an academic perspective, it provides a foundation for further research on construction delay factors in Malaysia, offering insights that contribute to the current body of knowledge on project management and construction performance. From an industry perspective, the study's findings can help construction practitioners understand the key factors that lead to delays and develop more effective planning and risk mitigation strategies. This knowledge is valuable not only for high-rise developments but also for a wide range of construction projects, helping companies anticipate potential issues and reduce the likelihood of future delays.

Construction risk management plays a crucial role in identifying potential risks throughout the construction lifecycle and implementing suitable mitigation measures. By anticipating and responding to risks, project teams can minimise negative impacts, safeguard project objectives and enhance overall performance. Construction risk management theory provides the theoretical basis for identifying and assessing delay factors in construction projects, as guided by the project management body of knowledge (PMBOK: PMI, 2021) and ISO 31000 (ISO, 2018) risk management standards (Takagi *et al.*, 2024). The theory emphasises a systematic process of identifying risks from the beginning of a project through its completion, assessing their probability and consequences, and determining appropriate strategies for avoidance, mitigation, transfer or acceptance. The risk management cycle also requires continuous monitoring and review to ensure risks are managed effectively and project goals are achieved. Within this theoretical structure, construction-related risks such as financial, schedule, technical, legal, regulatory and environmental directly correspond to factors that cause delays. These risks affect crucial project dimensions, including cost control, time management, quality and safety. In this study, risk is conceptualised as any event or condition that may hinder timely progress, disrupt workflow or increase project cost, thereby functioning as a delay factor.

Previous research consistently highlights that delays are prevalent in construction projects worldwide, with their severity and duration varying significantly (Latif *et al.*, 2023; Selcuk *et al.*, 2024). Delays may arise from internal or external sources and are typically attributed to financial constraints, material shortages, ineffective planning, poor communication, contractor-related issues, consultant inefficiencies, owner-related delays or environmental conditions. Studies conducted across different countries have identified overlapping causes, such as late payments, inadequate site management, design changes, a lack of skilled labour, slow decision-making and adverse weather conditions (Abdul Nabi and El-adaway, 2022; Viles *et al.*, 2020). These findings illustrate that while factors differ across contexts, understanding the specific causes in a given project is essential in managing and reducing delays.

Different categories of delays, such as non-excusable, excusable, compensable or concurrent, determine whether additional time, cost or liability applies (Kumar and Kumari, 2025). Responsibility for delays varies and may involve contractors, consultants, owners, suppliers or external agencies. Because the causes of delay differ from one project to another, empirical studies emphasise the need for project-specific investigation to understand the most critical contributing factors (Doloi *et al.*, 2012; Zidane and Andersen, 2018). This research therefore examines delay factors specific to the high-rise commercial development project in KL to determine their significance and relationship to project performance.

The conceptual framework for this study was developed by integrating insights from construction risk management theory and previous empirical studies. It identifies five major independent variables, namely financial constraints, contractual changes, site-related matters, communication issues and external factors, as the key factors influencing delays in the project. These five variables represent the core categories of risks that commonly affect construction projects. The framework illustrates how each factor is theoretically linked to project delay and provides a structured basis for the research methodology, guiding data collection, analysis and interpretation. By grounding the study in established risk management principles, the framework connects theoretical concepts with the practical realities of high-rise construction projects. Figure 14.1 illustrates the conceptual framework outlining the relationships between delay factors and construction project performance.

Drawing on prior literature and the conceptual framework, the study formulates five hypotheses corresponding to the identified delay factors. These hypotheses examine how financial constraints, contractual changes, site-related matters, communication issues and external factors influence project delays. By testing these relationships, the study seeks to identify which factors have the greatest impact and how they contribute to overall project performance.

Figure 14.1 Conceptual framework

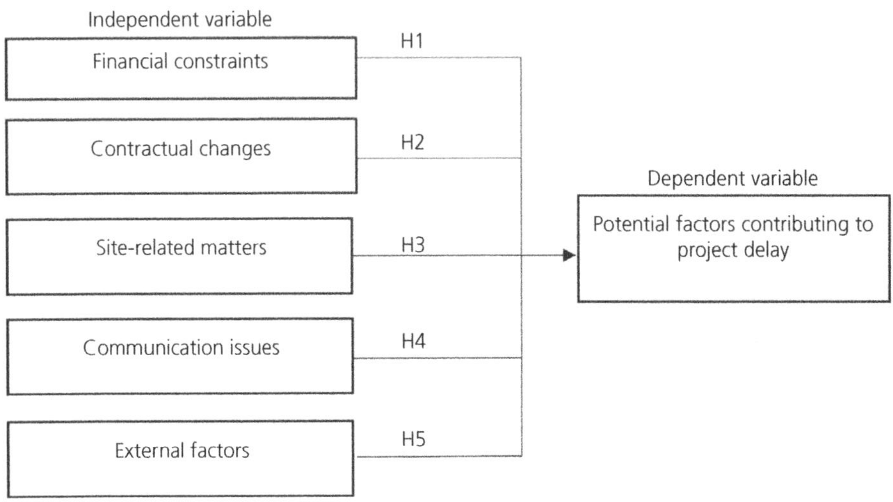

14.2. Methodology

14.2.1 Research design

In this study, the research questions aim to identify the key factors contributing to delays in construction projects. A quantitative research approach was selected because it efficiently collects data from large samples, yielding measurable and reliable results (Sekaran and Bougie, 2016). Quantitative research involves collecting and statistically analysing numerical data to answer specific questions, making it well-suited to describing patterns and testing relationships (Fischer *et al.*, 2023). Thus, the study adopts both descriptive and explanatory research designs.

Data collection was conducted in two phases. The first phase involved administering a questionnaire to project stakeholders to capture their perceptions of delay factors. The second phase drew on findings from previous studies to support the development of the conceptual framework and strengthen the interpretation of the results. Through this combined process, the study identified and examined the causes and effects of delays within construction projects.

14.2.2 Population and sampling procedures

Sampling involves selecting suitable individuals, events or objects for data collection, and the target population must be appropriately defined to ensure that survey responses effectively address the research questions (Sekaran and Bougie, 2016). For this study, the population comprises all companies and personnel contractually involved in the project, including the developer, project management consultant, main contractor, subcontractors and professional consultants such as architects, quantity surveyors, landscape specialists and mechanical, electrical, civil and structural engineers. The overall population consists of approximately 3000 employees directly engaged in or responsible for work on the project.

A probability sampling method was adopted, specifically simple random sampling, as it gives every member of the population an equal chance of being selected and ensures that the sample is representative of the broader population (Creswell, 2021). Because researchers often face constraints such as limited time, cost and accessibility, it is not always feasible to collect data from an entire population (Asfoor *et al.*, 2021). Therefore, selecting a subset of the population is both practical and sufficient.

To determine an appropriate sample size, this study follows Roscoe's (1975) guidelines and the Taro Yamane formula (Udeh, 2024). For a total population of 3000, a minimum of 97 respondents at a 10% precision level is considered adequate to produce reliable, accurate results (Sekaran and Bougie, 2016). The final sample consists of personnel at both managerial and non-executive levels who are directly involved in the project.

14.2.3 Data collection method

Data were collected primarily through self-administered questionnaires, with secondary sources such as reports, journals, articles and books. The questionnaire served as the main data-gathering instrument, and its development required attention to clarity, simplicity and ease of understanding for respondents. Items were adapted from earlier studies related to construction delays and modified to suit this research. The questionnaire consisted of closed-ended questions, allowing respondents to select answers from predetermined choices.

A five-point Likert scale was used to capture respondents' perceptions, ranging from strongly disagree (1) to strongly agree (5) (Jaymin-Sanchaniya *et al.*, 2024). The instrument was structured into three sections: respondent background and experience; project performance information; and delay-related factors. The questionnaire was distributed through email and Google Forms to stakeholders directly involved in the project, including the developer, main contractor, consultants and subcontractors. The data collected constitute the primary source of information for this study. Survey responses enabled the identification of the most significant delay factors and the stakeholders contributing most to them. By analysing these perceptions, the study identified the most critical causes of delay for the project.

14.2.4 Operationalisation and measurement

In this study, the independent variables represent the core factors expected to influence the dependent variable. These variables include financial constraints, contractual changes, site-related matters, communication issues and external factors. Financial constraints include unstable funding, irregular progress payments and sudden changes in material prices, all of which can disrupt workflows and slow project progress. Contractual changes involve commercial, technical or design modifications that require lengthy discussions, approvals and coordination between project stakeholders, often resulting in time extensions and cost overruns. Site-related matters include discrepancies in contract documents, delays in material supply, shortages of quality materials and an insufficiently skilled workforce. Communication issues arise from misinterpretation, ineffective coordination or breakdown in communication between clients, contractors, subcontractors, consultants and suppliers. External factors are influences beyond the control of project stakeholders, such as adverse weather conditions, natural disasters, regulatory changes, political instability and economic fluctuations.

No mediating variables were included in this research, as the study focuses directly on the relationship between the independent and dependent variables. The dependent variable, potential factors of project delay, represents the main outcome of interest. The purpose of the study was to identify the extent to which each independent variable contributes to project delays and to understand the nature of these influences within the context of the high-rise commercial development project.

14.2.5 Data analysis techniques

In identifying the factors contributing to construction delays, both descriptive and inferential statistical techniques were used to analyse the collected data. These methods enable the organisation, interpretation and drawing of meaningful conclusions about the relationships between variables. Descriptive statistics provide an overall view of the data, particularly the demographic background of the project stakeholders captured in Section 1 of the questionnaire. The quantitative data gathered online through Google Forms were analysed to describe respondents' characteristics and attitudes toward the delay factors, using measures such as mean, median, mode and standard deviation (SD).

The statistical package for the social sciences (SPSS) was the primary tool used for data analysis. It was used to summarise data using frequency distributions and cross-tabulations and to examine relationships between variables. In addition to descriptive analysis, inferential statistics were applied to generalise findings from the sample to the population. Techniques such as correlation analysis, regression analysis and hypothesis testing allowed the significance of observed patterns in the data to be determined and the theoretical relationships proposed in the study to be tested.

14.3. Results and analysis

14.3.1 Demographic profile

Figure 14.2 presents the demographic distribution of respondents involved in the construction project. The largest proportion is consultants at 26.9%, including project management, architect, civil and structural, and mechanical and electrical consultants, indicating their substantial influence on project execution. Subcontractors account for 22.2%, underscoring their critical operational role, while local authorities and government agencies also show notable involvement, consistent with their regulatory responsibilities for overseeing project compliance. The main contractor accounts for 13.9%, a significant share that reflects its central role and direct involvement in project delays. This demographic distribution is essential for interpreting the study's findings, as the stakeholders' roles and backgrounds shape their experiences, perceptions and insights into the causes of project delays.

14.3.2 Reliability analysis

Reliability analysis was conducted to assess the internal consistency of all constructs in the study, using Cronbach's alpha (α) as the primary indicator. A coefficient above 0.70 is

Figure 14.2 Demographic profile of project stakeholders

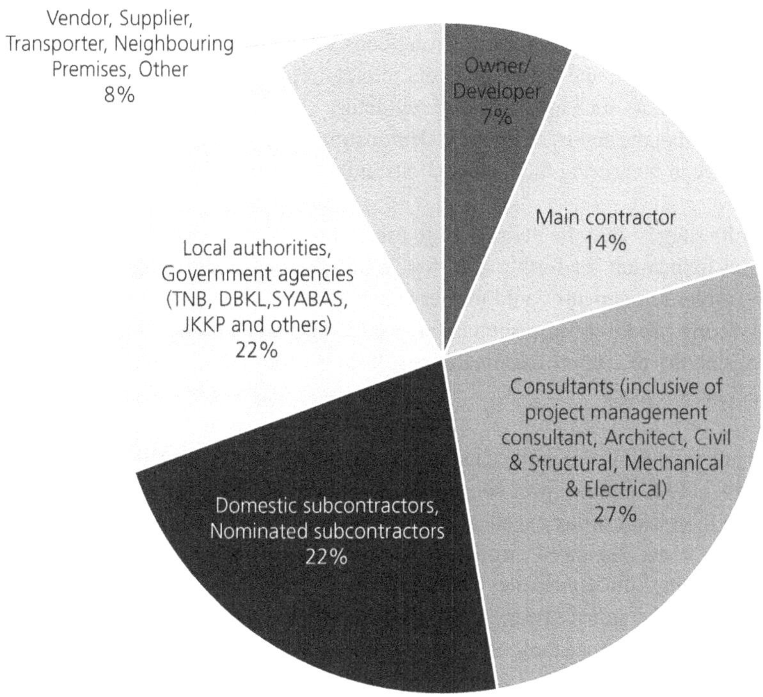

generally considered acceptable (Dobrucali *et al.*, 2024). The results show that all five constructs exceed this threshold, indicating strong reliability (see Table 14.1). Financial constraints recorded an alpha of 0.872, contractual changes 0.865, site-related matters 0.848 and both communication issues and external factors 0.861. These values demonstrate that the items within each construct consistently measure their respective concepts and are therefore suitable for further analysis. Each factor reflects stable and dependable measurement:

Table 14.1 Reliability assessment of project delays

Variable (factor delays)	Cronbach's alpha, α	No. of items
Financial constraints	0.872	6
Contractual changes	0.865	6
Site-related matters	0.848	6
Communication issues	0.861	6
External factors	0.861	6

financial constraints show high reliability in assessing monetary issues that contribute to delays; contractual changes reliably capture the effects of design modifications and approval processes; site-related matters consistently measure operational challenges at the construction site; communication issues show strong consistency in identifying problems related to stakeholder interaction; and external factors reliably represent influences outside the control of project participants such as weather or regulatory changes. Overall, the Cronbach's alpha results confirm that the survey instrument demonstrates good to excellent internal consistency and is effective in measuring the factors contributing to project delays.

These results suggest that the items used to assess the delay factors are dependable and provide a solid foundation for further analysis. While the constructs exhibit good to excellent reliability, minor refinements could further enhance accuracy and reduce measurement error. Ensuring strong reliability is essential for producing valid findings and strengthening the instrument's ability to support meaningful conclusions about factors influencing organisational and project performance.

Discriminant validity was examined using corrected item–total correlation analysis as shown in Tables 14.2–14.5. Across most constructs, correlation coefficients exceeded the minimum threshold of 0.40, confirming excellent discriminant validity. Within the financial constraints construct, all items displayed strong correlations with the overall scale, indicating that each item meaningfully contributes to measuring financial-related delays. A similar pattern emerged in the contractual changes construct, in which all items correlated well with the total scale, confirming that they collectively represent the intended factor. For site-related

Table 14.2 Item total statistics – financial constraints

	Scale mean if item deleted	Scale variance if item deleted	Corrected item total correlation	Cronbach's alpha if item deleted
Changes in cost or material prices	18.62	13.565	0.763	0.836
Problem between contractors and subcontractors regarding payment	19.04	11.961	0.776	0.832
Delay in progress payments from the client to the contractor and subcontractor/suppliers	18.35	11.969	0.759	0.836
Financial difficulties for the relevant parties involved	18.35	15.202	0.534	0.872
Insufficient funds by the client	18.35	14.193	0.588	0.865
Extension and changes to the contract affected the overall budget	18.40	15.008	0.682	0.854

The corrected item–total correlation coefficients for the financial constraint items are all greater than 0.40, indicating excellent discriminant validity.

Table 14.3 Item total statistics – communication issues

	Scale mean if item deleted	Scale variance if item deleted	Corrected item–total correlation	Cronbach's alpha if item deleted
Delay in the client's finalisation of the contract award	18.08	12.339	0.636	0.850
A change in contractual specifications slows project execution	18.02	16.037	0.403	0.879
Slow preparation and delayed finalisation of the additional work package	18.07	11.022	0.848	0.804
Inadequate project feasibility study by the owner/client	18.67	12.075	0.717	0.832
Delay in design freeze	18.05	13.054	0.727	0.831
Slow finalising consultants' onboarding	17.72	13.922	0.675	0.843

One item ('Delay and slow decision-making under supervision') returned a corrected item–total correlation of 0.293, falling below the 0.40 threshold; the remaining items exceeded this criterion.

Table 14.4 Item total statistics – site-related matters

	Scale mean if item deleted	Scale variance if item deleted	Corrected item–total correlation	Cronbach's alpha if item deleted
Delay in site possession by the substructure contractor	18.39	13.174	0.834	0.788
Ineffective planning and scheduling of the projects	18.02	14.261	0.640	0.823
Construction mistakes and defective work	18.53	15.504	0.496	0.846
Failure of machinery/equipment	18.50	13.449	0.687	0.813
Slow mobilisation at the site has delayed consultants' onboarding	18.31	10.554	0.752	0.809
Shortage and delay in material delivery/supply to the site	18.20	15.267	0.472	0.850

The corrected item–total correlation coefficients for the communication issues items exceed 0.40, indicating excellent discriminant validity.

Table 14.5 Item total statistics – external factors

	Scale mean if item deleted	Scale variance if item deleted	Corrected item–total correlation	Cronbach's alpha if item deleted
Poor communication between the parties involved	18.22	13.688	0.687	0.833
Contractor's poor coordination with relevant parties in the project	18.31	11.560	0.713	0.831
Slow response and poor site inspection	18.31	13.321	0.728	0.825
Delay and slow decision-making under supervision	18.39	16.595	0.293	0.888
Conflicts between parties	18.19	12.694	0.753	0.819
Lack of communication between parties	18.07	12.742	0.768	0.816

The corrected item–total correlation coefficients for the external factors items are all greater than 0.40, indicating excellent discriminant validity.

matters and communication issues, the corrected correlations likewise exceeded the required level, demonstrating that the items effectively capture operational and communication-related problems. External factors largely met the discriminant validity criteria; however, one item ('Delay and slow decision-making under supervision', $r = 0.293$) fell below the 0.40 threshold. These findings collectively indicate that the measurement tool is both reliable and valid, strengthening the overall quality of the research instrument used to analyse project delay factors.

14.3.3 Descriptive analysis

Descriptive analysis provides a clear overview of the raw characteristics of the data and helps summarise the responses before proceeding to inferential analysis. The results in Table 14.6 show that financial constraints recorded the highest mean (3.70, SD = 0.73), indicating that it is perceived as the most influential factor contributing to project delays. Issues such as fluctuating material prices, payment delays along the contractual chain, financial difficulties among stakeholders, and insufficient client funding appear to affect project progress significantly.

Contractual changes recorded a mean of 3.62 (SD = 0.71), while site-related matters averaged 3.67 (SD = 0.73). Communication issues (3.65, SD = 0.72) and external factors (3.69, SD = 0.74) were also rated highly, indicating general agreement that all five factors contribute meaningfully to construction delays. Contractual changes showed the widest response range, suggesting more diverse views among respondents, while external factors recorded the highest standard deviation, reflecting slightly greater variability in perceptions of external

Table 14.6 Descriptive statistics of factor delays

	n	Minimum	Maximum	Mean	SD
Financial constraints	108	2.17	4.50	3.7037	0.7293
Contractual changes	108	2.17	4.57	3.6204	0.71362
Site-related matters	108	2.17	4.50	3.6651	0.72933
Communication issues	108	2.00	4.50	3.6497	0.72306
External factors	108	1.83	4.50	3.6867	0.74316
Valid *n* (listwise)	108				

influences. Overall, all five factors are viewed as significant contributors to project delays, with financial constraints emerging as the most critical on average. The results also show moderate variability across respondents, indicating consistent but not uniform perceptions of how these factors affect project performance.

In regression analysis, an acceptable histogram should indicate that the residuals are approximately normally distributed, supporting the validity of the model. Figure 14.3 demonstrates

Figure 14.3 Regression standardised residuals

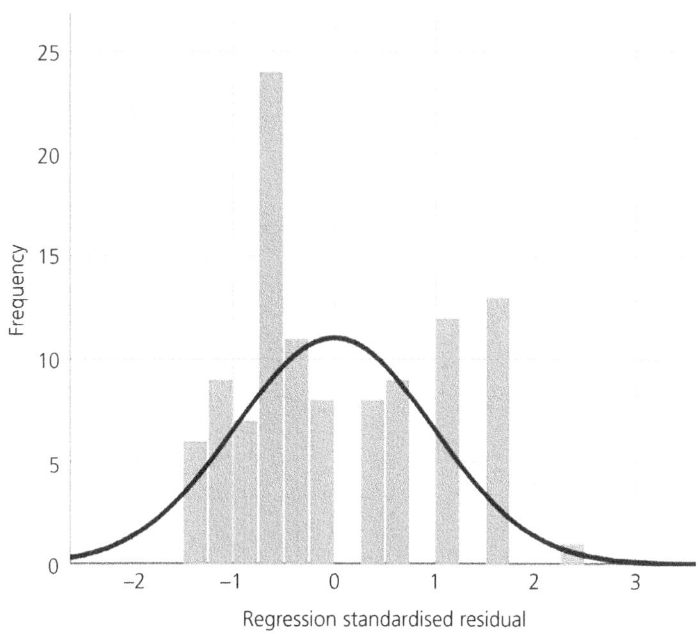

a reasonably bell-shaped pattern, indicating that the residuals follow a normal distribution. The mean of the residuals is close to zero, suggesting that the model does not systematically overestimate or underestimate project delay factors. The standard deviation of the residuals is 0.976, indicating that the errors are clustered near the mean and that the model fit is satisfactory. With a sample size of 108, the analysis is adequately powered, and the residual distribution between –2 and 3 is typical for real-world construction data. Overall, Figure 14.3 supports the assumptions of normality and constant variance required for linear regression, indicating that the model is appropriate for interpreting and predicting project delays. Minor deviations from perfect normality, however, suggest that further refinement could strengthen the model's performance.

The Q-Q plots for all variables, shown in Figures 14.4–14.8, indicate slight negative skewness and longer tails at the lower end of the distributions. In the case of project delays, a few

Figure 14.4 Normal Q-Q plot of project delay

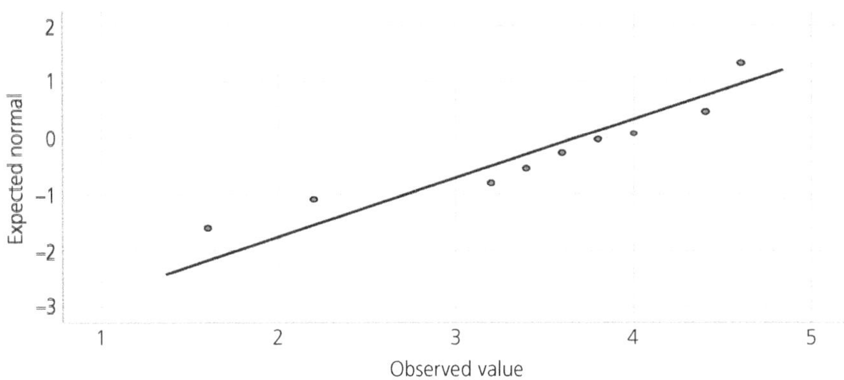

Figure 14.5 Normal Q-Q plot of financial constraints

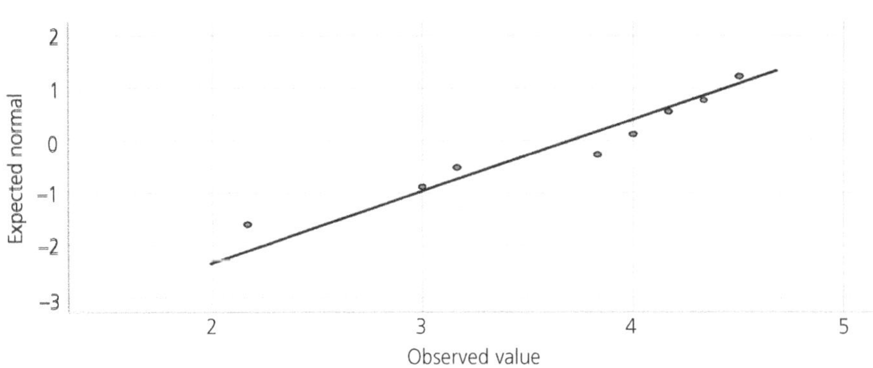

Figure 14.6 Normal Q-Q plot of contractual changes

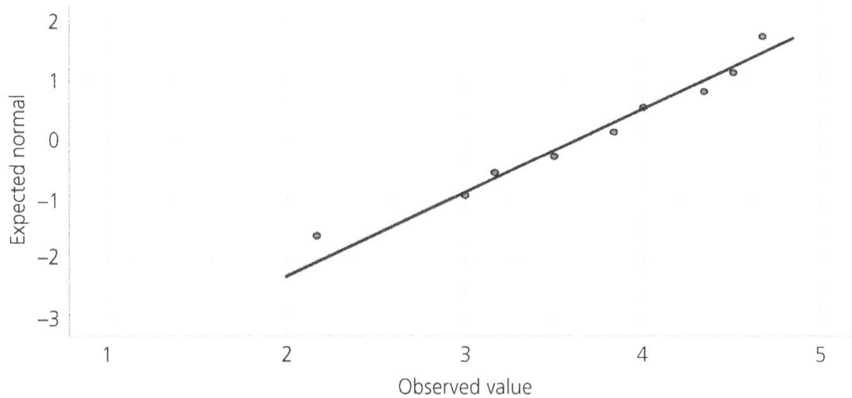

Figure 14.7 Normal Q-Q plot of site-related matters

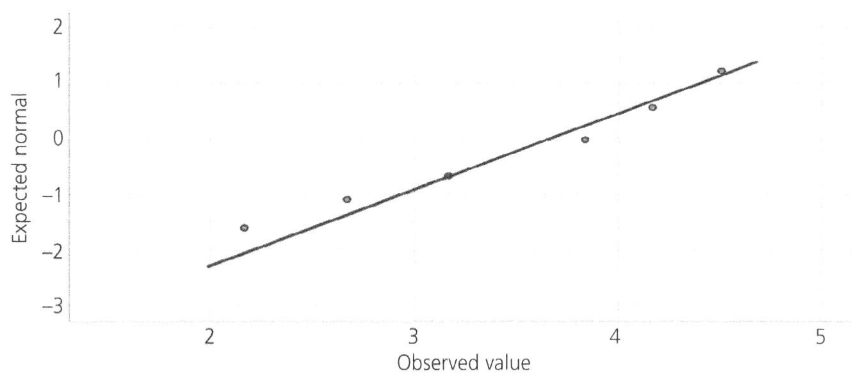

mild outliers are present, although none are severe enough to distort the data. Across financial constraints, contractual changes, site-related matters and communication issues, no significant outliers are observed; however, all plots deviate from the expected straight-line pattern, confirming that the data do not follow a normal distribution. Despite this non-normality, the sample size is sufficiently large to justify the use of parametric statistical techniques, which remain robust under such conditions.

14.3.4 Normality analysis

Table 14.7 reports the normality test results used to assess the distribution of the study variables. Meanwhile, Table 14.8 presents the test results for mean differences across the

Figure 14.8 Normal Q-Q plot of communication issues

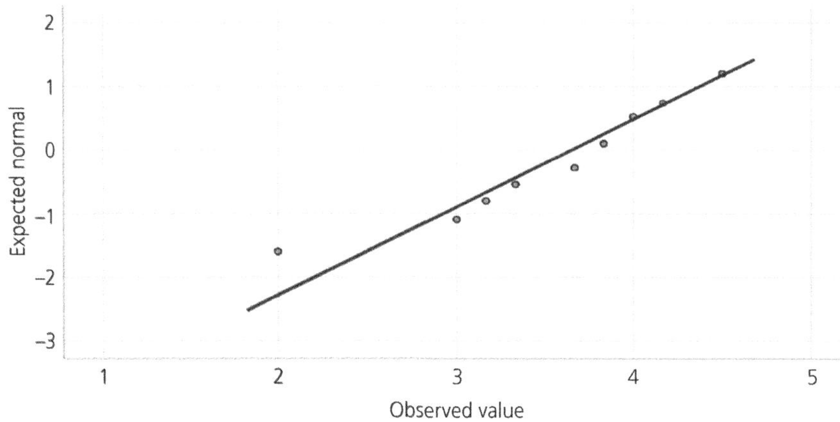

Table 14.7 Normality test results for study variables

	Kolmogorov–Smirnov[a]			Shapiro–Wilk		
	Statistic	*df*	Sig.	Statistic	*df*	Sig.
Financial constraints	0.237	108	0.000	0.855	108	0.000
Contractual changes	0.164	108	0.000	0.926	108	0.000
Site-related matters	0.258	108	0.000	0.869	108	0.000
Communication issues	0.176	108	0.000	0.872	108	0.000
External factors	0.216	108	0.000	0.815	108	0.000
Project delay	0.237	108	0.000	0.818	108	0.000

[a]Lilliefors significance correction

Table 14.8 Test for mean difference

	Sum of squares	*df*	Mean square	*F*	Sig.
Between groups	4.942	2	2.471	2.783	0.066
Within groups	93.213	105	0.888		
Total	98.155	107			

identified groups. The Shapiro–Wilk and Kolmogorov–Smirnov tests indicate that the individual variables do not follow a normal distribution ($p < 0.001$). However, in multiple regression analysis, the assumption of normality applies to the distribution of residuals rather than to the independent or dependent variables themselves. Given the sample size ($n = 108$) and the robustness of regression analysis to moderate deviations from normality, the analysis proceeded after confirming that the regression residuals were approximately normally distributed, as indicated by histogram and Q-Q plot inspection (Templeton and Blank, 2025).

The analysis of variance (ANOVA) results provide no significant evidence against the null hypothesis ($F(2, 105) = 2.78, p > 0.05$), indicating that the average project delay rate does not differ significantly across contract value categories.

14.3.5 Correlation analysis

Correlation analysis was conducted to determine the level of association between various factors expected to cause project delays. Correlation analysis helps measure the direction and magnitude of linear association. Table 14.9 presents the Pearson correlation analysis examining the relationships between potential delay factors and overall project delay.

The summary highlights five potential factors contributing to project delays within the high-rise commercial development. The analysis shows that communication issues ($r = 0.720$, $p < 0\ .01$) and financial constraints ($r = 0.680$, $p < 0.01$) demonstrate the strongest positive relationships with project delays. These are followed by contractual changes ($r = 0.650$, $p < 0.01$), site-related matters ($r = 0.590$, $p < 0.01$) and external factors ($r = 0.410$, $p < 0.01$), all of which are significantly associated with delays at the bivariate level. Among these, communication issues exhibit the most substantial relationship with project delay, while external factors show the weakest.

Table 14.9 Relationship between potential factors and project delays

Variable	Financial constraints	Contractual changes	Site-related matters	Communication issues	External factors	Project delay
Financial constraints	1.000					
Contractual changes	0.500[a]	1.000				
Site-related matters	0.520[a]	0.490[a]	1.000			
Communication issues	0.550[a]	0.510[a]	0.580[a]	1.000		
External factors	0.450[a]	0.480[a]	0.420[a]	0.440[a]	1.000	
Project delay	0.680[a]	0.650[a]	0.590[a]	0.720[a]	0.410[a]	1.000

[a]Correlation is significant at the 0.01 level (2-tailed).

The findings further indicate that communication issues and financial constraints show the strongest bivariate associations with project delay; the independent predictive contribution of each factor is examined in the regression analysis that follows. Multiple regression analysis confirms that these internal project management factors are statistically significant predictors, whereas external factors do not maintain an independent significant contribution in the multivariate model. Consequently, these primary drivers are the most critical factors associated with delays in project completion. The linearity and homoscedasticity assumptions were tested and found to hold, indicating that the variables are linearly related and a multiple linear regression model is appropriately applied to the data.

14.3.6 Regression analysis

Although the Kolmogorov–Smirnov and Shapiro–Wilk tests indicated that the observed variables were not normally distributed ($p < 0.001$), multiple regression analysis was still conducted. The normality assumption in linear regression concerns the distribution of residuals, not the distribution of the independent variables. Inspection of the histogram and Q-Q plot of standardised residuals demonstrated approximate normality. Additionally, with a sample size of 108, regression analysis is considered robust to moderate violations of normality. Therefore, the application of multiple regression is statistically justified. The overall significance of the regression model is confirmed by the ANOVA results presented in Table 14.10.

Table 14.10 presents the results of the multiple regression analysis identifying the main factors contributing to construction project delays. The regression model is statistically significant ($F(5, 102) = 23.31, p < 0.001$) indicating that the set of predictors collectively explains a significant proportion of the variance in project delay.

Based on the regression and total sum of squares values (52.340/98.155), the model explains approximately 53.3% of the variance in project delays ($R^2 = 0.533$). This indicates moderate to strong explanatory power, suggesting that financial constraints, contractual changes, site-related matters, communication issues and external factors jointly contribute meaningfully to construction delays.

The significant F-statistic indicates that at least one independent variable provides substantial predictive value. While the model does not account for all variation in project delay,

Table 14.10 ANOVA results for the multiple regression model

ANOVA					
Model	Sum of squares	*df*	Mean square	*F*	Sig.
Regression	52.340	5	10.468	23.31	0.000
Residual	45.815	102	0.449		
Total	98.155	107			

Note: dependent variable: project delay; predictors: (constant), external factors, site-related matters, contractual changes, financial constraints, communication issues.

explaining over half of the variance indicates that these five factors represent key determinants of delay within the studied project context. Overall, the regression results demonstrate a robust, practically meaningful model for understanding the drivers of construction project delays.

Table 14.11 presents the regression coefficients that identify the main factors contributing to construction project delays. The results indicate that financial constraints (B = 0.214, p = 0.007), contractual changes (B = 0.312, p < 0.001), site-related matters (B = 0.187, p = 0.013) and communication issues (B = 0.348, p < 0.001) are statistically significant predictors of project delay. This suggests that increases in these factors are associated with longer project delays.

Among the significant predictors, communication issues (β = 0.331) demonstrate the strongest standardised effect, followed by contractual changes (β = 0.295) and financial constraints (β = 0.228), indicating their relatively greater contribution to delay when controlling for other variables. Site-related matters also show a meaningful but comparatively smaller effect (β = 0.176). In contrast, external factors (B = 0.096, p = 0.167) are not statistically significant, suggesting that their independent contribution to project delay is limited within this model.

Collinearity statistics indicate no multicollinearity concerns, with tolerance values above 0.5 and variance inflation factor (VIF) values below 2.0. Furthermore, inspection of the scatterplot of regression-standardised residuals against predicted values, as shown in Figure 14.9, indicates that the assumptions of linearity and homoscedasticity are reasonably satisfied, as the residuals appear randomly distributed with no clear patterns.

Overall, the findings demonstrate that internal project management factors, particularly communication and contractual issues, play a more critical role in influencing construction project delays than external factors.

Table 14.11 Regression coefficients for factors contributing to project delay

Coefficients							
Model	Unstandardised coefficients		Standardised coefficient	t	Sig.	Collinearity statistics	
	B	Std. error	β			Tolerance	VIF
1 (Constant)	0.512	0.241		2.12	0.036		
Financial constraints	0.214	0.078	0.228	2.74	0.007	0.624	1.60
Contractual changes	0.312	0.085	0.295	3.67	0.000	0.558	1.79
Site-related matters	0.187	0.074	0.176	2.53	0.013	0.642	1.56
Communication issues	0.348	0.081	0.331	4.29	0.000	0.531	1.88
External factors	0.096	0.069	0.088	1.39	0.167	0.701	1.43

Note: dependent variable: project delay.

Figure 14.9 Regression standardised residual against regression standardised predicted value

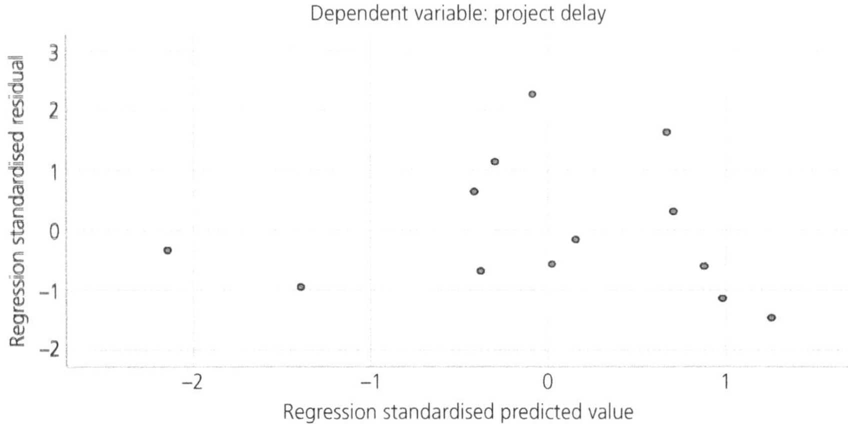

14.3.7 Hypothesis testing

Overall, the study partially supports the proposed hypotheses. Correlation analysis indicates that all five independent variables (financial constraints, contractual changes, site-related matters, communication issues and external factors) are positively and significantly associated with project delay at the bivariate level ($p < 0.01$). Communication issues recorded the strongest correlation with project delay ($r = 0.720$), followed by financial constraints ($r = 0.680$), contractual changes ($r = 0.650$) and site-related matters ($r = 0.590$). In contrast, external factors showed the weakest association ($r = 0.410$).

However, multiple regression analysis provides a more refined understanding of their independent effects. The results indicate that financial constraints ($B = 0.214$, $p = 0.007$), contractual changes ($B = 0.312$, $p < 0.001$), site-related matters ($B = 0.187$, $p = 0.013$) and communication issues ($B = 0.348$, $p < 0.001$) significantly predict project delay. Among these, communication issues ($\beta = 0.331$) emerged as the strongest predictor, followed by contractual changes ($\beta = 0.295$) and financial constraints ($\beta = 0.228$).

In contrast, external factors ($B = 0.096$, $p = 0.167$) were not statistically significant in the regression model, indicating that their independent contribution to project delay is limited when controlling for other variables. Therefore, hypotheses relating to financial constraints, contractual changes, site-related matters and communication issues are supported, whereas the hypothesis concerning external factors is not supported in the multivariate analysis.

14.4. Practical insights

The findings of this study offer several important practical insights for construction practitioners, project managers, developers and policy makers involved in high-rise construction projects in Malaysia. First, the results clearly indicate that communication issues and contractual changes are the most significant predictors of project delays. This highlights the need for project teams to prioritise effective communication mechanisms, including clear reporting structures, timely decision-making processes and regular coordination meetings among clients, consultants, contractors and subcontractors. Improving information flow and reducing misunderstandings can significantly minimise onsite disruptions and rework.

Second, site-related challenges such as material shortages, delayed mobilisation and inadequate site supervision require more proactive planning and monitoring. Project managers should strengthen site management practices by improving logistics coordination, ensuring timely material procurement and maintaining adequate levels of skilled labour availability. The adoption of digital project management tools for real-time progress tracking and site reporting can further enhance control over site operations.

Third, financial constraints are a statistically significant direct predictor of project delays ($\beta = 0.228$, $\rho = 0.007$), ranking third among the significant predictors. Controlling upstream issues such as design finalisation and variation management can help reduce financial stress, but financial risk management should also be addressed directly.

Fourth, contractual changes and discrepancies significantly affected project timelines. Early design freeze, clearer contract documentation and faster approval processes are essential to limit variations and avoid cascading delays. Stakeholders should establish clear procedures for managing variations and ensure accountability across contractual parties.

Finally, while external factors were perceived as contributors to delay, their relatively weaker influence suggests that many delays are within the control of project stakeholders. By embedding structured risk management practices aligned with PMBOK and ISO 31000 standards, construction organisations can better anticipate, mitigate and respond to delay risks. Overall, the study underscores that effective communication, disciplined contractual control and proactive financial management are critical to improving time performance and reducing delays in complex construction projects.

14.5. Conclusion and recommendations

14.5.1 Limitations

The study acknowledges several limitations that may affect the generalisability of its findings. First, although 108 responses were collected, the validity and robustness of the results could be enhanced with a larger sample. A broader sample, including more stakeholders from various high-rise construction projects across Malaysia, might reveal additional dimensions not captured in this study, leading to more comprehensive and generalisable conclusions. Second, the research focuses solely on the high-rise commercial development project in KL, which limits the applicability of the findings to other construction projects that may differ in scale, complexity, management structures and stakeholder dynamics. As such, the effectiveness of the identified factors in predicting delays may not hold for projects in different contexts or facing different challenges. Third, the use of self-reported survey data introduces the possibility of response bias, as participants may unintentionally overstate or understate the impact of certain delay factors based on personal perceptions or experiences. This subjectivity may distort the accuracy of the results, particularly since increased costs or perceived delays do not always reflect the project's actual conditions. Together, these limitations suggest that the findings should be interpreted with caution and further validated through research with more diverse samples and objective data sources.

14.5.2 Recommendations and future research

Future research should include comparative studies across multiple construction projects in Malaysia. Such an approach would help validate the study's findings and determine whether the identified delay factors are universally applicable or project-specific. Expanding the research to include various types of construction projects would allow the same delay indicators to be examined across contexts, thereby enhancing the external validity and generalisability of the results.

Additionally, influential variables such as technological challenges, political factors or cultural aspects may not have been included in this study. Future investigations should consider these and other possible determinants to provide a more comprehensive explanation of project delays. Exploring a broader range of potential influences may reveal important interactions and deepen understanding of recurring delay patterns in the construction industry.

Future studies may also benefit from adopting a mixed-methods research design that integrates both quantitative and qualitative approaches. While quantitative tools, such as structured questionnaires, can assess the frequency and severity of project delays, qualitative methods, such as interviews, focus groups and stakeholder discussions, can offer insights into the underlying causes of these delays. Such a dual approach would not only shed light on the factors contributing to project extensions but also support the development of more targeted and effective strategies to mitigate delay issues.

REFERENCES

Abdul Nabi M and El-adaway III (2022) Understanding disputes in modular construction projects: key common causes and their associations. *Journal of Construction Engineering and Management* **148(1)**: 04021184.

Asfoor HMA, Al-Jandeel AAT, Igorevich KK and Ivanovna LA (2021) Control of time, cost and quality of construction project management. In *Proceedings of the International Conference on Energy and Green Computing*, 9–10 December 2021, Meknes, Morocco, *E3S Web of Conferences*, vol. 336.

Bryde DJ and Robinson L (2005) Client versus contractor perspectives on project success criteria. *International Journal of Project Management* **23(8)**: 622–629.

Creswell JW (2021) *A Concise Introduction to Mixed Methods Research*. Sage, Thousand Oaks, CA, USA.

Dobrucali E, Demirkesen S, Sadikoglu E, Zhang C and Damci A (2024) Investigating the impact of emerging technologies on construction safety performance. *Engineering, Construction and Architectural Management* **31(3)**: 1322–1347.

Doloi H, Sawhney A, Iyer KC and Rentala S (2012) Analysing factors affecting delays in Indian construction projects. *International Journal of Project Management* **30(4)**: 479–489.

Fischer HE, Boone WJ and Neumann K (2023) Quantitative research designs and approaches. In *Handbook of Research on Science Education*. Routledge, Abingdon, UK, pp. 28–59.

Hansen AS, Svejvig P and Hansen LK (2022) Revisiting Shenhar and Dvir's diamond model: do we need an upgrade? In *Research on Project, Programme and Portfolio Management: Projects as an Arena for Self-Organizing*. Springer, Cham, Switzerland, pp. 175–190.

ISO (International Organization for Standardization) (2018) ISO 31000: Risk Management – Guidelines. ISO, Geneva, Switzerland.

Jaymin-Sanchaniya R, Thomson D, Kundzina A and Geipele I (2024) Effective project management practices in construction industry: quantitative study. *Engineering for Rural Development* **23**: 85–92.

Kumar M and Kumari S (2025) Causes of delays in road construction projects: a systematic review. *Journal of Financial Management of Property and Construction* **30(2)**: 256–294.

Latif M, Saleem M and Cheema SM (2023) Analyzing the key factors contributing to project delays in the construction industry: a comprehensive study. *Journal of Development and Social Sciences* **4(3)**: 936–944.

Naji KK, Gunduz M and Falamarzi MH (2022) Assessment of construction project contractor selection success factors considering their interconnections. *KSCE Journal of Civil Engineering* **26(9)**: 3677–3690.

PMI (Project Management Institute) (2021) *A Guide to the Project Management Body of Knowledge (PMBOK® Guide)*, 7th edn. PMI, Newtown Square, PA, USA.

Sekaran U and Bougie R (2016) *Research Methods for Business: A Skill-Building Approach*, 7th edn. Wiley, New York, NY, USA.

Sanni-Anibire MO, Mohamad Zin R and Olatunji SO (2022) Causes of delay in the global construction industry: a meta analytical review. *International Journal of Construction Management* **22(8)**: 1395–1407.

Selcuk O, Turkoglu H, Polat G and Hajdu M (2024) An integrative literature review on the causes of delays in construction projects: evidence from developing countries. *International Journal of Construction Management* **24(6)**: 610–622.

Sovacool BK and Ryu H (2025) Beyond economies of scale: learning from construction cost overrun risks and time delays in global energy infrastructure projects. *Energy Research & Social Science* **123**: 104057.

Takagi N, Varajão J and Ventura T (2024) Implementing success management on government-to-government projects: an integrated perspective with the PMBOK guide. *International Journal of Managing Projects in Business* **17(1)**: 153–171.

Tariq J and Gardezi SSS (2023) Study the delays and conflicts for construction projects and their mutual relationship: a review. *Ain Shams Engineering Journal* **14(1)**: 101815.

Templeton GF and Blank DB (2025) Understanding non-normality in business, finance, and accounting research. *Review of Quantitative Finance and Accounting*: 1–37.

Udeh GI (2024) The impact of project evaluation and analysis on project success: a case study of UGI Technologies, Lagos, Nigeria. *Systemic Analytics* **2(2)**: 256–278.

Viles E, Rudeli NC and Santilli A (2020) Causes of delay in construction projects: a quantitative analysis. *Engineering, Construction and Architectural Management* **27(4)**: 917–935.
Welde M and Bukkestein I (2022) Over time or on time? A study of delays in large government projects. *Procedia Computer Science* 196: 772–781.
Zidane YJT and Andersen B (2018) The top 10 universal delay factors in construction projects. *International Journal of Managing Projects in Business* **11(3)**: 650–672.

Conclusion

Advancing Project Management: Frameworks, Digital Transformation, Sustainability and Governance

Mohammed Reyasudin Basir Khan
ISBN 978-1-83742-799-4
https://doi.org/10.1108/978-1-83742-798-720261018

Chapter 15
Future directions in project management innovation

Mohammed Reyasudin Basir Khan

Tun Razak Graduate School, Universiti Tun Abdul Razak (UNIRAZAK), Kuala Lumpur, Malaysia

15.1. Introduction

Project management has expanded from a primarily operational discipline focused on planning and control into a strategic organisational capability focused on governance, digitalisation, stakeholder value and long-term resilience. Across sectors, projects have become the principal mechanism for implementing digital transformation, modernising infrastructure, achieving sustainability commitments and strengthening organisational competitiveness. This broadened mandate has increased the expectations placed on project managers and project organisations. The contemporary project manager is increasingly required to operate at the intersection of technology, people, institutions and strategy.

The chapters in this volume provide convergent evidence that project performance is shaped by more than technical scheduling and budgeting. Recurring challenges include stakeholder misalignment, communication breakdowns, uneven adoption of digital tools, governance gaps, issues with risk visibility and sustainability trade-offs. These challenges have been documented globally, especially in complex infrastructure and multi-stakeholder projects (Flyvbjerg and Gardner, 2023; Merrow, 2011). The book's empirical contexts, while often located in Malaysia, reflect patterns observed internationally: institutional fragmentation, challenges in contractor coordination, digitally mediated work, regulatory scrutiny and growing requirements for sustainability reporting.

This concluding chapter is designed to be stand-alone and to perform three functions.

- It consolidates the book's empirical and conceptual contributions within four domains: methods and frameworks; digital transformation; sustainability and strategy; and governance and risk.
- It develops implications for theory, practice and policy, connecting the empirical insights to established foundations such as the project management body of knowledge

(PMBOK), the technology acceptance model (TAM), the resource-based view (RBV), contingency theory and governance scholarship (Barney, 1991; Davis, 1989; Donaldson, 2001; PMI, 2021; Too and Weaver, 2014).
- It proposes a forward-looking agenda for project management innovation, identifying capability pathways likely to define the next decade of project delivery.

15.2. From methods to adaptive frameworks

15.2.1 Methodological discipline remains necessary but not sufficient

Across the empirical studies, methodological structure consistently supported clarity, accountability and coordination. In construction and infrastructure settings, formal planning tools and documentation practices helped coordinate multiple contractors and stakeholders. In engineering, procurement and construction (EPC) environments, standardised stage-gate logic, design review practices and procurement coordination supported control over technical dependencies. In manufacturing contexts, structured improvement methodologies such as Lean and Six Sigma enabled measurable reductions in variation, rework and defects, improving cycle time and quality outcomes.

These observations align with the long-standing proposition that structured processes reduce uncertainty and create repeatability in project execution (Kerzner, 2017). PMBOK's process orientation remains useful as a common language for integration, scope, schedule, cost, quality, resource, communications, risk, procurement and stakeholder management (PMI, 2021). However, the book's evidence shows that PMBOK-style procedural discipline is insufficient when applied without adaptation to context, capability and governance conditions.

15.2.2 Why adaptation is becoming central to modern frameworks

The empirical chapters collectively show that contemporary projects face

- greater uncertainty in supply chains, permitting and stakeholder behaviour
- higher interdependence between disciplines and organisations
- more stringent oversight and compliance expectations
- increasing digital mediation of planning, reporting and collaboration
- rising sustainability requirements in design and delivery decisions.

In such environments, purely linear models can become brittle. A core message across the studies is that frameworks must be adaptive, combining structured discipline with contextual tailoring. This aligns with contingency theory, which holds that effectiveness depends on fit between organisational practices and environmental conditions (Donaldson, 2001). It also aligns with project scholarship, which emphasises complexity, emergence and systems thinking in project environments (Baccarini, 1996; Geraldi *et al.*, 2011).

15.2.3 Communication as a methodological capability not a 'soft add-on'

A strong cross-cutting finding is that communication is not peripheral. It is a functional determinant of whether frameworks work in practice. Multiple studies across construction and infrastructure contexts show how incomplete information flow, unclear responsibility boundaries and inconsistent reporting contribute to delays and rework. This echoes broader evidence that project communication failures frequently become schedule and cost risks (Kerzner, 2017; PMI, 2021).

A key implication is that methodological frameworks should embed communication protocols explicitly, including

- escalation paths for technical and commercial issues
- disciplined meeting rhythms and decision logging
- standardised documentation and change-control records
- stakeholder mapping and expectation alignment routines.

15.2.4 Measurement, estimation and operational realism

Several empirical contexts in the book emphasise the reliability of estimation and the realism of planning. Cost estimation models, schedule-tracking tools and progress-reporting practices were repeatedly shown to influence management credibility and stakeholder confidence. This aligns with the project control literature, which highlights the importance of reliable baselines and transparent reporting for trust and governance (Flyvbjerg and Gardner, 2023; Merrow, 2011).

The overarching conclusion is that project methods must evolve toward adaptive, evidence-based frameworks where planning is not merely predictive but continuously updated through monitoring, risk feedback and digital data.

15.3. Digital transformation as a capability enabler

15.3.1 Digital tools change information flow but adoption is sociotechnical

Digital transformation appears to enable improved visibility, coordination and auditability in empirical studies. In transportation and infrastructure contexts, project management tools supported scheduling, documentation and progress monitoring. In engineering and EPC settings, digital systems enabled design coordination and procurement tracking. In manufacturing contexts, data-driven process control supported consistent improvement cycles.

However, the empirical evidence also shows that digital investment does not automatically produce performance gains. Digital transformation is best understood as a sociotechnical change process in which technology interacts with users, roles, routines and culture. This aligns with research emphasising that digital transformation succeeds when organisations change processes and behaviours alongside technology (Carvalho *et al.*, 2022; Vial, 2019).

15.3.2 TAM remains relevant but project settings add new determinants

The book's findings strongly align with TAM: perceived usefulness and perceived ease of use influence both acceptance and actual use (Davis, 1989). Where tools were viewed as cumbersome, duplicative or misaligned with workflows, adoption was limited. Where tools clearly reduced reporting burden, improved coordination or supported decision-making, users were more willing to engage.

Yet project environments introduce additional acceptance factors:

- role clarity around who updates which data and when
- data governance concerning accuracy, version control and approvals
- training and onboarding for contractors and new team members
- leadership signalling about whether digital reporting is mandatory
- interoperability across systems and organisations.

This suggests that future research should extend TAM in project environments by incorporating governance and workflow integration variables (Cai *et al.*, 2023).

15.3.3 Digital transformation as a governance amplifier

Across sectors, digital tools strengthen governance by producing traceability, consistent reporting and audit trails. These capabilities are particularly valuable where projects involve public investment, regulated industries or security-sensitive constraints. Governance scholarship emphasises that oversight depends on information quality, decision-making clarity and accountability mechanisms (Too and Weaver, 2014). Digital systems can support these conditions when they are integrated into formal governance processes rather than treated as optional tools.

15.3.4 Sectoral maturity differences and the problem of fragmented adoption

The empirical chapters show uneven digital maturity. Larger organisations and infrastructure operators often have greater capacity for system investment and training. Smaller contractors may struggle with cost, skills and organisational readiness. Fragmented adoption creates data silos and inconsistent reporting, which reduces the governance and performance benefits of digital systems.

A practical implication is the need for minimum digital operating standards within project ecosystems, including common data definitions, reporting formats and update cycles. This supports interoperability and reduces friction across multi-stakeholder delivery settings (Carvalho *et al.*, 2022; Vial, 2019).

15.4. Sustainability and strategic value creation in projects

15.4.1 Sustainability is shifting from compliance to strategy

Across relevant empirical chapters, sustainability appears both as a technical requirement and a strategic orientation. Green building frameworks, resource-efficiency decisions and sustainability-linked quality practices demonstrate that sustainability increasingly shapes project objectives and stakeholder expectations. This aligns with the triple bottom line perspective, which integrates environmental and social outcomes with economic performance (Elkington, 1997).

15.4.2 Practical barriers to sustainability integration

The empirical studies highlight barriers commonly reported internationally:

- cost and schedule pressures that deprioritise sustainability features
- insufficient sustainability expertise in project teams

- limited access to sustainable materials and suppliers
- weak measurement systems for sustainability performance
- fragmented responsibility across design and delivery stakeholders.

These barriers reinforce the argument that sustainability requires capability development and governance alignment. Sustainability in projects is most effective when supported by explicit metrics, procurement strategies and stakeholder agreements (Soares *et al.*, 2024).

15.4.3 Lifecycle thinking and early integration

A recurring lesson is that sustainability achieves stronger outcomes when integrated early, during feasibility, concept design and procurement planning. Late-stage sustainability changes often generate rework or cost escalation. Lifecycle thinking holds that early design decisions strongly shape long-term environmental and operational impacts (Silvius and Schipper, 2014).

15.4.4 Sustainability, risk and reputation

Sustainability also intersects with risk management. Environmental non-compliance, stakeholder opposition, reputational harm and future regulatory tightening represent material risks. Projects that treat sustainability as strategic risk management rather than optional branding are better positioned for long-term value creation.

15.5. Governance, risk and the foundations of execution quality

15.5.1 Governance defines decision rights and accountability

Across the empirical evidence, governance quality repeatedly emerges as a differentiator between projects that remain controllable and those that drift into delay, rework and dispute. Governance includes:

- decision-rights allocation
- reporting cadence and transparency
- escalation mechanisms
- assurance and audit processes
- stakeholder accountability structures.

This aligns with the view that governance is not an administrative overhead but a value-creating structure that enables alignment and control in complex environments (Too and Weaver, 2014).

15.5.2 Risk management as a continuous practice

Risk management practices varied widely across cases. In some settings, formal registers and structured monitoring were in place. In others, risk management was informal. Across sectors, schedule delays and cost escalation frequently reflected unmanaged coordination risk, contractor performance risk, documentation risk and approval risk. These patterns align with broader evidence on megaprojects, which shows that risks compound when governance and oversight are weak (Flyvbjerg and Gardner, 2023).

15.5.3 Communication breakdown as a governance failure

The book's empirical evidence repeatedly shows that communication failures function as governance failures. When responsibilities for reporting, approvals and change control are unclear, the project becomes vulnerable to ambiguity-driven delay and conflict. This supports the argument that governance frameworks must include communication standards as enforceable routines, rather than optional good practices.

15.5.4 Digital governance and auditability

Digital tools strengthen governance when embedded into oversight routines. Audit trails, version control and real-time dashboards can support transparency and improve decision speed. However, digital governance requires standards for data ownership, update responsibility and approval workflows; tools can create a false sense of confidence.

15.6. Implications for theory, practice and policy

15.6.1 Implications for theory

- *PMBOK and contemporary extensions*: the findings reinforce PMBOK's relevance but suggest the need for stronger integration of sustainability governance, digital reporting norms and capability maturity. PMBOK's seventh edition shifts toward a focus on principles to support this broader evolution (PMI, 2021).
- *TAM extensions for project environments*: evidence validates TAM but indicates that project environments require expanded constructs such as workflow integration, governance enforcement and multi-stakeholder interoperability (Davis, 1989; Venkatesh and Davis, 2000).
- *RBV and capability maturity*: the book illustrates that project performance depends on capabilities such as leadership, communication discipline, digital readiness and governance routines. These resources align with RBV's emphasis on capability-based advantage (Barney, 1991).
- *Contingency theory and sectoral fit*: sector variation across defence, construction, manufacturing and infrastructure contexts supports the contingency perspective that practices must fit environmental and institutional constraints (Donaldson, 2001).

15.6.2 Implications for practice

- Treat frameworks as adaptive systems with built-in learning loops not static templates.
- Institutionalise communication routines and decision logging as part of core controls.
- Ensure digital transformation includes training, workflow redesign and governance enforcement.
- Integrate sustainability early and treat it as a strategic risk and value dimension.
- Strengthen governance with clear decision rights, escalation pathways and auditability.

15.6.3 Implications for policy

- Develop sector-wide digital reporting standards for public infrastructure and regulated projects.
- Strengthen sustainability incentives and require measurable sustainability reporting for large projects.
- Improve governance requirements in public procurement, including audit trails and transparent progress reporting.
- Encourage capability development through professional standards and certification pathways.

15.7. An implementation roadmap for project management innovation

While the preceding sections outline the theoretical, practical and policy implications of project management innovation, organisations often struggle to translate such insights into actionable change. Innovation in project management is rarely achieved through isolated initiatives or short-term interventions. Instead, it requires a structured and phased approach that aligns organisational capabilities, governance mechanisms and strategic objectives. Drawing on the empirical evidence throughout the book, this section proposes an implementation roadmap for organisations to operationalise project management innovation systematically and sustainably.

15.7.1 Establishing strategic alignment and leadership commitment

The first stage of any innovation roadmap involves establishing strategic clarity and leadership commitment. Across the empirical chapters, projects that demonstrated strong performance consistently benefited from visible senior leadership support. Leadership commitment influences resource allocation, priority setting and cultural acceptance of change. Without this commitment, even well-designed frameworks and digital tools remain underutilised.

Organisations should begin by articulating how project management contributes to strategic objectives such as digital transformation, sustainability performance, infrastructure reliability or operational efficiency. This articulation helps position project management as a strategic capability rather than a purely operational function. Senior leaders should formally endorse project management standards, digital reporting requirements and governance structures, signalling that compliance and continuous improvement are organisational priorities.

15.7.2 Assessing organisational capability maturity

The second stage involves diagnosing the current project management capability. The empirical evidence suggests that many organisations adopt advanced tools or frameworks without understanding their readiness to support them. Capability assessment should therefore precede major transformation initiatives.

A comprehensive capability assessment typically includes:

- evaluation of existing project management frameworks and their consistency of application
- assessment of digital readiness, including system integration and data quality
- review of governance structures, decision rights and escalation mechanisms
- analysis of risk management practices and documentation discipline
- assessment of human capabilities, including training, experience and leadership skills.

Such assessments allow organisations to identify capability gaps and prioritise interventions. They also support a staged approach to innovation, reducing the risk of overinvestment or change fatigue.

15.7.3 Strengthening core frameworks and processes

Once capability gaps are identified, organisations should focus on strengthening foundational project management processes. The empirical chapters demonstrate that innovation is most effective when built on a stable methodological base. This includes clear planning standards, cost-estimation practices, progress-tracking routines and change-control mechanisms.

Rather than introducing entirely new frameworks, organisations may adapt existing standards such as PMBOK to reflect contextual requirements. Adaptation may involve incorporating sustainability criteria into planning templates, embedding communication protocols into governance routines or aligning risk registers with digital reporting systems. This approach reinforces consistency while allowing flexibility.

Importantly, process strengthening should be accompanied by documentation standards and performance indicators. Clear documentation reduces ambiguity, supports knowledge retention and enhances auditability, particularly in regulated or public-sector project environments.

15.7.4 Embedding digital tools into daily practice

Digital transformation should be approached as an enabler of improved project discipline rather than as an end in itself. The empirical findings show that digital tools deliver value only when embedded into daily workflows and governance routines.

Organisations should prioritise

- integration of digital tools with existing project processes
- clear definition of roles and responsibilities for data entry and validation
- mandatory use of digital systems for reporting, approvals and change tracking
- structured training programmes tailored to different project roles
- continuous user support to address adoption barriers.

Embedding digital tools into formal governance processes ensures consistent data quality and usage across projects. It also enhances transparency and reduces reliance on informal or fragmented reporting methods.

15.7.5 Integrating sustainability into project lifecycle management

The roadmap should explicitly incorporate sustainability as a core dimension of project management. The empirical evidence demonstrates that sustainability initiatives are most effective when integrated early in the project lifecycle rather than treated as add-on requirements.

Organisations can operationalise sustainability by

- including environmental and social criteria in project selection decisions
- embedding sustainability indicators into planning and cost estimation tools
- aligning procurement strategies with sustainability objectives
- incorporating sustainability risks into risk management frameworks
- evaluating project performance using sustainability metrics alongside time, cost and quality.

By institutionalising sustainability practices, organisations reduce long-term risk and enhance stakeholder trust.

15.7.6 Reinforcing governance and risk oversight

Strong governance is essential for sustaining innovation. The roadmap should therefore include explicit mechanisms to reinforce governance. Empirical chapters repeatedly showed that unclear decision rights and weak oversight undermine project performance.

Effective governance reinforcement may include

- clearly defined decision-making authorities at different project stages
- regular governance reviews supported by digital dashboards
- independent assurance or audit mechanisms for large or high-risk projects
- formal escalation pathways for unresolved issues
- alignment between governance structures and organisational accountability frameworks.

Risk management should be integrated into governance routines rather than treated as a separate activity. Continuous risk identification, monitoring and mitigation improve execution quality and resilience.

15.7.7 Developing human-centred capabilities

Innovation in project management ultimately depends on people. Across all sectors examined in the book, human capability emerged as a decisive factor influencing framework adoption, digital usage, sustainability integration and governance effectiveness.

Organisations should invest in

- continuous professional development for project managers
- training in communication, negotiation and stakeholder management
- leadership development programmes focused on adaptive and ethical decision-making
- knowledge-sharing mechanisms to retain organisational learning
- cross-functional collaboration to reduce silos.

Human-centred capability development ensures that technical and digital innovations translate into sustained performance improvement.

15.7.8 Institutionalising continuous learning and improvement

The final stage of the roadmap involves embedding continuous learning into organisational routines. Project management innovation is not a one-time transformation but an ongoing process. Lessons learned, post-project reviews and performance analytics should inform continuous refinement of frameworks, tools and governance practices.

Organisations that institutionalise learning are better positioned to adapt to emerging challenges such as technological disruption, regulatory change and evolving stakeholder expectations. Continuous improvement also supports the progression of capability maturity over time.

15.8. Future research and innovation pathways

15.8.1 AI and predictive project intelligence

Research should examine how AI supports forecasting, risk detection and adaptive scheduling, while addressing governance, transparency and accountability concerns (Flyvbjerg and Gardner, 2023).

15.8.2 Hybrid collaboration and distributed delivery

Future work should explore governance, trust formation and communication structures in hybrid project teams, including digital fatigue, knowledge retention and virtual stakeholder engagement.

15.8.3 Sustainability as a performance dimension

Research should develop sustainability metrics integrated with project controls, including lifecycle costs, carbon accounting, circular-economy practices and environmental, social and governance (ESG)-aligned governance.

15.8.4 Integrated governance ecosystems

Future innovation will require governance platforms linking reporting, compliance, risk and stakeholder communications through digital audit trails and real-time dashboards.

15.8.5 Human-centred leadership and behavioural project science

More research is needed on decision-making under uncertainty, change management during digital adoption, stakeholder psychology and leadership capability in complex sociotechnical environments.

15.9. Concluding integration

The collective evidence in this book supports a clear conclusion: project management innovation is an integrated capability agenda. Methods provide structure, digital tools enable visibility, sustainability frames long-term value and governance ensures disciplined execution. The future of project management will be shaped by how effectively organisations align these domains into coherent operating systems (see Figure 15.1).

Figure 15.1 Integrated pathways for project management innovation

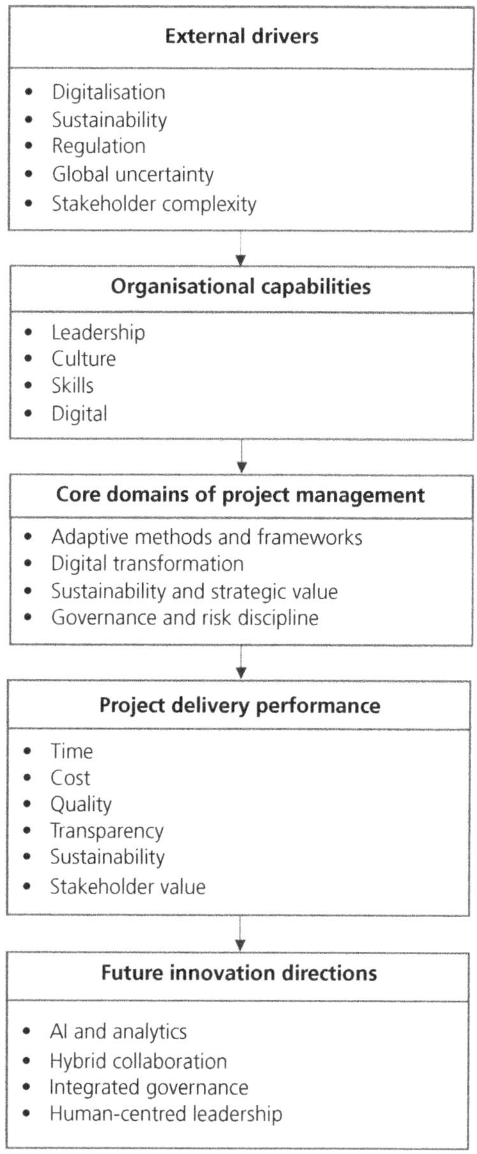

In this integrative view, innovation is best understood as a capability system rather than an isolated intervention. Organisations that treat project management as an evolving system of methods, technology, governance and human capabilities are more likely to deliver value reliably in uncertain environments.

REFERENCES

Baccarini D (1996) The concept of project complexity: a review. *International Journal of Project Management* **14(4)**: 201–204.

Barney J (1991) Firm resources and sustained competitive advantage. *Journal of Management* **17(1)**: 99–120.

Cai J, Li Z, Dou Y, Li T and Yuan M (2023) Understanding adoption of high off-site construction level technologies in construction based on the TAM and TTF. *Engineering, Construction and Architectural Management* **30(10)**: 4978–5006.

Carvalho MM, Desouza KC and Patah LA (2022) Digital transformation: a review and research agenda. *Technological Forecasting and Social Change* **174**: 121209.

Davis FD (1989) Perceived usefulness, perceived ease of use, and user acceptance of information technology. *MIS Quarterly* **13(3)**: 319–340.

Donaldson L (2001) *The Contingency Theory of Organizations*. Sage, London, UK.

Elkington J (1997) *Cannibals With Forks: The Triple Bottom Line of 21st Century Business*. Capstone, Oxford, UK.

Flyvbjerg B and Gardner D (2023) *How big things get done: The surprising factors that determine the fate of every project, from home renovations to space exploration and everything in between*. Crown Currency.

Geraldi J, Maylor H and Williams T (2011) Now, let's make it really complex (complicated): a systematic review of the complexities of projects. *International Journal of Operations & Production Management* **31(9)**: 966–990.

Kerzner H (2017) *Project Management: A Systems Approach to Planning, Scheduling, and Controlling*. Wiley, Hoboken, NJ, USA.

Merrow EW (2011) *Industrial Megaprojects: Concepts, Strategies, and Practices for Success*. Wiley, Hoboken, NJ, USA.

PMI (Project Management Institute) (2021) *A Guide to the Project Management Body of Knowledge (PMBOK® Guide)*, 7th edn. PMI, Newtown Square, PA, USA.

Silvius AJG and Schipper R (2014) Sustainability in project management: a literature review and impact analysis. *Social Business* **4(1)**: 63–96.

Soares I, Fernandes G and Santos JM (2024) Sustainability in project management practices. *Sustainability* **16(10)**: 4275.

Too EG and Weaver P (2014) The management of project management: a conceptual framework for project governance. *International Journal of Project Management* **32(8)**: 1382–1394.

Venkatesh V and Davis FD (2000) A theoretical extension of the technology acceptance model: four longitudinal field studies. *Management Science* **46(2)**: 186–204.

Vial G (2019) Understanding digital transformation: a review and a research agenda. *The Journal of Strategic Information Systems* **28(2)**: 118–144.

Advancing Project Management: Frameworks, Digital Transformation, Sustainability and Governance

Mohammed Reyasudin Basir Khan
ISBN 978-1-83742-799-4
https://doi.org/10.1108/978-1-83742-798-720261019

Index

Note: Italics in page numbers indicates figures.

academic institutions, 141
accountability
  expectations, 4–5
  governance defines, 287
adaptation becoming central to modern frameworks, 284
agency theory, 203
analysis of variance (ANOVA), 39, 44, 78, *78*, 91, 99, *99*, *135*, 171, *172*, 207, 210–213, 234, 254, 273
artificial intelligence (AI), 29, 52, 103, 109, 292
Association of Southeast Asian Nations (ASEAN), 103
assumption testing, 118
auditability, 288
average annual growth rate (AAGR), 123

Bartlett's test, 165, 168
behavioural project science, 292
blockchain, 29
box plot, 68, 71
building information modelling (BIM), 52, 88, 110, 156, 158, 224
Building Research Establishment Environmental Assessment Method (BREEAM), 51

central limit theorem, 17
central reference, 241
change management theory, 53
ChatGPT, 109, 112
climate change, 239
cloud platforms, 4, 156
coefficients analysis, 78–80, 171
coefficient regression, *100*
collaboration, 63
  tools, 229, 235
collinearity
  diagnostics, 80, *81*
  statistics, 275
communication, 202
  breakdown as governance failure, 288
  issues, 268
  as methodological capability not soft add-on, 284–285
  tools, 229, 235
comparative research, 121
complex projects, 206
construct validity, 165
construction
  risk management, 260
  sector, 124, 223, 239
construction delays
  conceptual framework, *262*
  correlation analysis, 273–274
  data analysis techniques, 264
  data collection method, 263
  demographic profile, 264, *265*
  descriptive analysis, 268–271, *269*
  hypothesis testing, 276
  item total statistics, *266–268*
  limitations, 277–278
  methodology, 262
  normality analysis, 271–273
  normality test results for study variables, *272*
  normal Q-Q plot of project delay, *270*
  normal Q-Q plot of financial constraints, *270*
  normal Q-Q plot of contractual changes, *271*

normal Q-Q plot of communication issues, *272*
normal Q-Q plot of site-related matters, *271*
operationalisation and measurement, 263–264
population and sampling procedures, 262–263
practical insights, 276–277
recommendations, 277
recommendations and future research, 278
regression analysis, 274–276
regression standardised residuals, *269*
reliability analysis, 264–268, *265*
research design, 262
results and analysis, 264
test for mean difference, *272*
construction industry, 125, 155
pressures, 158
Construction Industry Development Board (CIDB), 239
construction projects, 4, 125, 240, 260–261
delays, 259
management, 241
contemporary project management
contingency theory, 5
PMBOK, 5
resource-based view, 5
sustainability models, 5
TAM, 5
theoretical foundations of, 5
content validity, 165
contingency theory, 5, 12–14, 203, 241–242, 284
convenience sampling, 245
coordination, 202
corporate policies, 182–183, *189*, 190–191
correlation, 188, 194
coefficients, 41, 232
correlation analysis, 38, 41–42, 57, 72, 91, 97, 116, 133, *135*, 166, 169–171, 173, 213–214, 232, 247, 252, 273–274, 276
ANOVA results for multiple regression model, *274*
descriptive analysis for research variables, *169*
macroenvironmental factors, 214–215
organisational context, 215
Pearson correlation coefficient, 97–98
Pearson's correlation analysis, 74–77
Pearson's correlation test, *170*
project-specific factors, 215
relationship between potential factors and project delays, *273*
Scatter plots, *171*
Spearman's correlation analysis, 72–74
cost
cost–benefit analyses, 82
overruns, 199
reduction, 140
risk, 127, 130, 133–134
cost management, 52, 56, 58–59, 65, 205
descriptive results, *59*
expertise, 201
histogram, Q-Q plots and box plot for r, *66*
impact of GBC against, 72, 74–75
Kolmogorov–Smirnov and Shapiro–Wilk test results for, *65*
cost performance, 201
communication and coordination, 202
cost management expertise, 201
decision-making processes, 202
risk management culture, 202
technology adoption, 202
COVID-19 pandemic, 123
critical path method (CPM), 160
critical success factors theory (CSF theory), 13
Cronbach's alpha, 16, 41, *41*, 71, *71*, 117, 165, 167, 188, 228, 231, *231*, 247, 264
cross-sectional quantitative design, 245

data, 55, 144, 245
analysis, 15, 36, 247
analytics, 52
collection, 207, 257, 262
screening, 118
data analysis techniques, 38
construction delays, 264
digital project management systems, 116
engineering firms, 91–92
engineering management practices, 166
GBI, 57

LSS, 145
project cost performance, 207–208
project delivery perspective, 246–247
project management applications, 229
QMS, 38–39
data collection method, 145
construction delays, 263
defence projects, 15
digital project management systems, 114
engineering firms, 90
engineering management practices, 165
GBI, 56
infrastructure projects, 128
LSS, 145
project delivery perspective, 244–246
project management applications, 228
QMS, 37
sustainable project management, 186
decision makers, 48
decision rights, governance defines, 287
decision-making processes, 202
defence environments, 12
defence organizations, 11, 26–27
defence practitioners, 26
defence projects, 4, 11, 13, 23, 26
conceptual framework, *13*
data collection method, 15
demographic profile, 15–16, *16*
descriptive analysis, 18–21
hypothesis testing, 22–23, *22*
implications and strategic considerations, 27–28
limitations, 28–29
methodology, 14
normality analysis, 17–18, *18*
operationalisation and measurement, 15
population and sampling procedures, 14–15
practical insights, 23–27
recommendations, 27
recommendations and future research, 29
regression analysis, 21–22
reliability analysis, 16–17, *16–17*
research design, 14
response rate, 15
results and analysis, 15
Spearman's rank correlation, 21, *21*
standardised beta coefficients, *21*
survey items, mean and standard deviation, *19–20*
theoretical contributions, 28
training, technology and resource management, 28
define, measure, analyse, improve and control cycle (DMAIC cycle), 139, 142, 144
degrees of freedom, 78, 234
delays, 261
factors, 259, 266
demographic analysis, 40, 247
demographic profile, 39–40
demographic variables, 249
dependent variable (DV), 38, 91, 115, 127, *131–132*, *212*, 233, 248
descriptive analysis, 38, 40–41, 92, 130–133, 145–148, 166, 168–169, 194, 229
construction delays, 268–271
defence projects, 18–21
engineering firms, 96–97
Likert distribution across resource management variable, 191
of macroenvironmental factors, 210
project delivery perspective, 249–252
project management applications, 229–231
sustainable project management, 190–191
techniques, 116
descriptive statistics, 40, 57–58, 96, *191*, 207, 244, 264
cost management, 58–59
GBC impact on project management practices, 63–64, *64*
quality management, 61–62
of project environment, *249*
of project team, *250*
of project management process, *250*
of project risks, *251*
of project performance, *251*
stakeholder engagement, 62–63
time management, 59–61
of variables, *130*
detail engineering completion rating index system (DECRIS), 160
detrended Q-Q plot, 68–69

digital collaboration tools, 88
digital governance, 288
digital infrastructure, 7
digital project management
  academic contributions, 111
  assumption testing and data screening, 118
  challenges in PMS adoption, 112
  conceptual framework, 112–113, *113*
  data analysis techniques, 116
  data collection method, 114
  demographic profile, 116–117
  digital transformation and industry applications, 112
  hypotheses development, 113–114
  implications, 120
  limitations, 120–121
  methodology, 114
  multiple regression analysis, 119
  operationalisation and measurement, 115, *115*
  Pearson's correlation analysis, 118–119
  population and sampling procedures, 114
  practical insights, 119–120
  recommendations, 120
  recommendations and future research, 121
  reliability analysis, 117–118, *117*
  research design, 114
  results and analysis, 116
  theoretical contributions, 110–111
  theoretical foundation, 111
  tools, 12, 23
digital systems, 286
digital technologies, 121
digital tools, 52, 156, 161, 175
  embedding digital tools into daily practice, 290
  and platforms, 162
digital transformation, 4, 9, 29, 31, 112, 290
  adoption model, 7–8, *7*
  as capability enabler, 285
  digital tools change information flow but adoption sociotechnical, 285
  as governance amplifier, 286
  sectoral maturity differences and problem of fragmented adoption, 286
  TAM remains relevant but project settings add new determinants, 285–286
digitalisation, 158, 164
  functions, 165
discriminant validity, 266
distributed delivery, 292
documentation, 60

economic growth, 205, 210, 214
economic instability, 123
education, 187
effective communication, 162, 202
effective coordination, 202
effective engineering management, 157
effective planning, 112
effective project management processes, 243
effective resource, 33, 96
effective stakeholder
  engagement, 33, 112
  management, 113
electronics manufacturing service (EMS), 139
emerging technologies, 112
employee, 142
  engagement, 145, 148
employment status data, 129
energy
  organizations, 49
  projects, 33, 47–48
  sector, 34
  services, 40
engineering, 157
  companies, 86
  disciplines, 160
  management, 158
  performance, 156
  projects, 4, 85, 88
  quality, 161
engineering, procurement and construction (EPC), 9, 155, 157, *167*, 284
engineering firms, 85
  conceptual framework, *89*
  correlation analysis, 97–98
  data analysis techniques, 91–92
  data collection method, 90
  demographic profile, 92–94
  descriptive analysis, 96–97, *96*
  hypothesis testing, 99–101
  limitations, 102
  methodology, 89

multiple regression, 98–99
normality analysis, 95–96, *95*
operationalisation and measurement, 90–91
population and sampling procedures, 89–90
practical insights, 101–102
Q-Q plot, *96*
recommendations, 102
recommendations and future research, 102–104
reliability analysis, 94–95, *95*
research design, 89
respondent career field, *93*
respondents by company size, *94*
results and analysis, 92
skewness and kurtosis, 97
engineering management practices, 157, 166
conceptual framework, *163*
correlation analysis, 169–171
data analysis techniques, 166
data collection method, 165
demographic profile, 166–167
descriptive analysis, 168–169
hypothesis testing, 173–175
limitations, 176–177
methodology, 164
moderation analysis, *173*
multiple regression analysis, 171–173
operationalisation and measurement, 165–166
population and sampling procedures, 164–165
practical insights, 175–176
recommendations, 176
recommendations and future research, 177
reliability analysis, 167–168, *168*
research design, 164
response rate, 166
results and analysis, 166
theoretical contributions, 176
enterprise resource planning (ERP), 159, 224
environmental, social and governance (ESG), 121, 292
environmental factors, 240, 255
environmental responsibility, 4, 62
environmental sustainability-oriented projects, 64
ethnicity, 116
evidence-based decision-making, 33
exchange rate fluctuations, 200
execution quality
communication breakdown as governance failure, 288
digital governance and auditability, 288
governance, risk and foundations of, 287
governance defines decision rights and accountability, 287
risk management as continuous practice, 287
exploratory factor analysis (EFA), 165, 168, *168*
extensions of time (EOTs), 223

F-statistic, 135
Facebook, 128
financial incentives, 59
first hypothesis (H1), 127
flexibility, 159
fragmented adoption, sectoral maturity differences and problem of, 286
framing project management innovations
conceptual models for, 5–6
digital transformation adoption model, 7–8, *7*
governance and risk performance model, 8–9, *8*
project management innovation ecosystem, 6–7, *6*
frequency distribution, 38, 57–58
front-end engineering design (FEED), 155

gender, 247
geopolitical challenges (GR), 87–88, 90, 99
geopolitical conflicts, 86
geopolitical risk (GR), 88, 97
geopolitical tensions, 85
global organizations, 4
global supply chains, 3, 85
governance amplifier, digital transformation as, 286
governance defines decision rights and accountability, 287

governance expectations, 4–5
governance failure, communication breakdown as, 288
governance model, 8–9, *8*
government
  policies, 51
  projects, 124
green building adoption, 51
green building certifications (GBC), 52, 54, 56, 63, 82
  histogram, Q-Q plots and box plot for impact of, *70*
  impact of, 69–71
  impact of GBC against cost management, 72, 74–75
  impact of GBC against quality management, 76
  impact of GBC against stakeholder management, 74, 76–77
  impact of GBC against time management, 72–73
  impact on project management practices, 63–64, *64*
  Kolmogorov–Smirnov and Shapiro–Wilk test results for impact of, *70*
  systems, 52
Green Building Index (GBI), 51
  conceptual framework, *55*
  correlation analysis, *72–77*
  data analysis techniques, 57
  data collection method, 56
  descriptive statistics, 58–64
  frequency distribution analysis, 57–58
  implications and strategic considerations, 82
  limitations, 82
  methodology, 55
  multiple regression analysis of green building practices, 77–80
  normality analysis, 65–71
  percentage of respondents by role in organisation, *58*
  practical insights, 80–82
  recommendations, 82
  recommendations and future research, 82–83
  reliability analysis, 71
  research design, 55
  results and analysis, 57
  study population and sampling procedures, 55–56
  variables operationalisation, 56–57
green building practices, multiple regression analysis of, 77
  ANOVA, 78
  coefficients analysis, 78–80
  collinearity diagnostics, 80
  model summary and statistical fit, 77–78
green building projects, 82
  quality management in, 67–68
  stakeholder engagement in, 68–69
  time management in, 65–67
green certification process, 62
Green Technology Financing Scheme (GTFS), 51
green-certified buildings, 51
gross domestic product (GDP), 239

high-performance materials, 62
homoscedasticity, 118
human capital theory, 88
human resource (HR), 28, 90–91, 101
  capability, 160
  manager, 165
human resource challenges (HRC), 97, 99
human-centred capabilities, developing, 291–292
human-centred leadership, 292
hybrid collaboration, 292
hypothesis development, 205
  macroenvironmental factors, 205
  organisational context, 205–206
  project-specific factors, 206
hypothesis testing, 164, 166, 213–214
  construction delays, 276
  defence projects, 22–23
  engineering firms, 99–101
  engineering management practices, 173–175, *173–174*
  infrastructure projects, 135–136
  macroenvironmental factors, 214–215
  organisational context, 215

project delivery perspective, 255
project-specific factors, 215
sustainable project management, 192–193, *192*

independent variables (IV), 38, 42, *131–132*, 233, 248
industrial sectors, 156
industrialised building systems (IBS), 103
industry
applications, 112
experience, 92
involvement, 129
perspective, 125
practitioners, 141
sector, 92
inferential analysis, 38, 92
techniques, 116
inflation, 200
influential variables, 278
information systems (IS), 226
infrastructure development, 125, 136
infrastructure projects, 223
conceptual framework, *127*
correlation analysis, 133
data collection method, 128
demographic profile, 128–129
descriptive analysis, 130–133
hypothesis testing, 135–136
limitations, 136–137
methodology, 128
multiple regression analysis, 133–135, *134*
percentage of respondents by age group, *129*
population and sampling procedures, 128
practical insights, 136
recommendations, 136
recommendations and future research, 137
reliability analysis, 129–130, *130*
research design, 128
response rate, 128
results and analysis, 128
institutional theory, 53, 87, 203
integrated governance ecosystems, 292
integration, 160
integration management (IM), 172–173
interest rates, 200
changes, 214
International Federation of Consulting Engineers (FIDIC), 156
Internet of Things (IoT), 29, 52, 103

job role, 39

Kaiser-Meyer-Olkin test (KMO test), 165, 168
Kolmogorov–Smirnov tests (KS tests), 17, 57, 65, *65*, 67–68, 273–274
Kuala Lumpur (KL), 259
kurtosis, 97
values, 189

labour scarcity, 242
large-scale data collection, 37
leadership
development programmes, 28
support, 140, 145
Leadership in Energy and Environmental Design (LEED), 51, 181
leadership support and training (LST), 12, 15, 17–18, 21–22, *24*, 27
lean manufacturing, 141
lean practices, 159
Lean Six Sigma (LSS), 139, 141, 284
areas of cost reduction, *147*
challenges in, *147*
conceptual framework, *143*
cost reduction observed due to, *146*
data analysis techniques, 145
data collection method, 145
descriptive analysis, 145–148
future research and recommendations, 150
limitations, 150
measures to improve, *148*
methodology, 144
multiple regression analysis, 148–149
operationalisation and measurement, 145
Pearson's correlation analysis, 148
population and sampling procedures, 144
practical insights, 149–150
recommendations, 150
research design, 144

results and analysis, 145
tool usage in EMS companies, *146*
lean value stream mapping, 142
lifecycle assessments (LCA), 183
linearity, 118
LinkedIn, 128, 145
longitudinal approach, 257

machine learning (ML), 109, 112
ML-based tools, 161
macroenvironmental factors, 205, 214–215
affecting project cost performance, 200
descriptive analysis of, 210
economic growth, 201
exchange rate fluctuations, 200
inflation, 200
interest rates, 200
regulatory requirements, 201
macroenvironmental variables, 214
Malaysia
construction industry, 123
construction sector, 51
rail sector, 227
as representative global context, 10
Malaysian construction
context, 54
projects, 53
Malaysian context, 198
Malaysian energy
industry, 49
sector, 31, 33, 37
Malaysian engineering sector, 195
Malaysian industries, 3
Malaysian projects, 197, 202
environment, 197
material quality, 203
mean (m), 18, 38, 190
measurement scales, reliability of, 209
microcosm, 10
mobile-based PMS, 112
multicollinearity, 118
multiple linear regression analysis, 43–47, *47*
multiple regression, 91, 98–99
multiple regression analysis, 21, 38, 92, *135*, 145, *149*, *233*, 274, 276
ANOVA, *234*
coefficients, *234*
digital project management systems, 119
engineering management practices, 171–173
findings of Pearson correlation, *253*
of green building practices, 77–80
infrastructure projects, 133–135
LSS, 148–149
project delivery perspective, 252–255
project management applications, 233–235

national defence organization, 11
national rail operator (NRO), 227
non-parametric methods, 17
normal Q-Q plot, 68–69
normality analysis, 65
construction delays, 271–273
cost management, 65
defence projects, 17–18
engineering firms, 95–96
impact of GBC, 69–71
quality management in green building projects, 67–68
stakeholder engagement in green building projects, 68–69
sustainable project management, 188–190, *188*
time management in green building projects, 65–67
normality test, 57, 65, 69–70, 95, 118, 188, 252

organisational capability maturity, assessing, 289–290
organisational context, 201, 205–207
communication and coordination, 202
cost management expertise, 201
decision-making processes, 202
risk management culture, 202
technology adoption, 202
organisational effectiveness, 203
organisational factors, 121, 216
organisational leaders, 193
organisational learning, 183, 185, 189–190, *190*, 192–194
theory, 204

organisational policies, 201
organisations, 3, 32, 57, 149, 202, 289

p-values, 80
Paris Agreement, 181
partial least squares (PLS), 29
Pearson coefficient, 74, *133*
Pearson correlation coefficients, 92, 97–98, *98*, *192*, 133, 252, *254*
Pearson's correlation analysis, *42*, 74, *75–77*, 118–119, *118*, 145, 148, *148*, 169, 191–192, 207, 213, *232*
  impact of GBC against cost management, 74–75
  impact of GBC against quality management, 76
  impact of GBC against stakeholder management, 76–77
  impact against time management, 75–76
  skewness and kurtosis, *97*
  project management applications, 232–233
perceived ease of use (PEOU), 111
perceived usefulness (PU), 111
performance measurement theory, 54
pilot test, 145, 246
planning
  process, 118
  quality of planning process, 112
  and scheduling effectiveness, 203
policy
  and industry relevance, 53
  makers, 141, 193, 240
political instability, 86
poor communication, 202
practical contribution, 53
practical insights, 23–27
predictive project intelligence, 292
primary data, 15, 145, 186, 245
proactive risk management, 256
probability sampling method, 263
process
  efficiency, 140, 143, 148
  standardization, 31, 34, 38, 42–43
  training, 42
program evaluation and review technique (PERT), 160
project complexity, 202–203
project complexity and risk (PCR), 15, 17, 21–22, *24*, 26–27
project cost performance, 197–199
  ANOVA and framework validation, 210–213
  conceptual framework, *206*
  correlation analysis and hypothesis testing, 213–215, *213*
  data analysis techniques, 207–208
  data collection, 207
  demographic, *208*
  descriptive analysis of macroenvironmental factors, 210
  fluctuation in exchange rates, *211*
  government regulations and policies, *211*
  hypothesis development, 205–206
  industry sector, *209*
  literature review, 199
  macroenvironmental factors affecting, 200–201
  measurement and reliability, 207
  methodology, 206
  organisational context and cost performance, 201–202
  population and sample, 207
  practical insights, 215–216
  project cost performance and cost overruns, 199
  project-specific factors influencing cost performance, 202–203
  recommendations, 216–217
  regression standardised residual, 213
  reliability of measurement scales, 209
  research design, 206–207
  respondent profile, 208–209
  results and analysis, 208
  statistics, *210*
  summary of literature and research gaps, 203
  theoretical foundation, 203–205
project delays, 259
project delivery perspective
  conceptual framework, *243*
  correlation analysis, 252
  data analysis techniques, 246–247

data collection method, 244–246
demographic profile, 247–248
descriptive analysis, 249–252, *249*
hypothesis testing, 255
limitations, 256–257
methodology, 244
multiple regression analysis, 252–255
operationalisation and measurement, 246
population and sampling procedures, 244
practical insights, 256
recommendations, 256
recommendations and future research, 257
reliability analysis, 248–249
research design, 244
respondents' construction site location, *248*
response rate, 247
results and analysis, 247
project environment, 254
project execution models (PEM), 158
project governance, 10
project implementation, 155
project lifecycle, 13, 242
integrating sustainability into project lifecycle management, 291
theory, 13
project management, 3, 51, 110, 183, 198, 225, 283
background and context, 3–4
changing landscape of, 4
conceptual models for framing project management innovations, 5–9
digital transformation and technology adoption, 9
experience in, *230*
frameworks and methodological foundations, 9
governance, sectoral realities and risk, 9–10
governance and accountability expectations, 4–5
Malaysia as representative global context, 10
process, 243–244, 249, 254, 256
project management-specific frameworks, 241
purpose, 4
rising complexity and interdependence, 4
structure, 4
sustainability and environmental responsibility, 4
technological transformation, 4
thematic synthesis of content, 9
theoretical foundations of contemporary project management, 5
triangle, 242
project management applications
conceptual framework, *226*
data analysis techniques, 229
data collection method, 228
demographic profile, 229
descriptive analysis, 229–231
limitations, 236
mean, standard deviation, minimum and maximum, *230*
methodology, 227
multiple regression analysis, 233–235
operationalisation and measurement, 228
Pearson's correlation analysis, 232–233
population and sampling procedures, 227–228
practical insights, 235
questionnaire sections, *228*
recommendations, 235–236
recommendations and future research, 236
research design, 227
response rate, 229
results and analysis, 229
project management body of knowledge (PMBOK), 5, 34, 125, 156, 158, 228, 241, 260, 283–284
project management information systems (PMIS), 225
project management innovation, 292
adaptation becoming central to modern frameworks, 284
AI and predictive project intelligence, 292
assessing organisational capability maturity, 289–290
communication as methodological capability not 'soft add-on', 284–285

developing human-centred capabilities, 291–292
digital transformation as capability enabler, 285–286
ecosystem, 6–7, *6*
embedding digital tools into daily practice, 290
establishing strategic alignment and leadership commitment, 289
future research and innovation pathways, 292
governance, risk and foundations of execution quality, 287–288
human-centred leadership and behavioural project science, 292
hybrid collaboration and distributed delivery, 292
implementation roadmap for, 289
implications for theory, practice and policy, 288–289
institutionalising continuous learning and improvement292
integrated governance ecosystems, 292
integrated pathways for, *293*
integrating sustainability into project lifecycle management, 291
measurement, estimation and operational realism, 285
methodological discipline remains necessary but not sufficient, 284
from methods to adaptive frameworks, 284
reinforcing governance and risk oversight, 291
strengthening core frameworks and processes, 290
sustainability and strategic value creation in projects, 286–287
sustainability as performance dimension, 292
project management practices, 12, 31
GBC impact on, 63–64, *64*
project management software (PMS), 109–110
challenges in PMS adoption, 112
project management theory (PMT), 34, 158, 162–163
project management tools (PMTs), 18, 223–224
project managers, 193
project monitoring and reporting efficiency (PMRE), 13, 17, 21, *25*
project outcomes, 110, 112–113
project performance (PP), 4, 89, 97, 244
project risk(s), 123, 243, 252, 254
project success, 123, 229
measures, 231
theory, 125–126
project sustainability management (PSM), 183
project team, 243, 251
project-specific factors, 206, 215–216
influencing cost performance, 202
material quality and resource allocation, 203
planning and scheduling effectiveness, 203
project complexity, 202–203
scope clarity and stability, 203
purposive sampling, 128

Q-Q plot(s), 68, 131, 189, 232, 252, 270
qualitative approach, 89
qualitative methods, 103
quality, 155
deficiencies, 124
performance, 49
risk, 127, 130, 134, 136
quality assurance (QA), 160
quality control (QC), 160
quality management, 31–32, 52, 56, 61–62, 80, 164, 170, 173
descriptive results, *61*
histogram, Q-Q plots and box plot for, *68*
impact of GBC against, 76
in green building projects, 67–68
impact against, 73–74
Kolmogorov–Smirnov and Shapiro–Wilk test results for, *67*
quality management plans (QMPs), 161
quality management system (QMS), 31, 33, 80
conceptual framework, *35*
correlation analysis, 41–42
data analysis techniques, 38–39
data collection method, 37

demographic profile, 39–40
descriptive analysis, 40–41, *41*
limitations, 49
methodology, 36
multiple linear regression analysis, 43–47
operationalisation and measurement, 38
percentage respondents by job role, *39*
population and sampling procedures, 37
practical insights, 47–49
recommendations, 49
recommendations and future research, 49–50
reliability analysis, 41
research design, 36–37
results and analysis, 39
simple linear regression analysis, 43
Spearman's rank correlation, 42–43
quantitative approach, 143–144
quantitative data, 114, 227
quantitative methods, 185, 206
quantitative research, 164
approach, 245, 262
design, 55, 244
quantitative survey
designs, 36
research design, 14
quantitative tools, 278

radio-frequency identification (RFID), 103
railway infrastructure projects, 227
railway project
environments, 224
organizations, 235
random sampling, 164
real estate sector, 51
real-time data, 225
regression, 188, 194
analysis, 21–22, 57, 116, 133, 252, 269, 274–276
ANOVA results for multiple regression model, *274*
coefficients, 78, *79*, *274*
model, 77, *77*, 274
standardised residual against regression standardised predicted value, *276*
regulatory complexity, 242
regulatory systems, 3
related theories, 126
reliability, 92, 165, 207
of measurement scales, 209, *209*
testing, 57, 247
reliability analysis, 145, 209
construction delays, 264–268
defence projects, 16–17, *16–17*
digital project management systems, 117–118
engineering firms, 94–95
engineering management practices, 167–168
green building index, 71
infrastructure projects, 129–130
project delivery perspective, 248–249
quality management system, 41
sustainable project management, 188
research strategy, 227
resource allocation, 34, 36, 38, 85, 203
efficiency, 91, 215
systems, 224
resource availability (RA), 17–18, 21–22, *25*, 26–27
resource dependence theory, 203
resource efficiency, 97
resource management, 28, 170, 173, 182, 188, *189*, 190–192, *191*
resource-based view (RBV), 5, 13, 53, 284
risk
assessment, 161
factors influence, 123
performance model, 8–9, *8*
risk management, 11, 113, 118–120, 126, 128, 161, 175, 241–242
as continuous practice, 287
culture, 202
cycle, 260
theory, 87, 125
robust risk management, 164

safety-related risks, 123
sampling, 49
design, 128
Sarawak's economic development, 239

scenario-based modelling, 103
schedule management, 159
schedule risk, 130, 133–134
science, technology, engineering and mathematics (STEM), 103
scope clarity and stability, 203
second hypothesis (H2), 127
secondary data, 246
Senai Industrial Park, 110
Shapiro–Wilk tests (SW tests), 17, 57, 65, *65*, 67–68, 273–274
Silicon Valley of East, 90
simple linear regression analysis, 43, *44–46*
Six Sigma, 139, 142
skewness, 97
small and medium-sized enterprise–multinational corporation (SME–MNC), 103
small and medium-sized enterprises (SMEs), 82
software development, 110, 112
Spearman's correlation analysis, *43*, 72, *72–73*
  impact of GBC against cost management, 72
  impact of GBC against stakeholder management, 74
  impact of GBC against time management, 72–73
  impact against quality management, 73–74
Spearman's rank correlation, 15, 21, *21*, 42–43
Spearman's rho, 17, 42
stakeholder engagement, 34, 52, 56, 62–63, 82, 159, 170, 173, 184, 188, 190, *190*, 193
  descriptive results, *62–63*
  in green building projects, 68–69
  Kolmogorov–Smirnov and Shapiro–Wilk test results for, *69*
stakeholder management, 34, 36, 43, 113, 117–120
  impact of GBC against, 74, 76–77
stakeholder(s), 63
  dynamics, 121
  involvement, 63
  networks, 3
  satisfaction, 229
  theory, 34, 53, 183, 185
standard deviation (SD), 18, 38, 40, 58, 96, 130, 169, 190, 230, 247, 264, 270
standard operating procedure (SOP), 29, 125
statistical package for social sciences (SPSS), 15, 36, 91, 114, 131, 145, 165, 207, 228, 247, 264
statistical techniques, 91, 114
statistical validity, 165
strengths, weaknesses, opportunities and threats (SWOT), 125
structural equation modelling (SEM), 29, 188
structured data, 114
supply chain disruption (SCD), 88, 91, 97, 99
supply chain management theory (SCM theory), 88
supply chain professionals, 90
survey methods, 14
survey research, 245
sustainability, 4, 52, 181–182, 193
  integrating sustainability into project lifecycle management, 291
  lifecycle thinking and early integration, 287
  models, 5
  as performance dimension, 292
  practical barriers to sustainability integration, 286–287
  risk and reputation, 287
  shifting from compliance to strategy, 286
  and strategic value creation in projects, 286
  sustainability-focused practices, 63
  theory, 53
sustainability balanced scorecard (SBSC), 184
sustainable construction
  methods, 64
  projects, 61
sustainable development theory, 183, 185
sustainable project management (SPM), 82, 181, 184
  conceptual framework, *184*
  Cronbach's alpha across variables, *188*
  data collection method, 186
  demographic profile, 186–188
  descriptive analysis, 190–191
  framework, 186
  hypothesis testing, 192–193

limitations, 195
methodology, 185
normality analysis, 188–190
operationalisation and measurement, 186
Pearson correlation analysis, 191–192
practical insights, 193–194
questionnaire sections, *185*
recommendations, 194
recommendations and future research, 195
reliability analysis, 188
research design, 185–186
respondents' demographic profile, *187*
results and analysis, 186
theoretical contributions, 194–195
systems theory, 54, 241–242
systems thinking, 204

t-values, 80
team capabilities, 244
technological dimensions, 86
technological transformation, 4
technology, 28, 102, 223
technology acceptance model (TAM), 4–5, 12–13, 88, 110, 228, 284
remains relevant but project settings add new determinants, 285–286
technology adoption (TA), 9, 15, 17–18, 21–22, *23*, 27, 88, 91, 97, 99, 101, 110, 169, 172, 176–177, 202, 215–216
telegram, 128
theoretical contribution, 53
third hypothesis (H3), 127
time management, 32, 48, 52, 56, 59–61, 82
descriptive results, *60*
histogram, Q-Q plots and box plot for, *67*
impact of GBC against, 72–73
in green building projects, 65–67
impact against, 75–76
Kolmogorov–Smirnov and Shapiro–Wilk test results for, *66*
total quality management (TQM), 33
Toyota Production System, 139, 141
traditional project management, 181
train-the-trainer model, 28
training, 28, 33, 35, 38
programmes, 28
triple bottom line (TBL), 182
triple constraints model, 182

unified theory of acceptance and use of technology (UTAUT), 228
United Nations Sustainable Development Goals, 181
user satisfaction, 113

validity, 165
variance inflation factor (VIF), 80, 118, 275
virtual reality (VR), 103
ViTrox, 103

waste reduction, 148
WhatsApp, 90, 128
workplace organisation practices, 142

Printed and bound by CPI Group (UK) Ltd, Croydon, CR0 4YY
28/09/2026
14971126-0007